T0134829

Sustainable Civil Infrastructures

Editor-in-Chief

Hany Farouk Shehata, SSIGE, Soil-Interaction Group in Egypt SSIGE, Cairo, Egypt

Advisory Editors

Khalid M. ElZahaby, Housing and Building National Research Center, Giza, Egypt

Dar Hao Chen, Austin, TX, USA

Sustainable Civil Infrastructures (SUCI) is a series of peer-reviewed books and proceedings based on the best studies on emerging research from all fields related to sustainable infrastructures and aiming at improving our well-being and day-to-day lives. The infrastructures we are building today will shape our lives tomorrow. The complex and diverse nature of the impacts due to weather extremes on transportation and civil infrastructures can be seen in our roadways, bridges, and buildings. Extreme summer temperatures, droughts, flash floods, and rising numbers of freeze-thaw cycles pose challenges for civil infrastructure and can endanger public safety. We constantly hear how civil infrastructures need constant attention, preservation, and upgrading. Such improvements and developments would obviously benefit from our desired book series that provide sustainable engineering materials and designs. The economic impact is huge and much research has been conducted worldwide. The future holds many opportunities, not only for researchers in a given country, but also for the worldwide field engineers who apply and implement these technologies. We believe that no approach can succeed if it does not unite the efforts of various engineering disciplines from all over the world under one umbrella to offer a beacon of modern solutions to the global infrastructure. Experts from the various engineering disciplines around the globe will participate in this series, including: Geotechnical, Geological, Geoscience, Petroleum, Structural, Transportation, Bridge, Infrastructure, Energy, Architectural, Chemical and Materials, and other related Engineering disciplines.

SUCI series is now indexed in SCOPUS

and EI Compendex.

More information about this series at http://www.springer.com/series/15140

Ibrahim El Dimeery · Moustafa Baraka ·
Syed M. Ahmed · Amin Akhnoukh ·
Mona B. Anwar · Mahmoud El Khafif ·
Nagy Hanna · Amr T. Abdel Hamid
Editors

Design and Construction of Smart Cities

Toward Sustainable Community

 Springer

Editors
Ibrahim El Dimeery
Faculty of Post Graduate Studies
and Scientific Research
German University in Cairo
Cairo, Egypt

Syed M. Ahmed
Construction Management Department
East Carolina University
Greenville, NC, USA

Mona B. Anwar
Civil Engineering Program
German University in Cairo
Cairo, Egypt

Nagy Hanna
Civil Engineering Program
German University in Cairo
Cairo, Egypt

Moustafa Baraka
Civil Engineering Program
German University in Cairo
Cairo, Egypt

Amin Akhnoukh
Construction Management Department
East Carolina University
Greenville, NC, USA

Mahmoud El Khafif
Civil Engineering Program
German University in Cairo
Cairo, Egypt

Amr T. Abdel Hamid
Information and Electronics Technology
German University in Cairo
Cairo, Egypt

ISSN 2366-3405   ISSN 2366-3413   (electronic)
Sustainable Civil Infrastructures
ISBN 978-3-030-64216-7   ISBN 978-3-030-64217-4   (eBook)
https://doi.org/10.1007/978-3-030-64217-4

Conference Organizing Committee

General Conference Chair

Ibrahim El-Dimeery

Conference Chairs

Moustafa Baraka
Sayed M. Ahmed

Technical Program Chairs

Amr T. Abdel Hamid
Amin K. Akhnoukh
Mona Badr El-Din Anwar

Technical Committee Members

Moustafa Baraka
Nagy Fouad Hanna
Mona B. El-Din Anwar
Amin K. Akhnoukh
Ramy Shaltout
Amr Maher El-Nemr
Ehab Nour El-Din
Ahmed Maher El-Tair
Yara Basyoni
Mahmoud El-Khafif

Ahmed Abdel Sattar
Amr T. Abdel Hamid
Tallal El-Shabrawy
Ayman Hamdy Nassar
Mohamed Ehsan Ashour
Abdel Aziz El-Ganzory
Ibrahim Amin Lotfy
Maggie Mashaly
Wassim Alexan

Conference Organization Chairs

Engy A. Maher
Omar M. Shehata

Financial Chair

Eman Ahmed Hamdy Azab

Organizing Committee

Eng. Abdullah Amr
Eng. Ahmed El-Sayed Aredah
Eng. Ghada Amr Diab
Eng. Abanoub Mamdouh
Eng. Hind Abdel Warith
Eng. Mariam Mohamed
Eng. Menntallah Saleh
Eng. Minar Abbas El-Aasser
Eng. Youmna Atef
Eng. Nada Wael El-Mansy
Eng. Noha Demerdash Saber

Eng. Ahmed Hazem Hewedy
Eng. Alaa Farahat
Eng. Ahmed Amr Hamza
Eng. Abanoub Wasfy
Eng. Hana Medhat
Eng. Maha El-Feshawy
Eng. Mohamed Abdel Khalek
Eng. Monica Wasfy
Eng. Passant Ahmed Youssef
Eng. Menntualah Reyad Abdelazim

Contents

Seismic Performance Assessment of Commonly Used Structural Systems and Retrofitting Techniques Using Pushover Analysis

Alisar Baderkhan and Ibrahim Lotfy[✉]

Department of Civil and Material Engineering, German University in Cairo,
New Cairo City, Egypt
ibrahim.lotfy@guc.edu.eg

Abstract. Seismic performance is an integral factor in the design of low-to-mid-rise buildings. Thus the selection of an optimized structural system is an important element in the design and planning of smart cities. This paper aims to investigate the most commonly used structural systems in reinforced concrete low-to-mid-rise buildings in Egypt as well as the most commonly used seismic retrofitting techniques. The findings of this study can facilitate the selection of structural system of new RC buildings in a smart city context, or retrofitting of existing buildings. The study uses Pushover Analysis techniques for seismic assessment. It is a multi-step static non-linear analysis which involves pushing the structure laterally until failure. Pushover analysis is becoming the preferred tool for seismic performance and evaluation of frame buildings, and some modern seismic design codes. The main objective is to investigate the lateral behavior of two commonly used structural systems for short buildings (5 stories) and mid-rise buildings (9 stories) which are: Moment Resisting Frames, Shear Walls. Additionally, two commonly used seismic retrofitting techniques are also investigated: RC Structures Retrofitted with Steel Bracing and Buckling Restrained Steel Bracing. Three main performance indicators were evaluated for each case. The response modification factor which reflects the ductility and over-strength of the structure. The performance point which indicates the expected seismic performance level. And the inter-story drift ratios which can be used to evaluate the damage demand of the structure. The results indicate that for low-rise buildings both the moment resisting frame and the shear walls systems showed similar performance levels; however, for mid-rise buildings, the shear walls system exhibited better performance.

Keywords: Pushover analysis · Seismic performance · Moment resisting frames · Shear walls · Steel braced RC building · Seismic retrofitting · Smart cities

1 Introduction

Seismic performance is an integral factor in the design of low-to-mid-rise buildings. Thus the selection of an optimized structural system is an important element in the design and planning of smart cities. This paper aims to investigate the most commonly

I. El Dimeery et al. (Eds.): JIC Smart Cities 2019, SUCI, pp. 1–12, 2021.
https://doi.org/10.1007/978-3-030-64217-4_1

used structural systems in reinforced concrete low-to-mid-rise buildings in Egypt as well as the most commonly used seismic retrofitting techniques. The findings of this study can facilitate the selection of structural system of new RC buildings in a smart city context, or retrofitting of existing buildings.

Pushover analysis is a simple technique that has been used to estimate the seismic performance of structures. It is a static non-linear analysis with two types: force-controlled or displacement-controlled approaches. In force-controlled pushover analysis, the failure lateral load should be known and hence applied along the height of the structural building, however some numerical problems that affect the accuracy of results may occur since the target drifts may be associated with a very small positive or even a negative lateral stiffness. On the other hand, displacement-controlled pushover analysis implements a lateral load pattern which is a representation of the lateral earthquake forces along the height of the building. This load pattern is used to push the building laterally until a pre-defined displacement failure state is reached. The expected damage in the main structural members is represented using plastic hinges assumed at strategic locations along the lateral resisting members.

2   Past Research

Many studies were conducted to investigate lateral load effect and plastic hinge properties and it was concluded that inverted triangular lateral load and user defined hinges with FEMA-356 standards [1] would give accepted results and good accuracy [2–8]. Moreover evaluating the capacity spectrum method and displacement-based analysis was conducted by several studies which had shown good results [9–14]. Considering the four structural systems in this research, different previous studies were accomplished to evaluate the pushover analysis method, and it was concluded that it is recommended to consider the structural damage as a function of the lateral deformation only [15–24].

3   Research Objectives

The objective of this paper is to investigate the lateral behavior of four lateral load resisting systems commonly used in Egypt for short buildings (5 stories) and mid-rise buildings (9 stories). Two lateral load resisting systems for new construction which are concrete Moment Resisting Frames (MRF) and concrete Shear Walls (SW). The two other systems are for retrofitting of existing concrete buildings with either steel bracing (SB) or buckling restrained steel bracing (BRSB). The four systems were investigated using a finite element pushover analysis and three main performance indicators were studied; Response Modification Factor, Performance Point and Interstory Drift Ratios. The response modification factor reflects the ductility and over-strength of the structure, the performance point indicates the expected seismic performance level, and the interstory drift ratios can be used to estimate the ductility and damage demand of the structure.

The first two lateral load resisting systems represent the most common lateral loads resisting systems for new construction in Egypt. Retrofitting an existing concrete building with steel bracing is a common retrofitting scheme while using buckling restraint steel bracing is a relatively new, state-of-the-art, application that is still being researched and developed [24]. By using a nonlinear, multi-step static, pushover analysis, an accurate estimation of the three key performance indicators can be achieved.

The approach of this study is systemized as follow; first the description of the pushover analysis method and the different lateral systems under investigation is established. Second, the pushover process and methodology was validated with the similar studies from the current literature [15, 20, and 24]. Third, using the validated process and methodology, the buildings were designed according to the Egyptians code specifications such as period of vibration, seismic zone, properties of the material and moment of inertia of the structural members, and 3D finite element models were constructed to the pushover analysis. Fourth, the key performance indicators were evaluated; the response modification factor, the performance point and the interstory drift ratios. Finally, conclusions and comparisons were drawn using the outcomes of the analysis.

4 Finite Element Analysis

Eight, 3-dimensional, finite element models were constructed in order to assess the performance and draw comparisons between the different systems; MRF, SW, SB and BRSB. Four models considered five story-structures representing low-rise buildings while another four were nine story-structures as mid-rise building. All buildings were designed according to the Egyptian Code of Practice for concrete structures 203-2017 and SAP 2000 package was used to perform the analysis.

The following assumptions were used in the design;

– Third Seismic zone in Egypt
– Peak ground acceleration of 0.15 g
– Response Spectrum type 1 of ECP 201
– Soil class is between C to D
– Seismic Weigh is the dead load and 25% of the live load

Figure 1 and 2 present elevation views for the five and nine-story structures respectively. For all buildings, the plan was considered symmetric with dimension of 15×15 m; it is divided into four bays in each direction. Story height was fixed at three meters high.

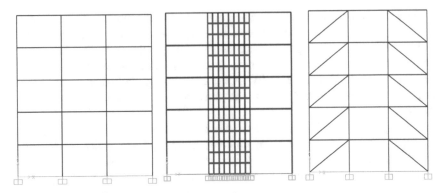

Fig. 1. Elevation views of the five-story structures for MRF (left), SW (middle) and SB & BRSB (right).

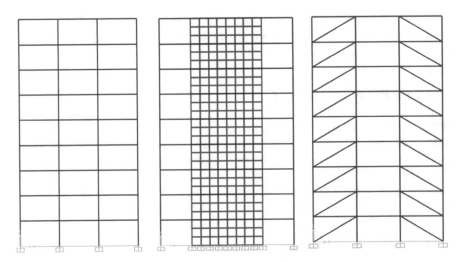

Fig. 2. Elevation views of the nine-story structures for MRF (left), SW (middle) and SB & BRSB (right).

5   Material Models, Non-linearity and Plasticity

A vital component of any non-linear analysis is the material definition and the plasticity considered. All concrete elements were defined using a non-linear concrete material model. Stirrups were assumed adequate to confine concrete with unconfined properties used where applicable. All steel elements were considered as elastic-perfectly plastic material with a clear yield stress in compression and tension. Buckling in compression was not considered in the material model but rather using hinges in specific prescribed location as explained in the following section.

In the MRF models, the concrete material model was assumed to be elastic with the plasticity defined in preselected hinges located at the beams and columns ends. The properties of the plastic hinges were considered as per the FEMA-356 standards [1]. Figure 3 presents a generalized plastic hinge backbone curve. Plastic hinge for beams were defined as moment only hinges while column hinges were defined as P-M; axial force and moment hinges. Yield moments and axial forces were considered accordingly using the reinforcement details and concrete dimensions obtained from the ECP 203 seismic design. Columns hinges were considered at different angle of interaction between the axial forces and moments.

In the SW models, a layered concrete fiber element was used to represent the shear wall element with the entire concrete and steel reinforcement material model assumed to be plastic. Finally, in the SB and BRSB models, plasticity was concentrated in hinges in the braces. Three hinges were used per brace; two at the ends and one the middle. For the SB model the steel braces were defined to yield in tension and buckle in compression; thus the hinges had an unsymmetrical backbone curve. However for the BRSB these unsymmetrical hinge backbones were reverted back to symmetrical as the braces will no longer buckle in compression.

An inverted triangle load pattern was assumed to perform a displacement-controlled pushover analysis which approximately represents the fundamental mode of vibration for the structures.

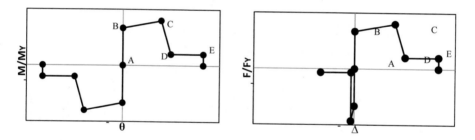

Fig. 3. Generalized plastic hinge backbone curves for bending (left) and axial (right).

6   Results

The pushover curves; base shear vs story drifts, were obtained for all models. To get a better understanding of the building behavior, data was normalized using the seismic weight of the structure and the overall height of the structure. The pushover curves can provide a practical and realistic measure of anticipated global performance of the structure in the event of an earthquake. Figures 4 and 5 present the generalized pushover curves for the five and nine-story models respectively.

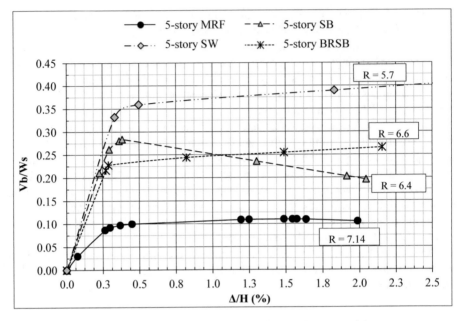

Fig. 4. Generalized pushover curve for the 5-story models.

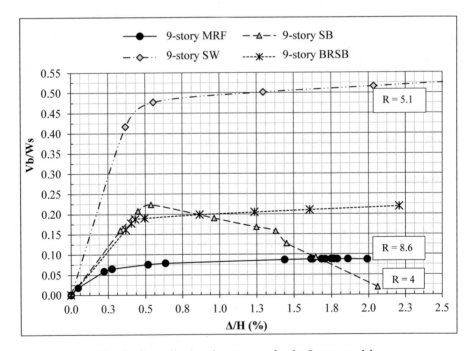

Fig. 5. Generalized pushover curve for the 9-story models.

Inspection of Fig. 4 and 5 reveals that all models behaved as expected with an almost linear initial stage followed by a clear yielding stage and a degraded stiffness post-yield. The SW system exhibited the largest initial stiffness followed by the bracing systems then the MRF system. It can be noticed that the post-yield stiffness for the SB system degraded faster than that of its BRSB counterpart which is attributed mainly to buckling. A larger difference in stiffness can be noticed between the SW system and the other system for the nine-story structure (Fig. 5) which is a result of a much larger concrete dimensions for the shear walls; the dimension increase was required by the ECP 203 to control the vibration period of the structure and get it within the range required by the code which is a conservative range.

The response modification factor was calculated for all the systems and is presented in Tables 1 and 2 along with the recommendation of the different design codes.

Table 1. Response modification factor "R" as calculated and as recommended by ECP 203.

System	Stories	Calculated response modification factor "R"	Response modification factor "R" recommended by ECP 203
MRF	5 (low-rise)	7.14 *	7 *
	9 (mid-rise)	8.6 *	7 *
SW	5 (low-rise)	5.7	5
	9 (mid-rise)	5.1	5
SB	5 (low-rise)	6.4	4.5
	9 (mid-rise)	4	4.5
BRSB	5 (low-rise)	6.6	4.5
	9 (mid-rise)	4.65	4.5

* Designed as frames with sufficient ductility as per ECP 203

Table 2. Comparison between the Response modification factor "R" calculated and recommended by the design codes.

System	Calculated response modification factor "R"	Response modification factor "R" recommended by ECP 203	Response modification factor "R" recommended by ASCE 7	Behavior factor "q" recommended by EuroCode 8
Ductility class	Sufficient ductility frames	Sufficient ductility frames	Special frames	Ductility class high
MRF	7.9	7	8	6.4
SW	5.4	5	6	4.8
SB	5.2	4.5	6	4
BRSB	5.6	4.5	6	4

Table 1 reveals that the specifications recommended by the ECP 203 are always more conservative than the actual calculated values for the response modification factor "R". The only exception was the nine-story SB system. This is likely due to the fact that the ECP 203 recommended values do not reflect the case studied in the paper which is concrete frame structure retrofitted with steel braces, as in ECP 203 the value of "R" is specified for steel structure retrofitted with steel braces. Table 2 presents a comparison of the different design codes with the calculated values for the response modification factor. It can be seen that the Eurocode 8 has the most stringent recommendation for the response modification factor; however these values can be increased in some cases using a ductility ratio.

Damage levels in the structure can be measured in a couple of ways. First a local damage indicator for each plastic hinge. This is achieved by inspecting the levels of plastic rotation in each individual hinge. The damage level is then classified according to the basic four damage levels which are: fully operational (FO), immediate occupancy (IO), life safety (LS) and collapse preventing (CP). The damage level in the structure can also be measure using the Interstory Drift Ratio (IDR). The IDR is the relative displacing between two consecutive floors divided by the story height. This ratio provides a direct indication of the level of straining actions expected to develop in the lateral elements during an earthquake event. These straining actions can be directly related to the expected damage demand in the lateral element of the story. IDR provides a practical and easy to understand measure of the anticipated damage state of the RC structures with three main damage states can be defined as: Repairable damage when IDR < 0.5%, Irreparable damage when IDR < 1.5%, and Severe damage when IDR < 2.5%.

Figure 6 and 7 present the plastic hinge progression for the MRF system throughout the damage states for the five and nine story buildings respectively.

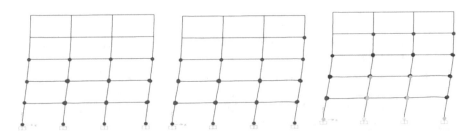

Fig. 6. Progression of the plastic hinge for the five-story MRF system; repairable damage state IDR < 0.5% (left), irreparable damage state IDR < 1.5% (middle), and severe damage state IDR < 2.5% (right)

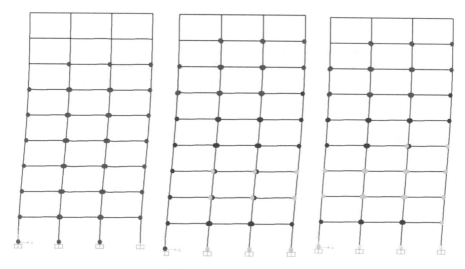

Fig. 7. Progression of the plastic hinge for the nine-story MRF system; repairable damage state IDR < 0.5% (left), irreparable damage state IDR < 1.5% (middle), and severe damage state IDR < 2.5% (right).

Figure 8 presents the "IDR" for all the systems evaluated along the height of the structure for the five and nine-story structures. The "IDR" the interstory drift ratio can be defined as the relative displacing between two consecutive floors divide by the story

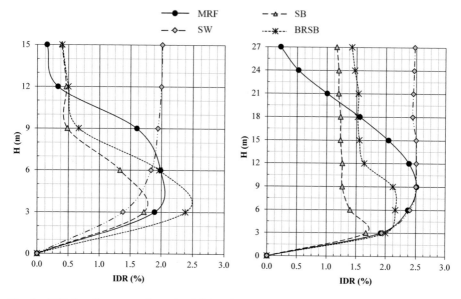

Fig. 8. IDR for all systems along the height of the structure for the five-story structures (left) and the nine-story structures (right).

height. It can be noticed that the maximum ductility demand occurs in the lower stories for both the five and nine-story structures.

To determine the performance point, the capacity method was used. It is acquired by intersecting the pushover curve with the design response spectrum. Figure 9 presents graphically how the performance point was obtained for the nine-story structures. Table 3 summarizes all the performance points for all the systems investigated.

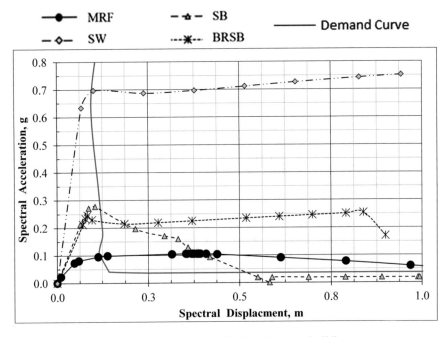

Fig. 9. Performance point for the nine-story buildings.

Table 3. Performance point for all the systems investigated.

	MRF		Braced		BRB		Shear wall	
	Base Shear, kN	Roof displacement, mm	Base shear, kN	Roof displacement, mm	Base shear, kN	Roof displacement, mm	Base shear, kN	Roof displacement, mm
5-story	1190	119	3190	85	2780	100	4640	90
	Life safety		Collapse prevention		Life safety		Life safety	
9-story	1620	133	4820	156	4180	151	12,430	144
	Life safety		Collapse prevention		Life safety		Immediate occupancy	

7 Summary and Conclusion

Seismic performance is an integral factor in the design of low-to-mid-rise buildings. Thus the selection of an optimized structural system is an important element in the design and planning of smart cities. In this paper, the seismic performance of four systems was investigated using pushover analysis. Each system was evaluated for a collapse prevention criterion at a top roof displacement. The SW system and the BRSB retrofitted system were successful in preventing collapse while sustaining significant permanent damage. However, the MRF and the SB retrofitted system were not successful in preventing collapse.

All the low-rise systems exhibited better seismic performance that the mid-rise systems with the exception of the shear wall system. This was due to a significant change in the shear wall geometry from low to mid-rise structure to accommodate building layout and maintain proper lateral stiffness to maintain the fundamental period in code recommended range.

The recommended values for the response modification factor "R" specified in the ECP 203 were conservative expect for one case; mid-rise building retrofitted with steel bracing. This was probably due to the fact that the code recommended values are specified for steel structures with bracing and not concrete structures retrofitted with steel bracing. A significant change in seismic weight and lateral behavior is present between both systems. Moreover, Eurocode 8 provides the most conservative recommendation for the response modification factor out of the ECP 203, ASCE7 and the findings of this study.

As Evident from the IDR, the maximum damage demand in all system occurred in the lower stories. This was confirmed by the progression of the plastic hinges and the location of collapsed hinges. This study confirms further the dangers of a common practice in Egypt with having softer lower stories for commercial use.

For low-rise buildings both the MRF and the shear wall systems showed better performance than their counterparts. However, for mid-rise buildings the shear wall system edged all others exhibiting better seismic performance. This was evaluated based on the damage state of the structure at its "performance point" which is the structure expected behavior as per the ECP 203 design response spectrum.

References

1. FEMA-356: Pre-standard and commentary for the seismic rehabilitation of buildings. Federal Emergency Management Agency, Washington DC (2000)
2. Jingjiang, S., Ono, T., Yangang, Z., et al.: Lateral load pattern in pushover analysis. Earthq. Eng. Eng. Vib. 2, 99 (2003)
3. Inel, M., Ozmen, H.: Effects of plastic hinge prosperities in non linear analysis of reinforced concrete buildings. Eng. Struct. 28(11), 1494–1502 (2006)
4. Alashker, Y., Nazar, S., Ismaiel, M.: Effects of building configuration on seismic performance of RC buildings by pushover analysis. Open J. Civ. Eng. 5, 203–213 (2015)
5. Kunadharaju, R., Cinitha, A., Iyer, N.: Seismic performance evaluation of existing RC buildings designed as per past codes of practice. Sadhana 37(Part 2), 281–297 (2012)

6. Hsieh, S., Deierlein, G.: Nonlinear analysis of three-dimensional steel frames with semi-rigid connections. Comput. Struct. **41**(5), 995–1009 (1991)
7. Albanesi, T., Biondi, S., Petrangeli, M.: Pushover analysis an energy based approach. In: 12th European Conference on Earthquake Engineering, p. 605. Elsevier, London (2002)
8. Lin, E., Pankaj, P.: Nonlinear static and dynamic analyses - the influence of material modeling in reinforced concrete frame structures. In: 13th World Conference on Earthquake Engineering, Vancouver BC, Canada, p. 430 (2004)
9. Mahaney, J., Paret, T., Kehoe, B., et al.: The capacity spectrum method for evaluating structural response during the Loma Prieta earthquake. In: U.S. Central United States Earthquake Consortium (CUSEC). Mitigation and Damage to the Built Environment, Memphis, Tennessee, US, pp. 501–510 (1993)
10. Kunnath, S., Valles-Mattox, R., Reinhorn, A.: Evaluation of seismic damageability of a typical RC in Midwest United States. In: 11th World Conference in Earthquake Engineering (11 WCEE), Acapulco, Mexico (1996)
11. Mwafy, A., Elnashai, A.: Static pushover versus dynamic collapse analysis of RC buildings. Eng. Struct. **23**(5), 407–424 (2001)
12. Chandler, A., Mendis, P.: Performance of reinforced concrete frames using force and displacement based seismic assessment methods. Eng. Struct. **22**(4), 352–363 (2000)
13. Elnashai, A.: Advanced inelastic static pushover analysis for earthquake applications. Struct. Eng. Mech. **12**(1), 51–70 (2001)
14. Hernandez-Montes, E., Kwon, O., Aschheim, A.: An energy-based formulation for first-and multiple-mode nonlinear static pushover analyses. J. Earthq. Eng. **8**(1), 69–88 (2004)
15. Fouad, A., EL-Esnawy, N.: Analysis of seismic capacity of RC frames with reduced stiffness in elevation. Int. J. Adv. Mech. Civ. Eng. **3**, 106–110 (2016)
16. Maske, A., Maske, N., Shiras, S.: Pushover analysis of reinforced concrete frame structures: a case study. Int. J. Adv. Technol. Eng. Sci. **2**(10), 118–128 (2014)
17. Vargas, Y., Beneit, L., Barbat, A., Hurtado, J.: Capacity, fragility and damage in reinforced concrete buildings: a probabilistic approach. Bull. Earthq. Eng. **11**(6), 2007–2032 (2013)
18. Chopra, A., Goel, R.: A modal pushover analysis procedure for estimating seismic demands for buildings. Earthq. Eng. Struct. Dyn. **31**, 561–582 (2002)
19. Rana, R., Jin, L., Zekioglu, A.: Pushover analysis of a 19 story concrete shear wall building. In: 13th World Conference on Earthquake Engineering, Vancouver BC, Canada, p. 133 (2004)
20. Hosseini, S., Bakar, S., Bagherinejad, K., Hosseinpour, E.: Pushover analysis of reinforced concrete building with vertical shear link steel braces. Malays. J. Civ. Eng. **27**(2), 169–179 (2015)
21. Veismoradi, S., Amiri, G., Darvishan, E.: Probabilistic seismic assessment of buckling restrained braces and yielding brace systems. Int. J. Steel Struct. **16**(3), 831–843 (2016)
22. Shin, J., Lee, K., Jeong, S., et al.: Experimental and analytical studies on buckling-restrained knee bracing systems with channel sections. Int. J. Steel Struct. **12**, 93–106 (2012)
23. Güneyisi, E., Muhyaddin, G.: Comparative response assessment of different frames with diagonal bracings under lateral loading. Arab. J. Sci. Eng. **39**, 3545–3558 (2014)
24. Ghowsi, A., Sahoo, D.: Seismic performance of buckling-restrained braced frames with varying beam-column connections. Int. J. Steel Struct. **13**, 607–621 (2013)

Modeling of a Reinforced Concrete Column Under Cyclic Shear Loads by a Plasticity-Damage Microplane Formulation

Mohamed Ali[1(✉)], Imadeddin Zreid[2], and Michael Kaliske[2]

[1] German University in Cairo, Cairo, Egypt
mohamedkhairy.96@gmail.com
[2] Institute for Stuctural Analysis, Technische Universität Dresden,
01062 Dresden, Germany
michael.kaliske@tu-dresden.de

Abstract. Modeling of nonlinear behavior and failure of concrete is a crucial aspect for reliable simulation of reinforced concrete structures. A new microplane formulation provide a way to model the behavior of concrete by defining constitutive model between stress and strain vectors on arbitrary oriented planes. A gradient enhanced element with coupled damage-plasticity microplane material is used to model the concrete behavior. The steel reinforcement will be simulated using one dimensional elements and a plastic material model. The incorporation of nonlinearities both in the concrete and in the steel reinforcement, as well as the use of a nonlocal formulation enable simulation of concrete under a wide range of loading situations as cyclic loading. Studies on model parameters and their identification will be performed to have an initial prediction about the effect of changing model parameters. A parametric study of a reinforced concrete column subjected to axial and cyclic shear loads is performed. The proposed microplane model shows a great capability of modeling cyclic loading. The damage split formulation can retrive the lost stiffness when changing from tension to compression stresses.

Keywords: Microplane formulation · Implicit gradient enhancement · Coupled plasticity-damage · Cyclic loading · Reinforced concrete

1 Introduction

Modeling of concrete behavior is considered one of the main civil engineering concerns. Due to its heterogeneous nature, concrete reacts differently under wide range of applied stresses and cyclic loading. The inelastic behavior of concrete in tension is a strong strain softening after the elastic limit and in compression is a small hardening after reaching the elastic limit then strain softening. Different behavior is encountered if concrete is under high or low confinement, as well as if concrete is reinforced with steel rebars, fibers, or textile. The Finite Element Method provides a good tool to simulate nonlinear inelastic behavior of concrete, in addition to failure modes and crack patterns. However, certain problems are encountered such as the difficulty to formulate general constitutive law for concrete that describes its distinct behavior, numerical complexity

© The Author(s), under exclusive license to Springer Nature Switzerland AG 2021
I. El Dimeery et al. (Eds.): JIC Smart Cities 2019, SUCI, pp. 13–21, 2021.
https://doi.org/10.1007/978-3-030-64217-4_2

of implementing this general law in FEM, and required time to solve nonlinear finite element analysis. The coupled plasticity-damage microplane model with implicit gradient enhancement [1] provides an alternative tool to model concrete nonlinear behavior. The coupled model can simulate concrete under cyclic loading. Smooth three-surface microplane yield function covers all possible range of applied stress that can affect concrete elastically and plastically. The damage is split into tension and compression in order to enable transition of stresses between tension and compression during cyclic loading. To this point, results are biased, as they are localized in one element regardless mesh size and become mesh dependent, which will cause divergence problems with complex geometries and loading conditions. The implicit gradient enhancement regularize the model by averaging equivalent strain over a volume surrounding each material points. The resulting nonlocal equivalent strain is considered as an extra degree of freedom. Therefore, the model is mesh independent, and enables performing various simulation without numerical instabilities.

2 Coupled Plasticity-Damage Microplane Model for Concrete

2.1 Microplane Theory

Constitutive laws of materials are established normally using strain and stress tensors, which means dealing with second or fourth order tensors. This is difficult with materials such as concrete to formulate a law that describe nonlinear inelastic behavior and microcracks that is already exist before loading between aggregates and mortar. Another alternative is describing constitutive law in terms of stress and strain vectors, not tensors, arbitrary oriented on various planes, thus this method is called microplane [2]. This method can describe the anisotropy of materials in simpler manner. Additionally, it can capture the oriented nature of the damage. The anisotropic behavior of materials, as concrete, and rocks, happens beyond elastic limit, in which cracks and plastic deformations evolved in different directions. The microplane method depends on material microstructure, in which planes are oriented according to loading and constraint conditions [3].

2.2 General Formulation

Plasticity is a permanent deformation of materials due to loading. Material stiffness is not affected. Plasticity is stress based, thus equations are implicit, and solution is iterative. Damage is the loss of material stiffness caused by initiation and propagation of microcracks and it is strain based, thus solution is direct without iteration. Previous models use damage and plasticity separately, although they are interconnected, in which plasticity is started, then damage evolved after small threshold that varies between tension and compression. Additionally, concrete strain softening is affected by material plasticity and degradation of stiffness. Subsequently, this model will use single inelastic surface to model this behavior. Therefore, the proposed model can be summarized in the following stress-strain formula

$$\sigma = \frac{3}{4\pi} \int_{\Omega} \left(1 - d^{mic}\right) \left[K^{mic} V\left(\epsilon_V - \epsilon_v^{pl}\right) + 2G^{mic} \boldsymbol{Dev}^T \cdot \left(\epsilon_D - \epsilon_D^{pl}\right)\right] d\Omega, \qquad (1)$$

where d^{mic} is the total damage ϵ_V and ϵ_D are the volumteric and deviatoric strains. pl represents plasticity. Projection tensors $\boldsymbol{V}$ and $\boldsymbol{Dev}$ are defined using vector n normal to each microplane as

$$\boldsymbol{V} = \frac{1}{3}\boldsymbol{1}, \quad \boldsymbol{Dev} = \boldsymbol{n} \cdot \boldsymbol{I}^{dev} = \boldsymbol{n} \cdot \boldsymbol{I}^{sym} - \frac{1}{3}\boldsymbol{n} \cdot \boldsymbol{1} \otimes \boldsymbol{1}, \qquad (2)$$

where $\boldsymbol{1}$ and $\boldsymbol{I}$ are the second and the fourth order tensors respectively. K^{mic} and G^{mic} are microplane bulk and shear moduli. The numerical integration is performed over surface of a sphere. it is computed over 42 microplanes, which can be reduced to 21 planes due to symmetry

$$\frac{3}{4\pi} \int_{\Omega} (\cdot) d\Omega = \sum_{mic=1}^{21} (\cdot) w^{mic}. \qquad (3)$$

The Drucker-Prager function with caps for tension and compression regions creates non smooth yield surfaces, which induce numerical instabilities at the corner regions. Thus, smooth cap models are introduced to avoid these problems and the resulting yield function in Fig. 1 according to [1] could be written as

$$f^{mic}\left(\sigma_D^e, \sigma_V^e, k\right) = \frac{3}{2}\sigma_D^e \cdot \sigma_D^e - f_1^2\left(\sigma_V^e, k\right) f_c\left(\sigma_V^e, k\right) f_t\left(\sigma_V^e, k\right), \qquad (4)$$

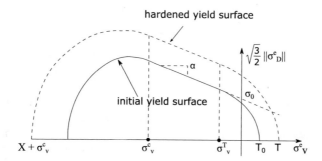

Fig. 1. Smooth three-surface microplane cap yield function

this function is evaluated within undamaged space. While, function f_1 is Drucker-prager yield function with hardening [4]

$$f_1 = \sigma_0 - \alpha\sigma_V^e + f_h(k), \qquad (5)$$

additionally, f_c and f_t are compression and tension caps. The modified cap function is produced by multiplying f_c with f_1^2 and f_t with f_1^2 for compression and tension respecively. Thus the total yield surface has a continuous derivative. The compression cap is given as

$$f_c = 1 - H_c \left(\sigma_V^C - \sigma_V^e \right) \frac{\left(\sigma_V^e - \sigma_V^C \right)^2}{X^2},$$ (6)

on the other hand the tension cap is given as

$$f_t = 1 - H_t \left(\sigma_V^e - \sigma_V^T \right) \frac{\left(\sigma_V^e - \sigma_V^T \right)^2}{\left(T - \sigma_V^T \right)}.$$ (7)

Damage of concrete is different in tension and compression. Strain softening in tension starts immediately after elastic limit, whereas in compression it is stiffer. Material stiffness is retrieved due to crack closure, when tension stresses is changed into compression stresses. Nevertheless, damage caused by compression remains in all cases. Therefore, damage is split into tension and compression damage as in [5]

$$1 - d^{mic} = \left(1 - d_c^{mic} \right) \left(1 - r_w d_t^{mic} \right),$$ (8)

where d_c^{mic} is the compression damage, d_t^{mic} is the tension damage. They have range from 0 to 1, in which 0 is undamaged and 1 is fully damaged. The damage laws are

$$d_t^{mic} = 1 - exp \left(-\beta_t \gamma_t^{mic} \right),$$ (9)

$$d_t^{mic} = 1 - exp \left(-\beta_t \gamma_t^{mic} \right).$$ (10)

Mesh dependency is a problem in FEM, especially in strain softening simulations. This may cause convergence problems and numerical instabilities, in which specific element size is required to converge. Results may be biased, as field variable is framed in specific elements without taking into consideration element size with respect to effect of applied loads on the whole geometry. This mesh dependent solution is a famous problem in the local plasticity models. The implicit gradient enhancement is a class of nonlocal methods that solves the mesh dependency of the plasticity-damage models. Nonlocal average of a local variable is computed by considering the nonlocal value as an extra degree of freedom [6]. The modified Helmholtz-type equation is introduced to describe the nonlocal field

$$\bar{\eta} - c\nabla^2 \bar{\eta}_m = \eta_m$$ (11)

A homogenized value for the equivalent strains η_t^{mic} and η_c^{mic} is considered as follows

$$\eta_m = \begin{bmatrix} \eta_{mt} \\ \eta_{mc} \end{bmatrix} = \begin{bmatrix} \frac{1}{4\pi} \int_\Omega \eta_t^{mic} d\Omega \\ \frac{1}{4\pi} \int_\Omega \eta_c^{mic} d\Omega \end{bmatrix}, \tag{12}$$

so that two extra degrees of freedom are required as mentioned before. For full regularization of plastic -damage model over-nonlocal formulation is developed [7], where the over-nonlocal variable $\widehat{\eta}^{mic}$ is given as follows

$$\widehat{\eta}_t^{mic} = m\bar{\eta}_{mt} + (1-m)\eta_t^{mic}, \tag{13}$$

$$\widehat{\eta}_c^{mic} = m\bar{\eta}_{mc} + (1-m)\eta_c^{mic}, \tag{14}$$

where material parameter m should be greater than 1 to have regularization.

2.3 Model Parameters

The model parameters can be identified by data from several experimental tests on the used material or by fitting results of the actual experiments using trial and error method. The elastic parameters are represented in modulus of elasticity E and Poisson's ratio v. The plastic parameters are uniaxial compressive strength f_{uc}, biaxial compressive strength f_{bc}, and uniaxial tensile strength f_{ut} of concrete. They specify the yield surface of the concrete. The initial yield stress σ_0 and the friction coefficient α are calculated as follows

$$\alpha = \frac{\sqrt{3}(f_{bc} - f_{uc})}{2f_{bc} - fuc}, \tag{15}$$

$$\sigma_0 = \left(\frac{1}{\sqrt{3}} - \frac{\alpha}{3} \right) f_{uc}, \tag{16}$$

the compression cap parameters σ_V^c, and R are determined using concrete triaxial test. They can be estimated as minimum value, if there is no triaxial test data available, as follows

$$\sigma_V^c = -\frac{2}{3} f_{bc}, \tag{17}$$

$$R = \frac{\sqrt{3}X_0}{f_{bc}}, \tag{18}$$

while the tension cap parameters are calculated in the model using the following empirical equations

$$\sigma_V^T = -\frac{f_{uc}}{3}, \tag{19}$$

$$T_0 = \frac{f_{ut}}{3}. \tag{20}$$

Cyclic tests are required identify damage parameters. Damage and hardening parameters are related, as they control softening and unloading slope of the material. D hardening material constant, β_c compression damage constant, and γ_{c0} compression damage threshold are estimated first by fitting uniaxial cyclic compression test. Then, R_t tension cap hardening constant, and β_t tension damage constant are estimated by fitting uniaxial cyclic tension test. The suggested values $R_t = 1$ and $\beta_c \approx 0.67\beta_t$ can be used as initial values if no uniaxial cyclic tension test data is available. Tension damage threshold γ_{t0} is kept equal 0, as softening in tension starts immediately after the elastic limit. In [1], a single element simulation is executed to identify the damage parameters using this approach.

Damage and nonlocal parameters are considered together as a material constant. The over nonlocal averaging parameter m could be considered as numerical parameter, in which it should have a value greater than 1 to regularize the solution [8]. Other parameter is c, which is the nonlocal interaction range parameter. There are many approaches to identify the gradient parameter c. One of them is using the results of homogeneous and nonhomogeneous tensile concrete test [9]. Damage parameters can be identified from homogeneous test as cracks are distributed and nonlocal parameter c can be identified from nonhomogeneous test.

3   Cyclic Shear Test of a Reinforced Concrete Column

A series of squat bridge columns, designed before 1971, are tested to failure and new design approaches are developed due to the need of earthquake resistant structures [10]. Test unit (R1A) is chosen to be simulated from several as built column model. The column was built on scale 1:2.5. it has rectangular footing that is firmed to stiff ground and top load stub to give boundary conditions as a real bridge column. Test setup was planned so that column is subjected to axial load and cyclic shear loads under reversed curvature. All loads are applied by hydraulic arm that is fixed to the load stub. Forces are transmitted from loading arm to column by four pins in the load stub. Axial force is applied using high strength rod against the ground and is transferred to column by a cross beam mounted on top of the load stub. Rigid body rotation is reduced using actuator to balance the applied loads. More details in Fig. 2. Concrete is modeled by the plasticity-damage model with three smooth caps yield function. Reinforcement is modeled by bi -linear isotropic hardening Von Mises plasticity model. The element used for concrete is 3D 8 nodes hexahedral element (CPT215), while for rebars is 3D discrete reinforcement (REINF264). Perfect bond is assumed between concrete and rebars. In Fig. 3, the total number of nodes, and elements are 8099, and 7557 respectively. The material parameters of concrete and steel are provided in Table 1. The loads applied are gravity, vertical (Pv = 513KN), horizontal loads.

In the Fig. 4, horizontal load-displacement diagram of the first three cycles of the experiment is plotted. After the first load cycle damage is observed in concrete due to tension. The unloading part of each cycle is quite predicted. Fig. 5, three small cycles and then monotonic loading till failure is applied. This is the same applied load by [11]. Horizontal load-displacement diagram is plotted. The three cycles will initiate damage. Then, monotonic loading is applied till failure.

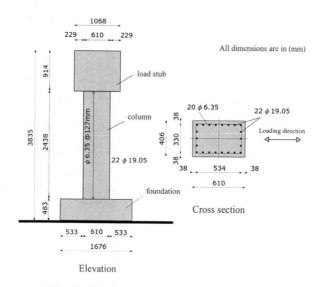

Fig. 2. Reinforced squat bridge column (R1A)

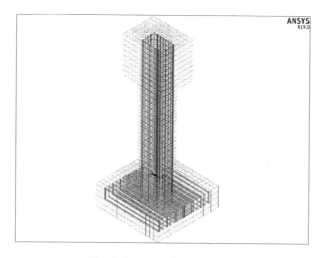

Fig. 3. FEM model of column

Table 1. Material parameters for the reinforced squat bridge column

Concrete					
Parameter	Value	Parameter	Value	Parameter	Value
$E(MPa)$	24132	$R(-)$	2	$\gamma_{co}(-)$	$5*10^{-5}$
$\nu(-)$	0.2	$D(MPa)$	$70*10^3$	$\beta_c(-)$	$4*10^3$
$\rho(kg/m^2)$	2400	$R_t(-)$	1	$c(mm^2)$	2500
$f_{uc}(MPa)$	37.92	$\gamma_{to}(-)$	0	$m(-)$	2.5
$\sigma_v^c(Mpa)$	−30	$\beta_t(-)$	$7*10^3$		

Steel					
Parameter	Value	Parameter	Value	Parameter	Value
$E(MPa)$	199948	$\rho(kg/m^2)$	7800	$f_{y\,stirr}(MPa)$	358.53
$\nu(-)$	0.2	$f_{y\,longi}(MPa)$	324.03		

In the two performed simulation, total damage is expected to be in upper and lower part of the column that is connected monolithically to load application head and foundation respectively. In Fig. 6, damage during different load steps is illustrated.

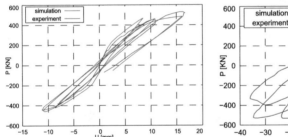

Fig. 4. Cyclic behavior of bridge column   **Fig. 5.** Envelope curve of bridge column

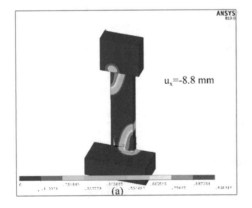
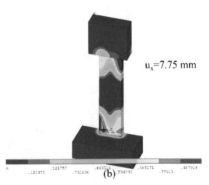

Fig. 6. Total damage due to cyclic loading applied to column

4 Conclusion

The gradient enhanced coupled plasticity-damage microplane model provides a robust approach to model concrete behavior. In this model, Drucker-Prager yield function with compression and tension caps form smooth three-surface yield function which cover all possible stress triaxialities. The damage model is split into tension and compression, as the initiation of damage is different in tension and compression in materials such as concrete and to account for the stiffness recovery during transition from tension to compression, due to crack closure. Implicit gradient enhancement is used to eliminate the mesh sensitivity caused by highly strain softening material as concrete. In this paper, reinforced squat bridge column is modeled. The applied cyclic load and the failure envelope is compared with the experimental results. The model parameters are identified, and the model shows a great capability to simulate these types of loading.

References

1. Zreid, I., Kaliske, M.: A gradient enhanced plasticity-damage microplane model for concrete. Comput. Mech. **62**, 1239–1257 (2018)
2. Bažant, Z.P., Oh, B.H.: Microplane model for progressive fracture of concrete and rock. J. Eng. Mech. **111**, 559–582 (1985)
3. Frigerio, A.: The microplane model for concrete in COMSOL. In: COMSOL Conference Europe 2011, Stuttgart (2011)
4. Zreid, I., Kaliske, M.: An implicit gradient formulation for microplane Drucker-Prager plasticity. Int. J. Plast **83**, 252–272 (2016)
5. Lee, J., Fenves, G.L.: Plastic-damage model for cyclic loading of concrete structures. J. Eng. Mech. **124**, 892–900 (1998)
6. Zreid, I., Kaliske, M.: Regularization of microplane damage models using an implicit gradient enhancement. Int. J. Solids Struct. **51**, 3480–3489 (2014)
7. Di Luzio, G.: A symmetric over-nonlocal microplane model M4 for fracture in concrete. Int. J. Solids Struct. **44**, 4418–4441 (2007)
8. Grassl, P., Jirásek, M.: Plastic model with non-local damage applied to concrete. Int. J. Numer. Anal. Meth. Geomech. **30**, 71–90 (2006)
9. Bazant, Z.P., Pijaudier-Cabot, G.: Measurement of characteristic length of nonlocal continuum. J. Eng. Mech. **115**, 755–767 (1989)
10. Priestley, M.J., Seible, F.: Steel jacket retrofitting of reinforced concrete bridge columns for enhanced shear strength-part 1: theoretical considerations and test design. ACI Struct. J. **91**, 394–405 (1994)
11. Hofstetter, B., Valentini, G.: Review and enhancement of 3D concrete models for large-scale numerical simulations of concrete structures. Int. J. Numer. Anal. Methods Geomech. **37**, 222–246 (2011)

Comparison Between BS EN 1998-1:2004, ECP 201-2011 and ASCE/SEI 7-16 for Code Requirements of Sky-Bridges Spanning Between Towers

Adel H. Elsaid[✉], Hatem AlShaikh, Yehia El-Ezaby,
and Charles Malek

Dar Al-Handasah, Cairo, Egypt
Adel.elsaid@dar.com

Abstract. The increasing urbanization coupled with the need to build smart and sustainable cities have led to the extensive construction of tall buildings, especially in highly populated countries. However, the wide use of tall buildings led to increasing the population of the cities at the ground plane which is the sole physical plane of connection. Introducing skybridges to link towers provides more levels for transportation. In addition, skybridges can also be used as escape routes in case of an emergency in one tower.

The connection between skybridges and towers may either be roller, pinned or fixed connections. In this study, a steel skybridge connecting 160 m tall reinforced concrete (RC) twin towers using pinned-roller connection will be investigated. The lateral load resisting system of the twin towers is RC cores. The seismic loads will be calculated using the Euro-Code 8 (BS EN 1998-1:2004), Egyptian Code of Practice (ECP 201-2011) and the American Society of Civil Engineers (ASCE/SEI 7-16). A comparison between the bridge seat requirements using the three codes will be investigated. Moreover, the outcomes of converting the connection between the twin towers and the steel bridge to pinned-pinned connection will be presented.

Keywords: Skybridges · Tall buildings · Seismic · Reinforced concrete

1 Introduction

Although the past decades have witnessed the construction of super tall buildings, the linked towers are still a rarity in design. Linking the towers using skybridges leads to easier transportation between towers especially if they are under the same owner-ship. Moreover, skybridges can bring added value to the property since they can be used as observation decks, recreation spaces and retails. The first modern skybridge connecting two towers at a height above ground is the National Congress Complex of Brasilia in Brazil (Taraldsen 2017).

The connection between towers and skybridges can be roller, pinned or rigid. Several skybridges were constructed using the roller connections including the Sky Habitat in Singapore, Nina Towers in Hong Kong, China and The National Congress

Complex of Brasilia in Brazil (McCall 2013). The Highlight Towers in Munich, Germany have an innovative structural design where two skybridges (at the 10th and 20th stories) can be disconnected and attached to the towers wherever needed (Emporis 2012). The famous skybridge linking the Petronas Towers in Kuala Lumpur, Malaysia, is considered a roller-connected skybridge with intermediate rotational pin support. The Linked Hybrid in Beijing, China is eight towers 22-stoery complex connected with skybridges. The skybridges are comprised of steel trusses supported on special "friction pendulum" seismic isolators (Nordenson 2009).

The Pinnacle @ Duxton is a large public housing project in Singapore that includes six skybridges, where of them connect the towers using pinned-pinned connections while the other three are connected using pinned-roller connections. On the other hand, several projects used rigid connections between the towers and the linking bridges. The Shanghai International Design Center in China, Suzhou "Gate to the East" building and China Central Television New Headquarters in Beijing are examples of towers connected by skybridges using rigid connections.

It can be seen that the construction of towers with skybridges is evolving; however, most of the international codes do not provide specific limitations and requirements for such critical construction. According to Luong, et al. (2012), the design of link structure shall include: (1) wind tunnel testing, (2) designing for vertical seismic, (3) ensuring the design has robustness and resilience to withstand rare seismic events, (4) comfort requirements under vibration and wind acceleration and (5) guaranteeing diaphragm action in critical floors/slabs. According to the Chinese code, buildings with high irregularity in plan and elevation, when rigidly connected by a skybridge, such as the Shanghai International Design Center (SIDC), require detailed study that usually includes refined structural analysis, scaled structural model test and large size member or joint test. For this particular irregular structure, the peer review committee recommended performing a shake table model test (Lu et al. 2009).

The primary objective of this study is to compare the pinned-roller bridge seat requirements using three different codes namely the Eurocode 8 (BS EN 1998-1:2004), the Egyptian Code of Practice (ECP 201, 2011) and the American Society of Civil Engineers (ASCE/SEI 7-16). Two 160 m high reinforced concrete (RC) twin towers and a steel skybridge will be used for this investigation. The outcomes of converting the connection between the twin towers and the steel bridge to pinned-pinned connection will be presented.

2 Description of the Building Under Study

This study is performed on a 160 m high RC twin towers connected by a structural steel skybridge at the 29th floor. The lateral load resisting system of each tower comprises central RC cores. The floor system is post-tensioned (PT) slab supported on the central cores, RC columns and peripheral beams as shown in Fig. 1. RC outriggers are introduced in the technical floor (at the mid-height of the tower) to reduce the tower's drift due to lateral loads. The skybridge consists of two structural steel trusses that support one floor with concrete on metal deck as shown in Fig. 2. The skybridge links the twin towers at the 29th floor by pinned-roller connection at the 29th floor.

Further, the twin towers and the skybridge are also studied for the case of having pinned-pinned connections at the 29th floor.

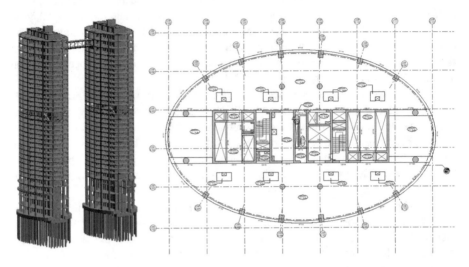

Fig. 1. 3D view and typical floor structural layout for the building under study

Fig. 2. 3D view of the skybridge linking the twin towers

3 Numerical Analysis

3.1 Finite Element Modeling

Two Finite element (FE) models were created using ETABS 17. The first model was created for a single tower and the reactions from the skybridge were assigned as loads on the supporting RC edge beam. This model simulated the skybridge connected to the towers by pinned-roller connections. The second model was created for the two towers and the linking skybridge which simulated the skybridge connected to the towers by pinned-pinned connections. For the two models, the slabs, core walls and coupling beams were modeled as shell elements, while the columns, beams and diagonal outrigger members were modeled as frame elements. Since the towers are supported on rigid rafts on piles, fixed base supports were assumed. Moreover, to simulate the pinned connection between the bridge beam and the towers, the flexural stiffness of the steel beam was released at the end connected to the tower. The 3D views of the created FE models are shown in Fig. 3. It should be mentioned that the structural design of the towers was performed using the Egyptian Code of Practice (ECP 203-2018).

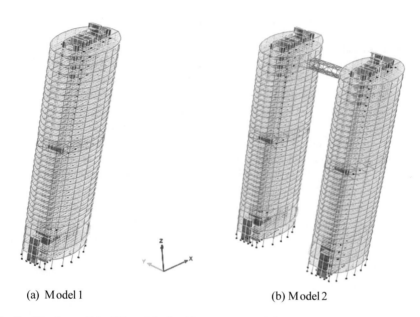

(a) Model 1   (b) Model 2

Fig. 3. 3D views of the FE models for (a) one tower and (b) two towers with the skybridge

3.2   Eigenvalue Analysis

An Eigenvalue analysis was performed for the two models to extract the dynamic characteristics of the structure in the two cases. The first three natural vibration periods for the two models are listed in Table 1.

Table 1. First three natural vibration periods (seconds) for the two models.

Vibration mode	Model 1		Model 2	
	Natural period (Sec)	Remark	Natural period (Sec)	Remark
1	5.855	Translation (Y-direction)	5.832	Translation (Y-direction)
2	4.010	Rotation	3.999	Rotation
3	3.951	Translation (X-direction)	3.968	Translation (X-direction)

As shown in Table 1, the first vibration mode is translational in Y-direction which is the direction with lower stiffness. It can also be concluded that linking the twin towers by the skybridge using the pinned-pinned connection at just one level (29th floor) did not change the dynamic behavior of the structure. This could be attributed to the fact that the two towers are an exact mirror image and consequently having similar dynamic behavior. Further, the bridge did not enhance the lateral stiffness of the two towers since the connection of the skybridge to the towers is pinned-pinned at one level which is less rigid than the fixed or pinned-pinned connections at two levels.

3.3   Lateral Drift and Skybridge Seat Calculations

In order to calculate the width of the skybridge seat (in case of pinned-roller connection), lateral seismic displacements were calculated using the ECP 201-2011. The design ground acceleration (a_g) is 0.15 g for Cairo, Egypt and the ground type is B as recommended by the soil investigation report. In order to evaluate the displacement values calculated using ECP 201-2011 when compared to international codes, seismic displacements were also calculated using BS EN 1998-1 and ASCE/SEI 7-16. The factors used in the seismic calculations using BS EN 1998-1 are similar to those of the ECP 201-2011. As for the ASCE/SEI 7-16, in order to estimate the risk-targeted Maximum Considered Earthquake (MCE_R) spectral response acceleration parameters, the following two methods were assessed and the most conservative one was used in this comparison. According to El-Kholy et al. (2018), the Maximum Considered Earthquake (MCE) spectral response acceleration parameters of ASCE/SEI 7-05 namely S_s and S_1 for Cairo, Egypt, are 0.4 g and 0.095 g, respectively. To convert theses values to risk-targeted, S_s and S_1 were multiplied by 1.1 and 1.3, respectively (ASCE/SEI 7-16, clause 21.2) so that the MCE_R S_s and S_1 are 0.44 g and 0.124 g, respectively. The other approach was to use the response spectrum based on the Uniform Building Code (UBC 97) to calculate S_1 and S_s, where S_{DS} (short-period spectral response acceleration) in ASCE/SEI 7-16 is equivalent to 2.5 Ca in UBC 97 and S_{D1} (1 s spectral response acceleration) in ASCE/SEI 7-16 is equivalent to the value of spectral acceleration at 1 s in UBC 97. Based on this approach, S_s and S_1 are 0.562 g and 0.225 g, respectively. The second approach was more conservative and hence it was considered in this study. It should be highlighted that the seismic calculations for the EC8 and ECP 201-2011 are based on 10% probability of exceedance in 50 years (EN1998-1:2004, clause 2.1) while those for ASCE/SEI 7-16 are based on 1% probability of collapse in 50 years (ASCE/SEI 7-16, clause 21.2).

Table 2 shows the outcomes of the seismic calculations. T_{used} is the fundamental period used in the seismic force calculations based on the codes' limitations and δ_Y is the in-elastic displacement in Y-direction at the skybridge level.

Table 2. Seismic in-elastic deformations calculated for the three codes.

Code	T_{used} (sec)	δ_Y (m)	Seismic joint (m)	Skybridge seat width (m)
ECP 201-2011	2.746 (X-direction) 2.746 (Y-direction)	0.43	0.61	2.22
BS EN 1998-1:2004	3.951 (X-direction) 5.855 (Y-direction)	0.43	0.61	2.22
ASCE/SEI 7-16	3.291 (X-direction) 3.291 (Y-direction)	0.82	1.64	4.28

The inelastic displacement of the tower at the skybridge level was calculated using the design response spectrum of BS EN 1998-1:2004 (q_d δe) and ECP 201-2011 (0.7 R δe) that does not exceed the those calculated using the elastic response spectrum. It could be observed that the seismic displacement calculated using the two codes are identical. This could be attributed to the fact that the elastic response spectrum of the ECP 201-2011 is similar to that of the BS EN 1998-1:2004. For the ASCE/SEI 7-16, the inelastic displacement was calculated as per clause 12.8.6 (equation 12.8-15), where, the elastic deformation is multiplied by the deflection amplification factor (C_d) and divided by the importance factor (I_e).

ASCE/SEI 7-16 specified a clause for members spanning between structures (clause 12.12.4) where:

1- The inelastic deformation is further amplified after multiplying by $1.5R/C_d$, where R is the response modification coefficient,
2- Considering additional deflection caused by diaphragm rotation including the torsional amplification factor,
3- Considering diaphragm deformations,
4- Assuming the two structures are moving in opposite directions and
5- Using the absolute sum of the inelastic displacements.

These requirements are not mentioned neither in the ECP 201-2011 nor the BS EN 1998-1:2004 where the seismic joints shall not be less than the square root of the sum of squares (SRSS) of the maximum horizontal inelastic displacement of the two buildings (EN 1998-1, clause 4.4.2.7 and ECP 201-2011, clause 8.8.2.7). Since the seismic joint for elements spanning between structures is not specified in the ECP 201-2011 and EN 1998-1, the SRSS of the maximum horizontal inelastic displacement of the two buildings was considered. Therefore, the seismic joint between the skybridge and the tower, at the roller side, calculated based on the ASCE/SEI 7-16 is the most conservative of the three codes. To account for the out-of-phase behavior of the two towers, the width of the skybridge seat shall be the sum of double the seismic joint and the bearing pad width as shown in Fig. 4.

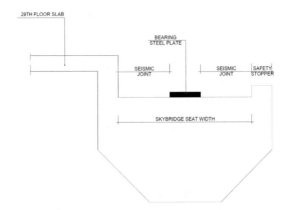

Fig. 4. Schematic sketch for the skybridge seat at the roller support

4 Results of Pinned-Pinned Skybridge Study

The finite analysis of Model 1 showed large inelastic displacements at the 29th floor which would lead to a large width required for the skybridge seat, especially if ASCE/SEI 7-16 approach is adopted. Therefore, the effect of connecting the skybridge to the towers using a pinned-pinned connection was studied. The eigenvalue analysis showed that in case the connection between the towers and skybridge is pinned-pinned at one floor, the skybridge is not highly contributing to the lateral stiffness of the towers.

In order to consider the worst design scenario, it was assumed that one tower would be finished and used prior to the other tower. Hence, the live, partitions and flooring loads were removed from one of the towers while keeping the other tower with all

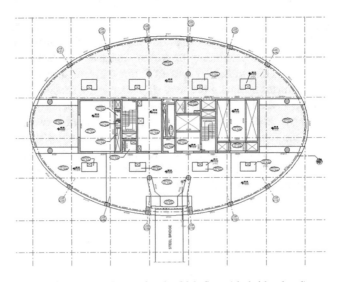

Fig. 5. Structural plan for the 29th floor (skybridge level)

loads assigned. This led to around 50% increase in the axial forces of the skybridge's lower chord. The axial load in the bridge's lower chord shall be transferred to the tower's cores through the 29th floor slab. This slab was designed to resist these high-tension forces by increasing its thickness and introducing adequate reinforcement and detailing. This is in addition to introducing two RC beams as shown in Fig. 5.

5 Conclusions

Based on the findings of the current study, the following outcomes may be drawn:-

1- The EN 1998-1 and ECP 201-2011 are not specifying certain requirements for the members spanning between structures similar to the ASCE/SEI 7-16.
2- For skybridges with pinned-roller connections, the width of the skybridge seat calculated using ASCE/SEI 7-16 is much more conservative than that calculated using BS EN 1998-1:2004 or ECP 201-2011.
3- For the ECP 201-2011 and EN 1998-1, the seat width of skybridges with pinned-roller connections shall assume the two structures are moving in opposite directions and using the absolute sum of the inelastic displacements for the two towers, not the SRSS.
4- The dynamic behavior of the twin towers with the skybridge connected using pinned-pinned connection at one level was very similar to the dynamic behavior of a single tower. This could be attributed to the fact that the two towers are an exact mirror image.
5- Studying the pinned-pinned skybridge by assuming that the two towers are dissimilar in loading (live load, partitions, and flooring) is essential to have conservative axial forces on the linking bridge.
6- The load path and magnitude of the axial forces transferred from the skybridge to the towers' slab shall be well defined and considered in the design.

References

American Society of Civil Engineers ASCE/SEI 7-16: Minimum Design Loads for Buildings and Other Structures, Reston, VA, ASCE (2016)
Egyptian Code of Practice ECP 201-2011: Calculating Loads and Forces in Structures, Cairo, Egypt (2013)
Egyptian Code of Practice ECP 203-2018: Design and Construction of Concrete Structures, Cairo, Egypt (2018)
El-Kholy, A.M., Sayed, H., Shaheen, A.A.: Comparison of Egyptian Code 2012 with Eurocode 8-2013, IBC 2015 and UBC 1997 for seismic analysis of residential shear-walls RC buildings in Egypt. Ain Shams Eng. J. **9**, 3425–4336 (2018)
Emporis: Highlight Munich Business Towers (2012). http://www.emporis.com/complex/highlight-munich-business-towers-munichgermany. Accessed 7 Aug 2019
ETABS, Version 17.0.1: ETABS Documentation. Computer and Structures Inc., Berkeley, CA, USA (2017)

Lu, X., Chen, L., Zho, Y., Huang, Z.: Shaking table model tests on a complex high-rise building with two towers of different height connected by trusses. Struct. Des. Tall Spec. Build. **18**, 765–788 (2009)

Luong, A., Kwok, M.: Finding structural solutions by connecting towers. Counc. Tall Build. Urban Habitat J. (CTBUH), (III), 26–31 (2012)

McCall, A.J.: Structural analysis and optimization of skyscrapers connected with skybridges and atria, Ph.D. thesis, Department of Civil and Environmental Engineering, Brigham Young University, USA, pp. 7–15 (2013)

Nordenson, G.a. A.: Linked Hybrid (2009). https://www.nordenson.com/projects/linked-hybrid. Accessed 7 Aug 2019

Taraldsen, T.: Linking skyscrapers – a conceptual study on skybridges' effects on the structural behavior of tall buildings subject to quasi-static wind loads. MSc thesis, Department of Structural Engineering, Faculty of Engineering Science and Technology, Norwegian University of Science and Technology, Trondheim, Norway (2017)

The European Standard EN 1998-1. Eurocode 8: Design of Structures for Earthquake Resistance - Part 1: General rules, seismic actions and rules for buildings, London, BSI (2004)

Analytical Fragility Curves for Non-Ductile Reinforced Concrete Buildings Retrofitted with Viscoelastic Dampers

Yasser S. Salem[(✉)], Guseppe Leminto, and Trung Tran

California State Polytechnic University, Pomona, CA, USA
ysalem@cpp.edu

Abstract. Non-Ductile reinforced concrete frames (NDRCF) structures are found to be vulnerable during major seismic events. Unfortunately, the majority of low-rise buildings around the world were built prior to 1976, when the current ductility requirements in modal building codes were not developed yet. Few of the smart cities have been looking for ways to become more resilient and resistant to earthquakes. In this study, the feasibility of retrofitting these structures, using viscoelastic dampers are evaluated. Analytical Fragility curves are developed to evaluate the benefit of using the viscoelastic dampers in reducing the risk of damages during major earthquakes. The methodology that was used to develop the curves is based on calculating the maximum story drift using nonlinear time history analysis for different ground motions with different intensities and frequencies. The fragility curves are developed from the adjustments of distribution function of the analysis results. The produced fragility curves highlight the advantages of using viscoelastic dampers in reducing the level of damage these structures can experience during earthquakes which can be an efficient strategy to increase the resilience of building in a smart city.

Keywords: Non-Ductile reinforced concrete frames · Visco-Elastic dampers · Analytical fragility curves

1 Introduction

Non-ductile concrete buildings were a prevalent construction type in highly seismic zones of the U.S. prior to enforcement of codes for ductile concrete in the mid-1970s. In California, no ductile concrete buildings were principally constructed between approximately 1890 (when elevators first enabled the construction of relatively tall buildings) and the mid 1970s (when improvements in building codes that reduce collapse risk were implemented (Comerio 1998). This type of construction is common internationally as well, and remains widespread in many developing countries. The poor seismic performance of non-ductile concrete buildings has been documented in many earthquakes in both developed and underdeveloped countries, including in recent events such as in New Zealand in 2011 (Smyrou et al. 2011).

The Uniform Building Code (UBC) was the national code adopted by most of the states in the USA during the time of this non-ductile concrete moment frame construction type. Much of the text related to the design and behavior of concrete structures within the UBC and other city codes is based on the American Concrete Institute (ACI) 318 document: Building Code Requirements for Structural Concrete and Commentary. All the ACI and UBC codes prior to 1976 had few requirements for ductile detailing, which makes all concrete moment frame buildings constructed prior to 1976 non-ductile moment resisting frames without seismic details. These buildings are vulnerable to numerous failure modes including: failure of column lap splices; strong beam/weak column failures; captive column failure; punching shear failures in flat plate slabs; and shear and axial load failure of columns with wide transverse reinforcement spacing (Faison et al. 2004). Few of the smart cities have been looking for ways to become more resilient and resistant to earthquakes. Considering the fact many of existing essential buildings such as hospitals, fire stations, schools and other critical buildings are non-ductile reinforced concrete buildings, the proposed fragility curves can help the decision makers in setting priorities to retrofit these building to increase the resilience of smart cities.

The objective of this study is thus to investigate the brace–damper system as a viable retrofitting method that would reduce the seismic demands and thus, the structure's vulnerability. Analytical fragility curves are developed to assess the feasibility of viscoelastic dampers in reducing the vulnerability risk of these structures (Fig. 1).

Fig. 1. Example of failure of non ductile reinforced concrete frames observed in the Alaska earthquake.

2   Retrofit Strategy

There are commonly two different approaches to mitigate the risk of damages of structures. The first approach attempt to increase the strength of the existing members by utilizing steel, jacketing, FRP jacketing, or increasing the size of existing cross section with a new reinforced concrete jacketing. The second approach is based on the concept of reducing the seismic demands on the building to a level where the frames can resist the demands with its existing strength. This can be achieved by adding viscoelastic dampers along the selected bay of each frame.

Figure 2 shows the acceleration response spectra of the N-S components of 1985 Mexico City Earthquake under various damping ratios. The reduction of the spectral acceleration with the increase of damping is readily seen in this graph. One important observation from this graph is acceleration reduction with the increase of damping is effective only in the range of 1.5 to 3.0 s of the natural period. This indicates that the viscoelastic damper concept has the potential to reduce the seismic demands to these structures to the point where the demands in the moment frames would be within the existing low strength level. The supplemental damping devices can also reduce the story drift which is an important metric in determining the risk of damage to framed structures during earthquakes.

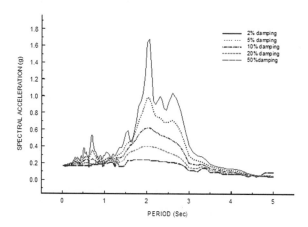

Fig. 2. Response Spectra Acceleration (SA) Mexico Earthquake September 1985.

3 Methodology

3.1 Mathematical Model

A typical six-story Archtype building was considered within the scope of this study. The building was designed according to UBC-1972 building code to represent a non-ductile reinforced concrete frame building ash shown in Fig. 3. The same building is retrofitted by adding a brace–damper system within the structure in the middle bay in each exterior elevations. Figure 4 shows a typical configuration of a viscoelastic damper. The structural system of the Archetype building consists of three bays of moment frames in each orthogonal direction. The building is composed of reinforced concrete frames. For simplicity, a symmetric floor plan was selected for this study. This symmetric plan allows the use of two-dimensional structural models. As mentioned earlier, the archetype represents a building with low level of strength and ductility. Thus, it represents all the shortcomings found in non-ductile reinforced concrete frames such as low ductility, strong beam-weak column, and low spacing of shear reinforcement.

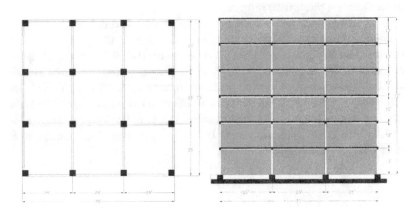

Fig. 3. Archetype building, non ductile reinforced concrete frames.

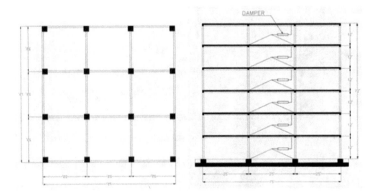

Fig. 4. Archetype building, retrofited with visco elastic dampers.

The archetype buildings were modeled based on the nonlinear behavior of materials and its damping characteristics. Pushover curves and inter-story drift ratio of the sample buildings at each step of pushover analysis were determined using SAP2000© software package. A time history analysis was carried out to determine the response of structure based off 22 far field, as well as near field ground motion which is selected from (PEER) ground motion database.

3.2   Analytical Model of Sample Building

To perform the push over and the nonlinear time history dynamic analysis, and to evaluate the vulnerability of the considered buildings, the building frames have been modeled using SAP 2000 software. The nonlinear or inelastic behavior of various structural members, including beams, columns, and bracing elements has been intro-duced to the software based on the FEMA 356 (2000) guidelines.

To account for the damping from supplemental damping devices, FEMA 356 specifies a damping modification factor to reduce the seismic effect (pseudolateral load in a given horizontal direction) on the structure. The damping modification factor comes from the estimated effective damping ratio βeff, which is expressed as follows for a structure with linear viscous dampers.

$$\beta_{eff} = \beta + \frac{\sum_j C_j \cos^2 \theta_j \phi_{rj}^2}{2\omega \sum_i \left(\frac{w_i}{g}\right) \phi_i^2} \tag{1}$$

where C_j is the damping constant of device j, θ_j is the angle of inclination of device j to the horizontal, φ_{rj} is the first mode relative displacement between the ends of device j in the horizontal direction, ω is the fundamental frequency of the rehabilitated building including the stiffness of the velocity dependent devices, w_i is the reactive weight of floor level i, and φ_i is the first mode displacement at floor level i.

The force-velocity relationship of the nonlinear damper is expressed as:

$$F = CV^\alpha \tag{2}$$

Where: F is the damping force, V is the velocity across the damper, and C is the damping constant of the damping device.

Using Eqs. 1 & 2, the approximate damping ratios of the system can be estimated and the supplemental damping devices can be sized to achieve the desired damping ratios.

3.3 Fragility Modeling

A lognormal distribution is used because all values are expected to be positive. The cumulative probability function. The probability of reaching or exceeding a limit state (LS) at a given earthquake intensity is given by Eq. 3.

$$P[ds|Sa] = \phi\left(\frac{1}{\beta_{ds}} \cdot \ln\left(\frac{S_a}{S_{a,ds}}\right)\right) \tag{3}$$

where Φ is the standard log-normal cumulative distribution function; S_a is the spectral acceleration amplitude (PGA in this case); S_a, ds is the median value of spectral acceleration (PGA) at which the building reaches the threshold of damage state, ds; and β_{ds} is the normalized composite log-normal standard deviation respectively (FEMA 2012). Demand spectra and capacity curves are described probabilistically by median properties and variability parameter, β_D and β_C, respectively (FEMA 2012).

$$S_{a,ds} = \delta_{R,S_{ds}} \alpha_2 \cdot h \tag{4}$$

It is noted that "Structural fragility is characterized in terms of spectral displacement and by equivalent-PGA fragility curves for buildings that are components of lifelines" (FEMA 2012). FEMA also explains that damage state medians are developed using functions to describe not only the drift at the damage state, but also variability in the capacity curve, demand spectrum, and the threshold of damage itself. First, damage state drift ratios are converted to spectral acceleration using Eq. 4, where δR, Sds is the drift ratio at the threshold of structural damage state, ds; α2 is the fraction of the building height at the location of pushover mode displacement; and h is the typical height, in inches, of the model building type of interest (FEMA 2012)

To account for variability, FEMA uses Eq. 5 where βSds is the log-normal standard deviation that describes the total variability for structural damage state, ds; βC is the log-normal standard deviation parameter that describes the variability of the capacity spectrum; βD is the log-normal standard deviation parameter that describes the variability of the demand spectrum; βM(Sds) is the log-normal standard deviation parameter that describes the uncertainty in the estimate of the median value of the threshold of structural damage state, ds (FEMA 2012).

$$\beta_{S_{ds}} = \sqrt{(\text{CONV}[\beta_D, \beta_C, S_d, S_{ds}])^2 + (\beta_M(S_{ds}))^2} \tag{5}$$

4   Results and Analysis

To assess the effectiveness of the retrofit using the viscoelastic dampers, two criteria were used in this study. Criterion 1 is the rotation of the concrete frames columns versus allowed limits of different performances levels as stated in FEMA 356 guidelines. Criterion 2 is the story drift between the ground level and the first floor which is the level that produced the maximum drift.

The fragility curves were developed for the three main performance levels prescribed in FEMA 356 which are immediate occupancy (IO), life safety (LS) and collapse prevention (CP). The 0.5 PGA was used as a point of reference to compare structure performances for all three criteria for the different performance levels. The 0.5 PGA represents the design basic earthquake (DBE) which represent an earthquake that has a 10% probability of exceedance in 50 years. For the column rotation of the concrete frames (criterion 1), Fig. 5 shows an example of one the fragility curves produced in this study. It can be seen that increasing the system damping reduced probability of failure.

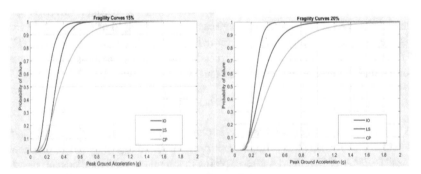

Fig. 5. Fragility curves for the archetype building retrofitted with visco elastic dampers

Table 1 summarizes the structure performance after retrofit based on criterion., increasing damping ratio from 2% to 25% reduced the probability of failure from 100% to 81% for the immediate occupancy level, 57% to 31% for the life safety performance level and 40% to 9% for the collapse prevention performance level.

Table 1. Probability of failure at different performance levels

Damping ratio	Performance level		
	Immediate occupancy	Life safety	Collapse prevention
2%	100	57	40
10%	100	50	36
15%	100	26	21
20%	85	13	6
25%	81	30	9

5   Conclusions

Fragility analysis has been widely used in performance-based design (PBD) over the last decades to assess the response of different structure systems. In the context of PBD, fragility assessment of non-ductile reinforced concrete frame building subjected to earthquake hazards is conducted in this study. This study revealed that the viscoelastic dampers were effective in reducing the risk of damages for concrete frame structures.

The fragility curve analysis shows clear improvements in each retrofit case. It is noticed that there is no significant improvement when the damping ratio of the system was increased from 15% to 25%. It was also noticed that the effectiveness of the supplemental damping devices are dependent on the structure natural frequency. Therefore, it is recommended to conduct a parametric study for structures with different heights and floor areas as that will produce response for structure with different dynamic properties allowing a general conclusion to be drawn over the effectiveness of this method of retrofit.

The produced fragility curves highlight the advantages of using viscoelastic dampers in reducing the level of damage these structures can experience during earthquakes which can be an efficient strategy to increase the resilience of vulnerable buildings in a smart city. The fragility curves can be a useful tool to assist the decision makers in setting priorities to retrofit these buildings especially the ones that are essential to the well-been of a society such as hospitals, fire stations and schools and that can improve the resilience of smart cities.

References

ATC 40: Seismic evaluation and retrofit of concrete buildings, Applied Technology Council, ATC., vol. 1 and 2, Redwood City (1996)

Altug Erberik, M.: Fragility-based assessment of typical mid-rise and low-rise RC buildings in Turkey. Eng. Struct. **30**, 1360–1374 (2008)

Comerio, M.C.: Disaster Hits Home: New Policy for Urban Housing Recovery. University of California Press, Berkeley (1998)

Baran, E., Mertol, H.C., Gunes, B.: Damage in reinforced-concrete buildings during the 2011 Van Turkey, Earthquakes. J. Perform. Constr. Facil. **28**(3), 466–479 (2014)

Erberik, M.A., Elnashai, S.A.: Seismic vulnerability of flat-slab structures, Technical Report DS-9 Project (Risk Assessment Modeling) Mid-America Earthquake Center, University of Illinois at Urbana-Champaign (2003)

Faison, H., Comartin, C., Elwood, K.: Housing report reinforced concrete moment frame building without seismic details. Report, Earthquake Engineering Research Institute (EERI) and International Association for Earthquake Engineering (IAEE) (2004)

Kirçil, M.S., Polat, Z.: Fragility analysis of mid-rise R/C frame buildings. Eng. Struct. **28**(9), 13351345 (2006)

Comerio, M.C., Anagnos, T.: Los Angeles inventory: implications for retrofit policies for nonductile concrete buildings. In: 15 World Conference on Earthquake Engineering, Lisbon (2012)

Maniyar, M.M., Khare, R.K., Dhabal, R.P.: Probabilistic seismic performance evaluation of nonseismic, RC frame buildings. Struct. Eng. Mech. **33**(6), 725–745 (2009)

Enhancing Progressive Collapse Resistance in Existing Buildings

Karim Hammad[1], Ibrahim Lotfy[1(✉)], and Mohamed Naiem[2]

[1] Department of Civil and Material Engineering, German University in Cairo, New Cairo City, Egypt
{karim.hammad,ibrahim.lotfy}@guc.edu.eg
[2] Department of Construction Engineering, American University in Cairo, New Cairo City, Egypt

Abstract. Progressive collapse is a chain reaction of structural element failures due to a relatively local structure damage. Different design codes began considering progressive collapse guidelines and specifications following the collapse of the Ronan Point building in 1996 and the world trade center in 2001. Researchers and government agencies were inclined to study means of enhancing progressive collapse resistance of structures in order to mitigate the destructive effect of accidents and/or reoccurring disasters. A common type of progressive collapse occurs as a results of a sudden column removal. The resulting dynamic straining actions can be resisted by three mechanisms; vierendeel action in framed structures, catenary action in the beams and membrane action in the slabs. This paper aims to enhance progressive collapse resistance of framed structures using retrofitting with steel plates. Using steel plates retrofitting to enhance progressive collapse resistance of existing structures can provide a clever and innovative solution for smart city applications. In this paper, the effect of the proposed scheme on progressive collapse resistance was investigated for multi-story reinforced concrete framed structures. A dynamic, non-linear, analysis following the guidelines of the ASCE 41 and the GSA was performed to accurately portray the behavior of the structures and assess their progressive collapse resistance in the event of sudden column removal. The findings of the parametric study conducted shows the potential of this method on the behavior of the retrofitted structures.

Keywords: Progressive collapse · Framed concrete structures · Column removal · Progressive collapse retrofitting · Smart cities

1 Introduction

The topic of progressive collapse has taken the attention of many researchers in recent years due to many structural collapse associated with accidents or attacks or failure of structural elements. One form of progressive collapse of buildings occurs when one or more vertical load carrying members (typically columns) are removed or collapse. After losing the original load path due to column removal, the load seeks an alternate path. The loads from the upper floors will be redistributed quickly; within a few milliseconds. As a result, all floors above the first floor will deflect identically and

I. El Dimeery et al. (Eds.): JIC Smart Cities 2019, SUCI, pp. 39–46, 2021.
https://doi.org/10.1007/978-3-030-64217-4_5

dynamically under uniform gravity loads to seek a new equilibrium path. The building will not collapse immediately due to spatial action of structure and catenary action of the frame beams or slabs.

There exists several alternative methods to enhance progressive collapse resisting capacity. Using rotational friction dampers is an example of such methods. The definition of building's capacity to resist progressive collapse is that the internal forces of the structure regulate spontaneous to prevent progressive collapse damage and then avoid the overall collapse when emergency or severe overloading of local lead to a sudden failure of some component. In this study a new method is proposed which is centered around retrofitting frames with steel plates to upgrade the hinge capacity against progressive collapse. It relies on the new composite beam structurally-sound hinges and its new-found load redistribution quality.

2 Past Research

Several researchers have performed studies to simulate the progressive collapse. More to the point, the following studies addressed mainly previous research on progressive collapse of 3D structures.

The progressive collapse of a typical ten-story reinforced concrete structures subjected to gravity loads was studied by Helmy [1] and designed according to ECP 203 [2]. The obtained results were compared to both GSA [3] and UFC guidelines [4]. The AEM was used in the analysis and nonlinear dynamic analysis was performed. The results were found to satisfy the GSA [3] limits and the structure was found to have low potential for progressive collapse, but when the results were compared to the UFC limits [4], the structure was found to have high potentials for progressive collapse for the cases of corner column and edge shear wall removal. Salem et al. [5] studied the progressive collapse of a multi-story reinforced concrete structure, designed according to the ACI 318-08 [6], due to removal of one or two interior columns. The aim of the study was to reach economic design of reinforced concrete structures to resist progressive collapse. The software used in this study is based on the AEM. Nonlinear dynamic analysis was performed together with the ability of software to predict cracks locations, elements separation, and elements collision. The results showed that the removal of one column only will not lead to progressive collapse, while the removal of two columns will lead to progressive collapse of a considerable part of the structure. It was also concluded that the AEM is capable of suggesting economical designs which can resist progressive collapse. Alrudaini and Hadi [7] studied a ten-story structure subjected to corner column removal from the ground floor using ANSYS 11. The results showed that the progressive collapse of the studied structure could be prevented by hanging the columns using vertical cables inside them, these cables will tie the columns with an upper steel frame on the top of the building. More recently, Elshaer [8] assessed a typical ten-story reinforced concrete building due to removal of corner, edge, and interior columns from the ground, fifth, eighth, and tenth floor. A nonlinear dynamic analysis was preformed using ELS. The obtained results showed that not considering the slab contribution gives inaccurate results and causes partial collapse of the structure. The cases, which were studied considering the slab contribution, did not collapse. Moreover, the elements rotation limits didn't exceed the allowable rotation limits given by the UFC guidelines [4].

3   Research Objectives

Resistance to progressive collapse is achieved either implicitly; by provisions of minimum levels of strength, continuity, and ductility or explicitly by:

1) Providing alternate load paths so local damage is absorbed and major collapse is averted; or
2) Providing sufficient strength to structural members that are critical to global stability [9].

To study the behavior of a reinforced concrete structure after the removal of one or more columns, several researchers studied the performance of the RC structures following the loss of a column via experimental and analytical approaches. The main objective of this study is to evaluate the effectiveness of the proposed steel plates retrofitting method in improving the progressive collapse resistance of existing structures. Moreover, studying the different factors and limitation affecting the performance of the proposed method.

The approach implemented in this study is a, non-linear, numerical simulation using a finite Element Model using SAP2000 modeling suite. The alternative load path method used by removing one or several major structural elements (i.e., introducing an initiating damage) and analyzing the remaining structure to determine if the initiating damage propagates. The representation of the progressive collapse of the structure followed these stages. First, the structure is loaded with dead and live loads (normal service load). Then structural elements are removed suddenly; this could be represented mathematically as a sudden change in the stiffness matrix of the structure. However in this study the "at rest" equilibrium state of the structure is simulated by applying the internal forces of the element to be removed to the supported structure. Initiating damage is simulated by a sudden removal of this internal force. The performance evaluation criteria for the nonlinear analysis procedures are based on plastic hinge rotations and displacement ductility ratios.

Additionally, different design parameters has been taken into consideration during the study to test different scenarios of progressive collapse applying different thicknesses of steel plates and evaluating their effectiveness. The performance has been evaluated based on plastic hinge rotation and displacement ductility ratios.

4   Finite Element Analysis

The objective of this analysis was to evaluate the effectiveness of the retrofitting method in upgrading the capacity of the building against progressive collapse.

The analysis was performed over a four floor 2D frame with 4 spans of 6 m each. The progressive collapse is initiated by removing the middle column. A nonlinear dynamic analysis using a time history function has been used to simulate the progressive collapse and to evaluate the retrofitting method suggested. Three different system where used for the retrofitting of the building. The first one was adding 2 m length steel plate on the bottom side of the beam covering its full width at the location of the hinges only at the first floor. The second system was similar to the first one but by repeating the steel plates in the second floor as well. Finally the last system was using 2 steel plates at the first floor at the location of the hinges in both top and bottom sides of the beam. The thickness of steel plates in each system varied from 5 mm to 20 mm.

A vital component of any non-linear analysis is the material definition and the plasticity considered. All steel elements were considered as elastic-perfectly plastic material with a clear yield stress in compression and tension. All concrete frames were modeled as linear-elastic. Plasticity was not considered in the material model but rather using hinges in specific prescribed location. To account for non-linearity preselected hinges located at the beams and columns ends was assumed. The properties of the plastic hinges were considered as per the GSA guidelines [3]. Plastic hinge backbone curves for beams were defined as moment only hinges while column hinges were defined as P-M; axial force and moment hinges. Yield moments and axial forces were considered accordingly using the reinforcement details, concrete dimensions and retrofitted steel plates. Columns hinges were considered at different angle of interaction between the axial forces and moments.

5   Results

A total of 12, nonlinear, time-history analyses were performed and the results were extracted from the models. Figures 1, 2, 3 and 4 show the results plotted against the time of column removal. Figure 1 presents the vertical displacement at the location of removed column in each model with respect to time from the column removal.

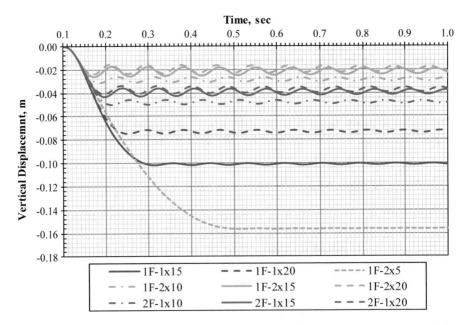

Fig. 1. Vertical displacement at the location of the removed column with respect to time for non-collapsed models.

Inspection of Fig. 1 reveals that all models with 5 mm plate have collapsed. The only exception is the model with 2 × 5 mm plate top and bottom which had larger vertical displacement as comparison to the any other non-collapsed model. Furthermore, models retrofitted with 15 and 20 mm thick plates were able to mitigate collapse in all cases when using one plate in the first floor only. By retrofitting the first and second floors frames with steel plates instead of only the first, it is observed that the maximum vertical displacement decreased and the plate thickness required for preventing the collapse of the building also decreased. Models retrofitted with 10 mm thick plate had a maximum displacement of around 50 mm. Finally, retrofitting frames with top and bottom steel plates yielded minimum vertical displacement at the location of the column removal compared to the rest of the models due to having a larger thickness of steel plates per floor and covering both sides of the beam. Figure 2 presents the vertical displacement at the location of removed column with respect to time for the models retrofitted with one steel plate at the bottom of the first floor frame.

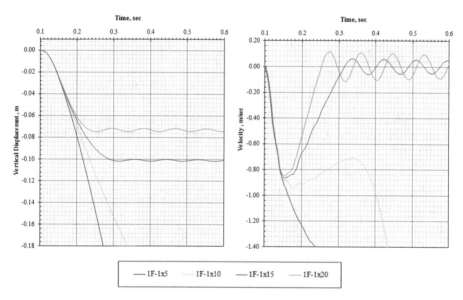

Fig. 2. Vertical displacement and velocity at the location of the column removal with respect to time for models retrofitted with one steel plate at the bottom of the first floor frame.

Inspection of Fig. 2 highlights the effect of increasing of the thickness of the retrofitting plate on the max deformation at the location of the column removal. The models retrofitted with 5 mm and 10 mm steel plates have collapsed and exhibited larger values of displacement and the velocity. Moreover, the velocity never returned to zero with time which indicate the displacement tends to infinity; i.e. collapse occurred. However, models retrofitted with 15 mm and 20 mm steel plates were able to prevent collapse with max displacement was 100 mm and 80 mm respectively at the location of the removed column. Considering the 12 m span length, these displacement values are deemed acceptable. Figure 3 presents the vertical displacement at the location of removed column with respect to time for the models retrofitted with one steel plate at the bottom of the first and second floor frames.

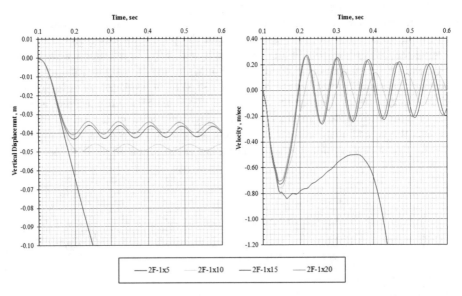

Fig. 3. Vertical displacement and velocity at the location of the column removal with respect to time for models retrofitted with one steel plate at the bottom of the first and second floor frames.

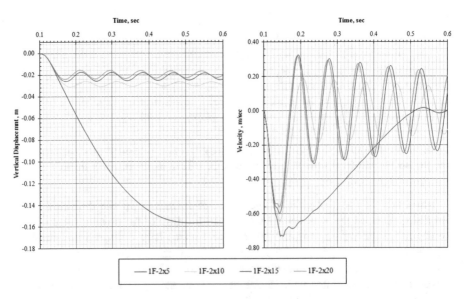

Fig. 4. Vertical displacement and velocity at the location of the column removal with respect to time for models retrofitted with two steel plates at the top and bottom of the first floor frame.

Figure 3 indicates that only one mode, retrofitted with a 5 mm thick steel plate, have been collapsed and experienced increasing value of displacement. The velocity history is curious as it clearly highlights the frame trying to stabilize the model and bring back the velocity to zero but did not have enough plasticity to do so. All the other models, retrofitted with 10 mm, 15 mm and 20 mm steel plates, prevented collapse and experienced equilibrium displacement of 50 mm, 40 mm and 38 mm respectively which consider acceptable. Figure 4 presents the vertical displacement at the location of removed column with respect to time for the models retrofitted with two steel plates at the top and bottom of the first floor frame. Inspection of Fig. 4 shows that no models have collapsed when retrofitted with top and bottom plates. Moreover, equilibrium displacement were less that other retrofitting techniques with 160 mm, 30 mm, 20 mm and 18 mm for 5 mm, 10 mm, 15 mm and 20 mm thick plates respectively.

6 Summary and Conclusion

Using retrofitting techniques to enhance progressive collapse resistance of existing structures can provide a clever and innovative solution for smart city applications. In this paper, a retrofitting technique centered around the use of steel plates to enhance the capacity of frames building against progressive collapse was evaluated. Three different retrofitting scenarios were investigated. A dynamic, non-linear, analysis following the guidelines of the ASCE 41 [6] and the GSA [3] was performed to accurately portray the behavior of the structures and assess their progressive collapse resistance in the event of sudden column removal. The behavior of the framed building was investigated in term of maximum vertical displacement at the location of the removed column. Damage and plasticity occurring in the frames were observed. Highlighted below are the finding and conclusions of this paper:

- From reviewing past literature it became apparent the importance of resisting progressive collapse in both existing and new buildings, retrofitting existing building against progressive collapse could be establish using different system.
- Retrofitting concrete framed structures with steel plates proved to enhance the hinges properties of the moment resisting frames and to delay and prolong the formation of plastic hinges. Thus it was able to prevent or at least delay progressive collapse and control the vertical deformations and velocities at the location of the removed column.
- The thickness of the steel plates used to retrofit the frames as well as its location effect dramatically the behavior of the building during a progressive collapse scenario.

References

1. Helmy, H.: Progressive collapse assessment of multistory reinforced concrete structures. Cairo University (2011)

2. ECP Committee 203: The Egyptian code for design and construction of concrete structures. Housing and Building Research Center, Giza, Egypt (2009)
3. GSA-2013: Progressive collapse analysis and design guidelines for new federal office buildings and major modernization projects. U.S. General Service Administration, Washington, DC (2013)
4. UFC: Design of buildings to resist progressive collapse. Unified Facilities Criteria, U.S Army Corps of Engineers (2009)
5. Salem, H.M., El-Fouly, A.K., Tagel-Din, H.S.: Toward an economic design of reinforced concrete structures against progressive collapse. Eng. Struct. **33**(12), 3341–3350 (2011). Corrected Proof. Accessed 28 July 2011. ISSN 0141-0296
6. ACI 318-05: Building code requirements for structural concrete and commentary. American Concrete Institute (ACI) (2005)
7. Alrudaini, T.M.S., Hadi, M.N.S.: A new design to prevent progressive collapse of reinforced concrete buildings. In: The 5th Civil Engineering Conference in the Asian Region and Australasian Structural Engineering Conference (2010)
8. Elshaer, A.Y.: Progressive collapse assessment of multistory reinforced concrete structures subjected to seismic actions. Master thesis, Cairo University, Egypt (2013)
9. Qian, K., Li, B.: Experimental study of drop-panel effects on response of reinforced concrete flat slabs after loss of corner column. ACI Struct. J. **110**(2), 319–329 (2013)

Damage Assessment and Sustainability of RC Building in New Cairo City Considering Probable Earthquake Scenarios

Abdelaziz Mehaseb Elganzory[1]([⊠]), Balthasar Novák[2],
and Ahmed Mohamed Yousry[3]

[1] German University in Cairo, Cairo, Egypt
abdelaziz.elganzory@guc.edu.eg
[2] Stuttgart University, Stuttgart, Germany
[3] Assuit Univerity, Assuit, Egypt

Abstract. Recent occurred earthquakes in Egypt have highlighted the need for more studies about risk assessment, sustainability of buildings and infrastructures, even in case of moderate earthquake events. This work investigates the topic of seismic risk assessment for urban areas located at the new Cairo and new capital cities. The focus has been done to quantitative damages and sustainability of representative reinforced concrete buildings. A new proposed model has been developed to evaluate the required physical damages. A new lateral load pattern during pushover method is proposed. It is found that, the results provided by the proposed lateral load pattern gives reliable results compared to those provided by time history estimates. The fragility curves for representative building models located at new Cairo and new capital cities are derived. Quantitative damage risk assessment is generally performed using the integration between the imposed seismic hazards, building fragility curves and exposure. This new proposed model can be used to evaluate damages for buildings located at new Cairo and new capital cities due to expected earthquakes. Similar models can be generated to evaluate damages in other regions in Egypt and worldwide.

Keywords: RC buildings · Sustainability · Damage assessment · Fragility curves · Seismic scenarios

1 Introduction

The current research concentrates on R.C buildings located at new Cairo and new capital cities. The selected buildings are designed and executed according to Egyptian code requirements with specific reinforcement details, which are different from other codes, and quality control is also different. Such differences mainly lead to; the derived fragility curves for selected buildings will be different from those counterparts located at other countries. Therefore it is recommended to generate a new model to assess the physical damages for R.C. buildings in the region under study. In the following sections, a full description will be given to the new developed damage assessment model. The model takes into consideration: occurrence of earthquake scenarios, selection of

I. El Dimeery et al. (Eds.): JIC Smart Cities 2019, SUCI, pp. 47–55, 2021.
https://doi.org/10.1007/978-3-030-64217-4_6

appropriate ground-motion prediction equations, determination of different buildings response parameters, derivation of analytical fragility curves for representative building models, and finally, evaluation of different structural damage distributions corresponding to the imposed earthquake excitation. In essence, the research provides a new proposed model to perform overall risk assessment for R.C. buildings located at new Cairo and new capital cities.

2   The Develop Model

$$\text{Risk} = \int P\,(\text{Hazard}) * \text{Consequences}\,(f(\text{Vulnerability} * \text{Explousure}))\,dt \qquad (1)$$

Seismic hazard is classified into two types: deterministic hazard and probabilistic one. In this research paper, deterministic seismic hazard assessment is selected. On behalf of deterministic seismic hazard assessment methodology, earthquake catalogue for new Cairo and new capital cities are collected and reviewed accurately to assume suitable earthquake scenarios. In general, for each region, specific attenuation equations are derived based on sufficient ground motion data which are recorded for several dozens of years. Unfortunately, for the region under study, no sufficient ground motion data are available; therefore it is suggested to use other-derived attenuation equations. In fact, the utilized equations are derived for other regions that are similar to the region under study. From these similar characteristics: the magnitudes of occurred earthquakes are of the moderate intensity and of shallow focal depth which ranges from 0–30 km beneath the ground surface. Using assumed earthquake scenarios of specific intensity combined with selected attenuation equation, the values of peak ground acceleration are determined underneath the selected buildings. It follows that; response spectrum is derived based on the soil category recommended by the Egyptian code [2]. In addition, SeismoArtif software [3] is utilized to generate a series of several accelerograms to match the derived response spectrum as shown in part A of the developed model (see Fig. 1).

Three representative building models from new Cairo and new capital cities are selected to reflect the majority of building stock located at the corresponding cities. The structural-characteristics of representative building models are modeled with finite element program SAP2000 [4]. It follows that, these buildings are exposed to several accelerograms through nonlinear time history analyses. Building responses in terms of top roof-displacements and base shears are determined. On the other hand, these buildings are exposed to a specific lateral load pattern through pushover analysis. Different damage limit states are determined based on chocen criteria as described in detail see Sect. 2.4.

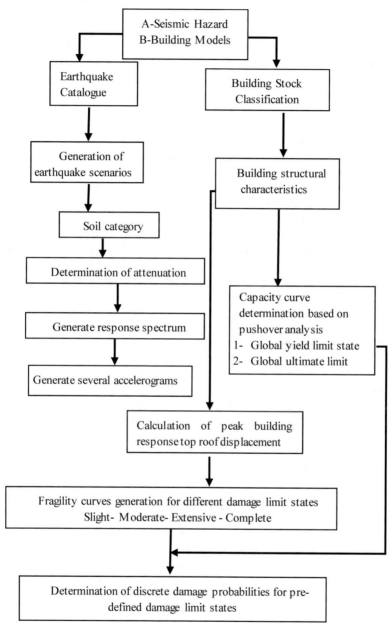

Fig. 1. Chart summarizing the developed model

Using formula provided by FEMA in advanced engineering building module (AEBM) [5], fragility curves are derived. The main input in the formula is different damage limit values for each building which are determined based on its capacity curve. Finally, to assess the expected damage distributions, it is needed to determine the target displacement. Numerous approaches are available to determine target displacement such as N2 method [6]. In this research, target displacement is obtained from average results of the nonlinear time history runs. Using target displacement value and fragility curves, discrete damage distributions are determined as shown in the developed chart (see Fig. 1).

2.1 Seismic Hazard

Deterministic seismic hazarld assessment approach (DSHA) is adopted to assume three earthquake scenarios. These scenarios are assumed based on the repetition of the historical events. Therefore, such scenarios were simply assumed arbitrarily at different locations at new Cairo city, to reflect the fact that Cairo is positioned within a broad seismic source zones, and may lead to, such events can occur anywhere within the source zone and the city.

Since Egypt did not have sufficient historical strong motion records, no attenuation relationships were derived specifically to the country. Therefore, other well-known derived attenuation relationships were utilized. In addition, most active zones surrounding Egypt are crustal and shallow earthquakes associated with surface ruptures. Riad et al. [7] decided to utilize the regional attenuation relationship derived by Campbell [8] for central United States to be used in Egypt as follows:

$$\ln(\text{PGA}) = 4.39 + 0.9222M - 1.27 \ \ln(r + 25.7) - 0.0018r \qquad (2)$$

Where, PGA is the peak ground is measured in terms of (cm/s^2) or gals, r is the epicentral distance, and M is magnitude measured on Richter scale. According to the Egyptian code, the ground shaking at a given location can be represented by an elastic response spectrum. Using software SeismoArtif [3] a three accelerograms are selected and scaled to match the corresponding derived response spectrum.

2.2 Building Stock and Structural Systems

In this research, three representative building models were analyzed. The first model is irregular 4-storey, 15 m height; RC solid and flat slab system. The building is selected to represent residential-villas region in new Cairo city. The assigned building is irregular in both elevation and plan. Therefore, a 3-D model is needed to reflect accurate dynamic characteristics of the building. The structural drawings and detailing of the three representative building are illustrated in [1], which presents plan, elevation, and cross sectional dimensions for the main structural elements of the buildings.

2.3 Derivation of Capacity Curves

A pushover analysis method is used to derive capacity curves for representative building models. Since the buildings have different structural characteristics in both orthogonal directions, pushover method is implemented for 3D structural model for both X and Y directions independently. Since 3-D structural model is performed, torsional moments and horizontal loading are applied at mass centres of each floor to adequately reflect horizontal displacements and torsion.

The method of modal combinations [1] is adopted while applying pushover analysis. In this procedure the spatial variation of the applied forces is determined from the relationship:

$$Fj = \Sigma \, \alpha n \, \Gamma n \, ma \, \phi n \, Sa \, (\xi n, Tn) \tag{3}$$

$$Mj = \Sigma \, \alpha n \, \Gamma n \, ma \, I \, \theta n \, Sa \, (\xi n, Tn) \tag{4}$$

Where, αn presents the modification factor which can be taken positive or negative; ϕn represents the mode shape vector related to the mode n, Sa represents the spectral acceleration corresponding to the period of the mode n. If the first two modes are taken into consideration, therefore, the following equation may be utilized:

$$Fj = \alpha_1 \, \Gamma_1 \, m\phi_1 \, Sa \, (\xi_1, T_1) \, 6 \, \alpha_2 \, \Gamma_2 \, m\phi_2 \, Sa \, (\xi_2, T_2) \tag{5}$$

$$Mj = \alpha_1 \, \Gamma_1 \, m \, I \, \theta_1 \, Sa \, (\xi_1, T_1) \, 6 \, \alpha_2 \, \Gamma_2 \, m\theta_2 \, Sa \, (\xi_2, T_2) \tag{6}$$

It means that, the performed procedure requires several pushover analyses where multiple combinations of the modal load patterns are adopted. To consider the best estimates of deformation and force demands, the envelope of the peak demand values are then used in the performance based-evaluation. Fig. 2 and Fig. 3 represent non-linear pushover curves for representative building no. 1.

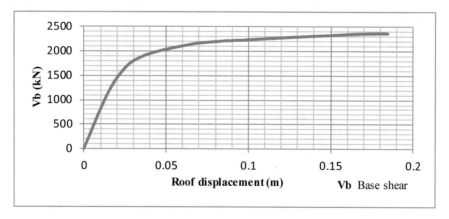

Fig. 2. Capacity, "pushover", curve for building no. 1"combination of the modes" as a lateral load pattern distribution

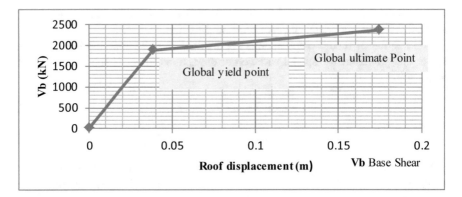

Fig. 3. Capacity, "pushover", joint plot for building no. 1"combination of the modes" as a lateral load pattern distribution

2.4   Derivation of Damage Limit States

Four damage limit states are determined in a quantitative way linked to the capacity curve as follows: slight damage is considered as the first occurrence of local yielding in the critical storey, moderate damage is considered as the mean yield value of ISD% in the critical storey, ultimate limit state is considered as the ISD% corresponds to the ultimate building base shear capacity, and finally, for the extensive damage limit state, some uncertainty is encountered in its definition, where the observation-based definition is hardly linked to analytical assessment, therefore, it is considered as the average value of ISD% between extensive and moderate damage limit states. Fig. 4 presents graphically the utilized methodology to determine quantitative damage limit states for selected representative building models as shown in Fig. 4 for representative building no. 2, it follows that; values of corresponding interstorey drift ratios to roof drifts are reported see Table 1.

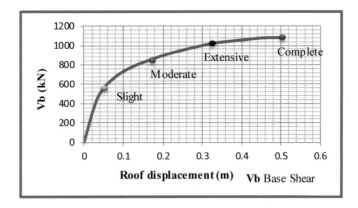

Fig. 4. Determination of four damage limit states for building no. 2

Table 1. Calculated damage limit states and corresponding drift ratios.

| Limit state | Interstory drift ratios (ISD%) | | Results |
	Building no. 1	Building no. 2	Building no. 3
Slight	0.41	0.415	0.55
Moderate	0.98	1.20	0.9
Extensive	1.77	1.94	1.70
Complete	2.57	2.87	2.32

The evaluated roof displacements and corresponding interstorey drift ratios are used to derive fragility curves for selected buildings, afterwards.

2.5   Derivation of Fragility Curves

The following equation is used to determine the corresponding fragility curves [5]:

$$P[ds/Sd] = \varphi[1/\beta ds] * \ln(Sd/Sd, ds) \tag{7}$$

Where, Sd, ds represents the median value of spectral displacement at which the building reaches threshold of specific damage limit state. Sd represents the imposed spectral displacement on the structure. βds represents standard deviation of the natural logarithm of spectral displacement at which the building reaches the threshold of the damage state. ds, and φ represents normal cumulative distribution function. Fig. 5 represents fragility curve for building no. 1.

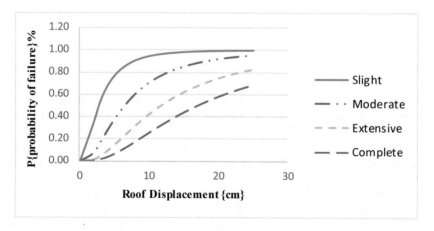

Fig. 5. Derived fragility curves for building no. 1

2.6   Risk Assessment and Seismic Performance Evaluation

In this paper, it is decided to assess target displacement from the computed nonlinear time history results that consider torsion effects and higher mode contributions. Also,

SAP 2000 [4] is used for elastic modal analysis, response spectrum, and nonlinear time historey methods of analysis. In addition, SeismoArtif [3] software is used to generate artificial accelerograms linked to real accelerograms. The chosen real accelerograms are Chi Chi, Fruili, and Imperial valley. It follows that, accelerograms are scaled and adjusted to be consistent with elastic response spectrum of the Egyptian code. Using target displacement and derived fragility curves, different structural damage distributions are determined.

2.6.1   Scenario no. 1 (Application of Intensity-Magnitude of 6 on Richter Scale)

First scenario of maximum magnitude 6 measured on Richter scale is assumed. It follows that, distance between the epicentre of the assumed scenario and the geocode of each representative building region is measured. Using the predefined-attenuation relationship [8], the maximum peak ground acceleration is determined. According to soil maps of the studied region and Egyptian code requirements, it is found that, C soil type is prevailing. Using the calculated maximum PGA (Peak Ground Acceleration) and soil type, elastic response spectrum is generated. Later on; several accelerograms are generated using SeismoArtif [3]. Also, response spectrum of generated accelerograms is generated to be consistent with Egyptian response spectrum.

Applying accelerograms to representative building models, target displacements are computed. Using the integration between the computed average value of top displacement and the fragility curves of the corresponding buildings, the probability of different damage limit states are determined. This process is repeated for different representative building models under different values of imposed seismic excitations, corresponding seismic risk values are obtained see [1] for all details.

3   Conclusion

An earthquake risk model related to structural damages of the building stock located at new Cairo and new capital cities in Egypt is performed. A simplified approach is used to assess seismic hazard in the investigated area. A new proposed lateral load pattern is adopted through the application of pushover analysis to capture the most accurate response of the analyzed buildings. Building capacity curves for related representative building models are derived. Fragility curves for the different representative building models are derived. These building models represent the majority of the building stock located at new Cairo and new capital cities. It follows that, several structural damage states are evaluated by combining obtained results from building responses and the derived fragility curves. It is found that, due to an adequate confinement for main structural components, the new constructed buildings in new Cairo and new capital cities shows a good responses for future probable earthquake scenarios.

References

1. Elganzory, A: Damage assessment and sustainability of RC buildings in new Cairo city considering probable earthquake scenarios. Ph.D. thesis. German University in Cairo, Egypt (2017)
2. ECP-Egyptian Code of Practice-201 Egyptian code of practice no. 201 for calculating loads and forces in structural work and masonry. National Research Center for Housing and Buildings, Egypt (2011)
3. Seismosoft Hompage. https://seismosoft.com/Seismosoft. Accessed 21 Nov 2019
4. Computers and Structures.INC. https://www.csiamerica.com/SAP2000. Accessed 21 Nov 2019
5. AEBM: Advanced Engineering Building ModuleTechnical and User Manual. FEMA, USA (1985)
6. Fajfar, P.: A nonlinear analysis method for performance-based seismic design. Earthq. Spectra **16**(3), 573–592 (2000)
7. Riad, P., Ghalib, M., El-Difrawy, M.A., Gamal, M.: A nonlinear analysis method for performance-based seismic design. Ann. Geol. Surv. Egypt **23**, 851–881 (2000)
8. Campbell, K.W.: Ground motion model for the central United States based on near-source acceleration data. Proc. Earthq. Earthq. Eng. **1**, 213–232 (1981)

Review of Seismic Provisions: Is Egypt Earthquake Safe?

Mohamed Afifi[1](✉) and Reem Ahmed[2]

[1] McGill University, Montreal, Canada
mohamed.afifi@mail.mcgill.ca
[2] Concordia University, Montreal, Canada

Abstract. A continuous improvement of any existing building code is a crucial responsibility of researchers, industry sponsors, and government officials. An integral part of any building code is its seismic provisions that need to be regularly updated as different aspects are revealed whenever an earthquake strikes. To give an example, major changes were introduced to the American code following the 1985 Northridge earthquake in California and similarly, provision alterations were applied to the Japanese code of practice following the 1995 Kobe earthquake. Likewise, the event of the 1992 Cairo earthquake drew major attention to enforcing earthquake resistant design the Egyptian code of building. Knowing that the Egyptian seismic provisions have not been majorly updated since last decade imposes a huge question of is it safe or is it overly conservative? And while major building codes are adding emphasis on the complex dynamic nonlinear analysis, the Egyptian provisions still utilize the traditional equivalent static load method as the main method of analysis. To answer these questions, a comparison of the Egyptian seismic provisions with its counterparts in the global code is presented in this study, followed by an application of the different provisions on the design of a steel building as a case study to verify the safety and feasibility of the current practice. Results reveal that the strict limits on drifts imposed by the Egyptian code, as well as the conservatism in calculating the seismic weight of the structure, yielded a structure with at least 23% more steel tonnage compared to structures designed according to other global building codes. Major steps need to be taken in order to optimize current code provisions to achieve the goal of building more sustainable cities.

Keywords: Building code · Dynamic analysis · Seismic · Steel structures

1 Introduction

Seismic events around the world draw the attention to structural flaws in buildings and other structures to resist earthquake loading. Assessment of failures is essentially carried out to identify failure modes and reasons, followed by reviews and updates to building codes and specifications. Historically, major updates to American seismic provisions were introduced following the 1985 Northridge earthquake in California and similarly, provision alterations were applied to the Japanese code of practice following the 1995 Kobe earthquake. In Egypt, the Cairo earthquake in 1992 (5.8 mb)

I. El Dimeery et al. (Eds.): JIC Smart Cities 2019, SUCI, pp. 56–64, 2021.
https://doi.org/10.1007/978-3-030-64217-4_7

which left over 9,000 buildings either completely or severely damaged and about 50,000 people homeless, was mainly due to the buildings were designed to resist only vertical loads and had insufficient lateral resistance [1]. Thus, the columns and beam column connections were found to have inadequate shear capacity, ductility, and confinement in plastic hinges [2, 3]. This event has drawn attention of officials in Egypt to the necessity of regularly updating the national building code to account for probabilistic loads like wind and earthquake, which were accounted for in subsequently published codes and specifications.

The design methods given by modern building codes guarantee acceptable safety level that depends on the probability of occurrence of the event. Global specifications allow designers to use various methods for seismic analysis staring from the simple equivalent static load analysis till the complex nonlinear dynamic analysis. Equivalent Static Load (ESL) is most popular among engineers for deign of buildings due to its simple methodology and lack of alternative methods [3] The most recent Egyptian code for load and forces, ECP2011 [4] and most of the global building codes depend on the conventional approach of equivalent static load analysis as the main method for evaluating seismic forces on symmetrical buildings. Hence, this paper aims at evaluating the ESL method of the Egyptian seismic provisos and how it compares to its international counterparts. In the last section of the paper, the design of the prototype structure is performed for each country and similarities and differences are highlighted.

2 Prototype Building

During the past two decades, the building environment in Egypt had extensively utilized medium-rise buildings having 6–12 stories, which is the maximum height allowed by the local authorities in most districts. These buildings are built with different configurations and structural systems having varying stiffness parameters that may have great influence on their seismic behaviour [3]. The structure plan view and the braced frame elevation are shown in Fig. 1. The building is an office building of the normal importance category. The building's layout is essentially five equal bays with a typical bay width of 9 m in both directions, and is representative of typical buildings in current practice in Egypt. The height of every story (column height) is taken equal to 5 m, as a normal height for office buildings. Beams are assumed on all grid lines, and the base columns are assumed to be fixed to the foundation.

The structure is assumed to be located at sites in Canada, United States and Egypt where similar seismic conditions and data prevail. Sites are located as follows: Montreal, QC, in Canada; Las Vegas, NV, in the U.S.; and Taba – South Sinai, in Egypt. For all three sites, the structure is assumed to be constructed on firm ground or very dense soil conditions, corresponding to site class C in USA and Canada with shear wave velocity between 360 and 760 m/s and site class B in Egypt with shear wave velocity between 360 and 800 m/s. In Egypt, the seismic input for design is preliminarily characterized by the maximum effective ground acceleration at the site. Taba is located in seismic zone 5B where a_g is equal to 0.30 g. This parameter can be compared to peak ground accelerations (PGA) specified in NBCC2015 [5] and ASCE 7–16 [6] for class C sites in Montreal and Las Vegas: 0.377 g and 0.298 g, respectively.

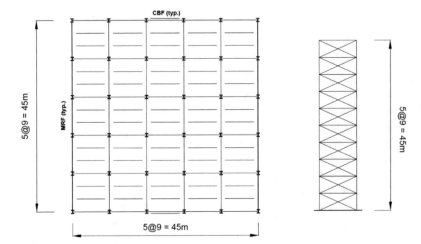

Fig. 1. Prototype structure.

3   Seismic Design Provisions in USA and Canada

This section briefly discusses the equivalent static force procedure of both the NBCC2015 [5] and ASCE7–16 [6]. Reports by Trembly et al. in 2015 [7] and Naqqash et al. in 2012 [8] were used extensively as basis for drafting this section of the paper. Stability requirements, P-delta effects as well as effects of accidental torsion were analysed but not indicated in the text for length limitations.

3.1   Canada: NBCC 2015 and CSA S16–14

In NBCC 2015, the minimum design base shear, V, is specified as:

$$V = \frac{S(T)M_V I_E W}{R_d R_o},$$

(1)

where S is the design spectrum, T is the period of the structure, M_v accounts for higher mode effects, I_E is the importance factor based on the use and occupancy, W is the seismic weight and R_d and R_o are the ductility and overstrength modification factors. The design spectrum for T > 0.2 is generally given as the product of spectral acceleration $S_a(T)$ and site coefficient F(T). Values of $S_a(T)$ are given in terms of uniform hazard spectral (UHS) ordinates, S_a, specified at periods 0.2, 0.5, 1.0, 2.0, 5.0 and 10 s for a return period of 2475 years or probability of exceedance of 2% in 50 years. The values for the chosen location are given in Table 1. F(T) is site coefficient that depend on the site class and soil type. Site class C corresponds to the reference ground type considered for the determination of S_a values and F is therefore equal to 1.0 at every period. Fundamental lateral period of vibration, T is dependent on height and type of Seismic Force Resisting System (SFRS) used, and is computed as shown in Fig. 2.

The M_v factor depends on the ratio S (0.2)/S (5.0) at the site, the period of the structure and the SFRS type. For moment frames and braced frames $M_v = 1.0$ in most situations except for ratios S (0.2)/S (5.0) greater than 40 in which case it may reach up to 1.03 and 1.07 for moment frames and braced frames, respectively. Importance factor (I_E) takes a value of 1.0, 1.3 or 1.5 for normal, high or post-disaster importance categories of structures. The seismic weight of the building, W, shall be determined to include the sum of dead loads, plus 25% of snow loads, plus 60% of storage loads and full contents of any tanks (Rogers, 2019). In the NBCC, R_d varies from 1.0 for the less ductile SFRSs to 5.0 for the most ductile systems. The factor R_o reflects the dependable overstrength present in the SFRS, depending on the difference between factored and nominal resistances and minimum level of strain hardening anticipated in tension [7], and varies between 1.5 and 1.0. For short period structures, the value of V from Eq. (1) should not exceed 2/3 the value computed at a period of 0.2 s. For steel frames with long periods, V must not be less than the value computed at a period of 2.0 s.

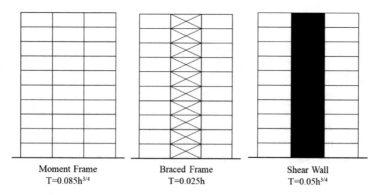

Moment Frame
$T=0.085h^{3/4}$

Braced Frame
$T=0.025h$

Shear Wall
$T=0.05h^{3/4}$

Fig. 2. NBCC2015 Fundamental period of vibration for different steel SFRSs.

$$Fx = (V - F_t)\left(\frac{W_X h_X}{\sum W_i h_i}\right), \text{ where } F_t = 0, \text{ for Ta} < 0.7\,\text{s}$$
$$= 0.07 T_a V \le 0.25, \text{ for } T_a \ge 0.7\,\text{s} \tag{2}$$

In this expression, F_t is a concentrated horizontal load applied at the top of the structure to account for higher mode effects. For buildings of the normal importance category, the design storey drifts Δx obtained from the displacements δx must not exceed the value of 0.025 h_{sx}, where hsx is the storey height at level x. In the NBCC, gravity dead (D) and live (L) are combined to earthquake loads (E) in load combination V: 1.0 D + 0.5 L + 1.0 E.

3.2   Usa: Asce 7–16 and Aisc341–16

According to ASCE7–16, the design base shear is specified as:

$$V = C_s W \tag{3}$$

Where W is the seismic weight and C_s is the seismic coefficient based on the value of the period. There are 3 ranges for the period values, short, intermediate and long, and for each range Cs is given by a different term. Minimum values are also specified that may govern long or intermediate ranges. Generally, C_s is function of either the short spectral acceleration (S_{DS}) or one-second spectral acceleration (S_{D1}), which are equal to 2/3 of the modified MCE$_R$ spectral values S_{MS} and S_{M1} specified for a certain location. For buildings or structures falling within the intermediate period range, C_s is specified as:

$$C_s = \frac{S_{D1}}{T(R/I_e)} \tag{4}$$

where R is the force modification factor that varies from 3.0 for steel SFRSs not designed or detailed for ductile response to 8.0 for the most ductile SFRSs; I_e is the importance factor ranging from 1.0 to 1.5 depending on risk category; and T is the fundamental dynamic period that can be obtained from dynamic analysis and cannot exceed $C_u T_a$, where $T_a = 0.0731 h_n^{0.75}$ and C_u varies from 1.4 in active seismic regions to 1.7 for low-seismic regions. The ASCE7–16 permits analysis using the equivalent static lateral force procedure if the height of the building is less than 48.8m and the period is less than 3.5Ts. The distribution of lateral forces can be computed by:

$$Fx = V\left(\frac{W_X h_x^k}{\sum W_i h_i^k}\right), \quad \text{where } k = 1.0 \text{ for } T < 0.5\,s$$
$$= 0.75 + 0.5T \le 2.5 \text{ for } T \ge 0.5\,s \tag{5}$$

Table 1. Spectral ordinates in the 2015 NBCC and ASCE 7–16

T(s)	S_a(g) NBCC2015	S_M(g) ASCE7–16
0.2	0.595	0.677
0.5	0.310	–
1.0	0.148	0.209
2.0	0.068	–
5.0	0.018	–
10	0.006	–

4 Seismic Design Provisions in Egypt – ECP2011

According to ECP2011, seismic design base shear, F_b, is specified as:

$$F_b = \gamma S_d(T_1)\lambda \frac{W}{g} \tag{6}$$

where γ is the importance factor taking values of 1.4,1.2,1.0 and 0.8 for post-disaster, high, normal and low importance categories, respectively. $S_d(T_1)$ is the ordinate of design spectrum at the fundamental period of vibration T_1; λ is the effective modal mass correction factor taking the value of 0.85 for $T \leq 2T_c$ (Upper limit of period of constant spectral acceleration), and n > 2 stories. W is the total weight of the building above the foundation level and g is the gravity acceleration = 9.81 m/s^2. The value of the fundamental period of vibration, T, is specified as follows:

$$T = C_t \times H^{3/4} \tag{7}$$

where C is a factor dependant on the structural system and material valued at 0.085 for steel moment resisting frames, 0.075 for concrete moment resisting frames and 0.050 for all other structures. H is the height of the structure above the foundation level. The ordinate of the design spectrum, $S_d(T_1)$, can generally be calculated from the following formula:

$$Sd(T_1) = a_g\gamma\frac{2.5}{R}\frac{T_c}{T} \geq 0.20\, a_g\gamma_1 \tag{8}$$

where a_g is the ground acceleration depending on the seismic zone for a return period of 475 years, S is the soil factor ranging from 1.0 to 1.6 depending on subsoil class, R is the reduction factor depending on structure type and material; there are 2 ductility levels for steel frames; Limited ductility (R = 5.0) and moderate ductility (R = 7.0), T_c is the upper limit of the period of the spectral acceleration chosen. The total base shear, Fb, shall be distributed among the different levels according to the following expression:

$$F_i = \left[\frac{Z_i W_i}{\sum_{j=1,n} Z_j W_j}\right].F_b \tag{9}$$

where F_i is the force acting horizontally on each floor i; F_b is the seismic base shear force obtained from Eq. 1; z_i and z_j are the heights of masses m_i and m_j, respectively; W_i and W_j are the weights of these masses; and n is the number of stories above the foundation level. Eq. 4 gives a linear shear distribution depending only on the storey height.

5   Seismic Design of Prototype Building

In this section, the seismic code provisions for the three countries are applied for the design of the 9-storey regular building structure described in Sect. 2. Key design parameters and results for the equivalent static force procedure are given in Table 2 for the three codes used. Gravity loads are assumed to be as follows; 3.6 kPa (DL), 1.0 (partitions), and 2.4 (LL). For this structure, lateral resistance is provided by two identical perimeter X braced frames (Fig. 1), having total height $h_n = 45$ m from ground level. In-plane torsion is ignored in this study; however, each braced frame is assumed to resist 50% of the applied lateral loads, including stability effects. Climatic loads like snow and wind loads were also ignored in the calculations.

5.1   Design Data

The braces of the structures are designed assuming a yield stress of $F_y = 300$ MPa. The modification factors for tension (ω) and compression (β) are taken equal to 1.4 and 1.1, respectively. In the analyses, the bracing members are assumed to have an equivalent cross-sectional area equal to 1.5 times the core cross-section area A_{sc}. This ratio is typical for braces detailed for high axial stiffness, when drift limits are expected to control the frame design. Beams and columns are assumed to be fabricated from ASTM A992 I-shaped members (US and Canada) with a F_y of 345 MPa, and grade C St-52 I-shaped members (Egypt) with a F_y of 360 MPa. Beams are non-composite and the frames are designed assuming that the beam-to-column connections are pinned.

5.2   Design Using the Equivalent Static Force Procedure

A linear modal dynamic analysis has been developed for the seismic design of the frames. Key design parameters and results for the equivalent static force procedure are given in Table 2 for the three codes used. For all three codes, the frame members were sized to satisfy minimum strength requirements and was then re-analyzed to obtain the fundamental period and the storey drifts. The procedure was iterative till convergence was reached and the final design values are presented in Table 2.

Table 2. Seismic design parameters and results - equivalent static force procedure (/building).

Parameter	NBCC2015	ASCE7-16	ECP2011
T(s)	1.125	1.271	0.868
Modification factor	$R_dR_0 = 3.9$	R = 6.0	R = 7.0
Seismic weight	W = 86 065	W = 85 705	W = 105 200
Base shear	V = 3045	V = 1565	F_b = 4287
Base shear ratio	V/W = 0.035	V/W = 0.018	F_b/W = 0.040
Design base shear	3540	2034	5015
Maximum drift ($/h_s$)	0.031	0.019	0.002
Steel tonnage	175	94	215

As can be observed form the table, values of both the seismic weight and base shear are significantly higher through the Egyptian code compared to its counterparts in the US or Canada. This is can be partially related to the shorter period computed with the Egyptian provisions (0.868 s) compared to 1.125 s and 1.271 s for the Canadian and US standards, respectively. The value of the reduction factor (R = 7.0) in the ECP2011 came in line with the value specified by the American code (R = 6.0), but its worth noting that majority of steel structural systems in Egypt take a reduction factor of either 5.0 or 7.0 based on anticipated ductility, regardless of types or geometry. The Canadian code specify lower reduction factor but this is compensated not applying the 2/3 factor applied to the spectral ordinates obtained. The Egyptian code came the most conservative in obtaining the seismic weight since a factor of 1.4 is applied to the dead load to obtain the ULS combination. The design base shears were obtained after applying the amplification factors specified by each code to satisfy the notional loads and the P-Δ effects. In the NBCC 2015 U_2 = 1.16 was applied to amplify the base shear, and the redundancy factor ρ = 1.3 was applied in the US case. The Egyptian code specifies an amplification factor of $1/(1 - θ)$ based on the ratio of gravity and lateral load applied at each storey, in this building θ = 0.23 and the amplification factor was 1.17 which came in agreement with the factor specified by the NBCC2015. The design base shear specified by the ECP2011 came 1.4 and 2.46 times the ones specified by the NBCC2015 and ASCE7–16, respectively. Satisfying the stringent drift limitations had a major impact on design, the final frame design in Egypt came the heaviest with 215t tonnes which is 23% heavier than Canadian frame (175t) and 128% heavier than the American frame (94t).

6   Conclusions

Seismic design provisions of the Egyptian code of practice were reviewed and compared to their counterparts in global design codes. The seismic design requirements were applied to a single-bay, Single X CBF used in a 9-storey office building located at three different sites with similar seismic and soil conditions. Earthquake effects were determined using the equivalent static force procedure method. The main conclusions of the study can be summarized as follows:

- Egyptian code has a single formula for computing the period of a structure that depends on structural system/material and height of building. However, the formula is very general and does not include much variation of systems or geometries, and was found to yield shorter period value when compared to other codes.
- Base shear values obtained from ECP2011 were found to be much higher than ones computed by other codes, mainly because of the conservatism in calculating the seismic weight taking in consideration entirety of dead load multiplied by factor of 1.4 in addition to a portion of the live loads.
- Force modification factors, base shear ratio, and stability requirement factors (P-Δ effects) of ECP2011 lie within similar ranges compared to NBCC2015 and ASCE7–16.

- ECP2011 imposes stricter drift limits and that has directly affected the design. The final frame design in Egypt came 23% heavier than Canadian frame and 128% heavier than the American frame.
- Using the equivalent static method of the ECP2011 has an overall similar methodology compared to other codes but has proven to be overly conservative and much resources can be saved if current provisions are further optimized.

References

1. NY Times. https://www.nytimes.com/1992/10/13/world/earthquake-in-egypt-kills-370-and-injures3300.html. Accessed 11 Nov 2019
2. Abdel Raheem, K.A., Abdel Raheem, S.E., Soghair, H.M., Ahmed, M.H.: Evaluation of seismic performance of multistory buildings designed according to Egyptian code. J. Eng. Sci. **38**(2), 381–402 (2011)
3. Abdel Raheem, K.A.: Evaluation of Egyptian code provisions for seismic design of moment-resisting-frame multi-story buildings. Int. J. Adv. Struct. Eng. **5**, 20 (2013)
4. ECP: ECP-201 Egyptian code for calculating loads and forces in structural work and masonry. Housing and Building National Research Center, Ministry of Housing - Utilities and Urban Planning, Cairo (2011)
5. NBCC: National building code of Canada, 14th edn. Canadian Commission on Building and Fire Codes, National Research Council of Canada (NRCC), Ottawa (2015)
6. ASCE: ASCE/SEI 7–16, Minimum Design Loads for Buildings and Other Structures, American Society of Civil Engineers (ASCE), Boca Raton, FL (2016)
7. Tremblay, R., Fahnestock, L., Herrera, R., Deghani, M.: Comparison of seismic design provisions in Canada, United States and Chile for buckling restrained braced frames. In: 8th International Conference on Advances in Steel Structures, Lisbon, Portugal (2015)
8. Naqash, M.T., De Matteis, G.: Seismic design of steel moment resisting frames European versus American practice. NED Univ. J. Res. **9**, 45–60 (2012)

A Case Study of the Application of BIM in China: Tianjin Chow Tai Fook Financial Center

Junshan Liu[✉], Dan Li, and Scott Kramer

Auburn University, Auburn, AL, USA
liujuns@auburn.edu

Abstract. Building Information Modeling (BIM) is widely used in the AEC industry. With BIM, project stakeholders can coordinate and manage the information using 3-dimential models throughout the project lifecycle. To apply BIM, innovative measures need to be implemented at the company and project level, and such measures may include the reengineering of the process and business structure, the establishment of hardware and software infrastructure, and training of their employees. China's construction market has continued to grow and its construction industry accounts for a large portion of China's gross domestic product (GDP). There are currently no BIM application standards or regulations for Chinese construction companies, however, the Chinese construction industry and the government have realized the benefits of BIM and are actively promoting the application of this technology. This research aims to gather data from the construction management team of the Tianjin Chou Tai Fook Finical Center project, standing 530 m tall as the eighth tallest building in the world in Tianjin, China, to identify the benefits, challenges, and experience for implementation of BIM in the project.

Keywords: BIM · China · Construction management · Case study

1 Introduction

Building Information Modeling (BIM) is currently one of the most popular topics in the construction industry. BIM is an IT approach applied to the AEC industry to provide a digital representation of infrastructure. Through this technology, users can coordinate and manage the information through graphical models and data. Information is saved, and users can disseminate and manage it throughout the project lifecycle (Autodesk Whitepaper, n.d.). BIM is a tool for changing the dynamics of the engineering and construction industry. BIM uses software to help plan, design, build and manage infrastructures and buildings (Autodesk 2019). This tool also helps the sector eliminate errors and increase productivity. Multiple software applications of BIM such as Revit, Navisworks, Tekla, and Bentley, etc., jointly promote the development of the construction industry. Table 1 shows a list of major BIM software developers.

To apply BIM, innovations need to be implemented at the company or the project level, and such measures may include the reengineering of the process and business structure, the establishment of hardware and software infrastructure, and the education

I. El Dimeery et al. (Eds.): JIC Smart Cities 2019, SUCI, pp. 65–72, 2021.
https://doi.org/10.1007/978-3-030-64217-4_8

Table 1. Major BIM software developers

Major BIM software developer	Country	Main function	Year started
Autodesk	U.S	BIM modeling/CM	1982
Bentley System	U.S.	BIM modeling	1984
Graphisoft	Hungary	BIM modeling	1982
Nemetschek SCIA	Germany	CM/Design-aid	1974
Synchro. Ltd (acquired by Bentley)	U.S.	CM	2001
Solibri (acquired by Nemetschek)	Germany	CM/Design-aid	1996
Tekla Corporation (acquired by Trimble)	U.S.	BIM modeling	1966

and employment of experts. Adopting BIM generating an additional cost for all parties involved in the projects. Owners and general contractors are mainly interested in the return on investment (ROI) of implementation of BIM (Azhar et al. 2012).

China's construction market has been growing rapidly, and the construction industry now accounts for a large portion of Gross Domestic Product (GDP). China's total construction volume is 50 billion square meters, with expansion expected to continue through 2050. And the annual new construction is half of the world's total construction area. (Hong et al. 2014). China also is one of the most important construction markets in the world. According to market size, this figure far exceeds 13 trillion RMB (around two trillion U.S. dollars). This includes substantial investments in the construction of both infrastructures and buildings. There is a great opportunity for technology and processes in China's construction industry to improve industry efficiency and safety (Dodge Data & Analytics 2015). China's construction market is huge, but there are currently problems such as short building life, low material utilization, and low industry productivity, which provides more opportunities for BIM development in China. Increasing industry productivity through the development of BIM technology and improving project delivery time will be a major topic in the construction industry in the coming years.

The Chinese construction industry and the government are actively promoting the application of BIM technology. BIM technology started late in China, but more and more companies are beginning to pay attention to BIM and use it. With the implementation and application of BIM technology, many barriers and unknown factors have been encountered, which creates both challenges and opportunities for China's construction companies. Various high-profile construction projects are being carried out across China, and BIM technology is increasingly being used through innovated processes. These processes can not only optimize the design but also improve construction productivity. In the next few years, enterprises with higher BIM technology application in China will grow by 108 percent, ranking among the top five countries and regions with the fastest growth of BIM application (Dodge Data & Analytics 2015).

There is currently no BIM application standards for Chinese construction companies. However, there are many construction companies trying to adopt BIM technology, and there are a lot of construction projects that are applying BIM technology right now. The experiences of these companies are valuable. By collecting and analyzing data from these projects, researchers can provide information and help Chinese companies adopt BIM technology and apply it to future projects.

The aim of this research is to investigate the implementation, benefits, risks, and challenges of the applications of BIM during the construction phase in the Tianjin Chow Tai Fook Financial Center (TCTFFC) project.

2 Methodology

Tianjin Chow Tai Fook Financial Center (TCTFFC) was developed by New World China Real Estate Co., LTD. It is a new landmark building in Binhai New Area, Tianjin, China. As shown in Fig. 1, its main building is 530 meters tall, which is known as "the diamond of the north". It is ranked as the tallest building in the north of the Yangtze River, the fourth tallest building in China and the eighth tallest building in the world in the category of "completed and capped super high-rise buildings". The total construction area of this project is about 390,000 square meters, which consists of four basement floors, five podium floors, and one 100-floor-tower. The general contractor of TCTFFC is China Construction Eighth Engineering Co., LTD. BIM technology was used by the project's construction team to integrate and manage the construction process of concrete structure, steel structure, MEP systems, curtain wall systems, and interior decoration.

Fig. 1. A rendering of the TCTFFC.

There were three main steps in this case study, as follows:

- Literature review.
- Analysis of the project files provided by the project BIM team.
- Online Interviews with the key personnel's of the project's BIM team.

However, due to the page limitation, this paper only presents a part of the research data gathered from the project files and its analysis.

3 Results

According to the general contractor, the design and construction drawings of TCTFFC provided by the architecture and engineering firm were all in 2D format, which means BIM technology was not implemented in the design phase. Due to the complexity of the project, communication and coordination between various parties are critical to success. Three of the major challenges that the owner and general contractor were facing during the construction of the project were:

- Complexity of the communication and coordination
- Complexity of the structure system
- Complexity of the curtain wall system

3.1 Project Challenges

3.1.1 The Complexity of the Project Communication and Coordination
The TCTFFC building has four floors underground and 100 floors above ground, which creates many working areas, as shown in two of construction photos of the project in Fig. 2. Subcontractors carried out construction on each floor and working areas according to the construction process, there were many working procedures in each trade, and shifting between trades was frequent. Managing and sharing construction documents, identifying construction tasks of each working area, organizing and managing subcontractors and crews, and determining and adjusting the construction schedule were just some of the problems faced by the general contractor on a daily basis. There were a lot of design errors and drawing corrections in this project. Making the construction documentation process be synergistic among many subcontractors is the key point of the management. The project used Revit on architectural and MEP modeling, and Tekla were used on structural steel modeling. Other trades used different modeling software and exchanged building information with Revit by using specific formats.

3.1.2 The Complexity of the Structure System and Its BIM Resolution
TCTFFC's tower core tube has a very complex structure system. After six times of descending upward, the overall plane area of the tower core tube reduces by nearly one half. The outer area of the tower increased gradually from the second floor, reaching the maximum at the sixteenth floor, and then decreasing gradually. The structure of sixteenth through fifty-first stories changed greatly, but the structure of fifty-first through eighty-eighth stories changed slightly. This unique structure brought great challenges to the layout of construction elevators and tower cranes. The structure system of the building is shown in Fig. 3.

Fig. 2. Project under construction (Source: skyscapercenter)

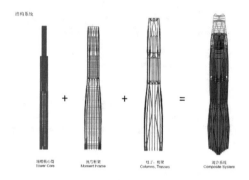

Fig. 3. Structure framing system of the project.

3.1.3   The Complexity of the Curtain Wall System

The TCTFFC main tower's curtain wall facade was composed of eight curved arcs, from the bottom to the top. The arc lengths of each floor were different, forming a hyperboloid structure of the external facade. The building presented a uniform streamline shape extending upward. In total, there were over 14,800 curtain wall units of the tower building. Among them, there were about 12,700 units of standard shape, 900 units of V-shaped orifices and 1,200 units of tower crowns. The total area of the curtain wall system was about 110,000 square meters. The variety of shapes and the unique design concept significantly increased the difficulty of curtain wall in design, fabrication, and installation. Ensuring the quality of curtain wall construction and the accuracy of component production was a major problem faced by the contractor of the project.

3.2   BIM Resolutions

To overcome these challenges and achieve project success, a BIM management team consisting of more than 40 BIM engineers had been created, as show in Fig. 4 for the team's organizational chart. The team leader was responsible for managing the team and directly reporting to the general contractor. The task of BIM modeling was divided into eight different independent categories since a single BIM file would be too large to handle. In order to facilitate the management, the model was divided into architecture, structure, mechanical, electrical, plumbing, fire protection, structural steel, and curtain wall system. The modeling files were stored in DWG, IFC, NWC, and RVT formats. The purpose of storing documents in various formats was for data exchange and compatibility. This chapter explains how this BIM team worked and how they resolved some of the challenges.

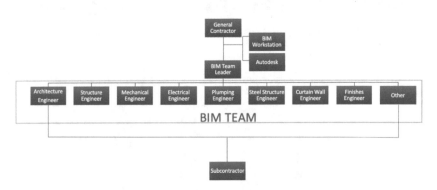

Fig. 4. Project's BIM team organizational chart.

3.2.1   BIM Management Platform

At the beginning of the project, aiming at the characteristics of complex construction drawings and subcontracting, the project team developed an implementation plan of BIM application and formulated the management process of BIM construction documentation. In order to meet the needs of managing the construction documents and changes, a BIM collaboration platform was also created, as shown in Fig. 5.

This workflow was based on the web-based BIM management platform, and it used BIM models as the carrier to carry out the construction documents and information exchange. This workflow had clear management level to ensure the efficiency of drawing corrections. And this information-based work flow improved the application efficiency of BIM models in the construction stage. Each trade had a team of engineers to work with models in which they would integrate and check the models for errors and omissions. Based on collision detection of both a single trade and the combined

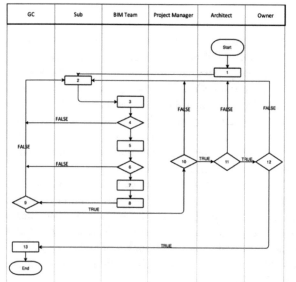

GC	Sub	BIM Team	Project Manager	Architect	Owner

Figure Legend:
1. Design Review/Clarification
2. Detail Design (Trade)
3. BIM Modeling
4. Clash Detection (Trade)
5. Comprehensive Arrangement (Trade)
6. Clash Detection (Project)
7. Comprehensive Arrangement (Project)
8. Construction Drawing (Project)
9. Approval
10. Approval
11. Approval
12. Approval
13. Construction Instruction

Fig. 5. BIM construction documentation management process

models, changes were proposed, and construction drawings were exported after the models were being adjusted and refined, and then signed by the project stakeholders.

In addition to the strict requirement of construction schedule from the owner, the design drawings issued in the early stage were full of problems, which led to the increase of the intensity and difficulty of the construction documentation process. Before the start of construction, the BIM team completed the modeling work in 28 days and submitted 131 comprehensive reports related to BIM works. Based on the BIM management platform, multiple models were created for each trade. These models were then combined, which led the management team to be able to identify problems and corrected them on time. An LOD 400 model was established according to different construction requirements in the modeling stage.

3.2.2   BIM and 3D LiDAR Scanning for Curtain Wall Design and Installation

In order to assist the shop drawing design, fabrication and installation of the complicated curtain wall system, both BIM technology and 3D laser scanning were used in this project. The 3D point cloud data of the building structure was captured by laser scanners. Point cloud data was processed and then used to create curtain wall models. As-built models of structural and architectural systems that had already been completed were used for the final design and fabrication of the curtain wall system; these as-built models were created by precise building data captured by laser scanners. Laser scanning had many benefits to the contractors in fabricating and installing curtain wall units more accurately and improved the efficiency of curtain wall system installation.

4 Conclusion

For a long time, the construction industry has been characterized by inefficiency. Traditional construction methods used 2D drawings to guide construction, which caused material waste and schedule delay. The development of BIM technology has created revolutionary changes to the construction industry. The application of BIM technology can reduce rework, material waste, and shorten construction schedule. BIM can help project owners make a timely decision and help contractors identify design errors in the early stage. BIM can also simulate the construction process, through which the construction party can identify other issues that may occur in the construction process and make feasible plans to resolve them.

China has one of the World's largest and most important construction markets. With China's economic development, the demand for high-rise buildings is also increasing. High-rise buildings are usually characterized by large scale, complicated working procedure, long construction period, and high cost. The advantages of BIM technology perfectly match the needs of design and construction of high-rise buildings. Therefore, the application of BIM technology in high-rise building construction is arguably the most popular trends in China's construction industry.

This study took Tianjin CTF Financial Center project as a case study. Through the collection of project data, this research discussed in details how the project utilized BIM technology in the construction phase. It was witnessed that the project obtained a lot of benefits by adopting BIM. As a result, the project was successfully completed on time, with minimum amount of rework and waste of materials. And the cost of the project was under the budget. Overall, the project has improved construction efficiency with BIM technology. Through this project, the BIM team's engineers learned valuable experience in the application of BIM technology in high-rise buildings. This is also an important benefit for the general contractor.

References

Autodesk: Autodesk Whitepaper (n.d.). http://www.laiserin.com/features/bim
Autodesk.com: What is BIM—Building Information Modeling—Autodesk (2019). https://www.autodesk.com/solutions/bim. Accessed 12 Feb 2019
Azhar, S., Khalfan, M., Maqsood, T.: Building information modelling (BIM): now and beyond. Constr. Econ. Build. **12**(4), 15–28 (2012). https://doi.org/10.5130/AJCEB.v12i4.3032
Dodge Data & Analytics: The business value of BIM in China. Dodge Data & Analytics Research & Analytics, Bedford (2015)
Dodge Data & Analytics: The business value of BIM for construction major global markets. Dodge Data & Analytics Research & Analytics, Bedford (2013)
Hong, L., Zhou, N., Fridley, D., Feng, W., Khanna, N.: Modeling China's building floor-area growth and the implications for building materials and energy demand, p. 12 (2014)

Identification of Wastes in Construction Projects: Case Study of Porto Sokhna Island Project

Islam El-Sayed[1], Ahmed Abaza[1], Alyaa Kamel[1],
and Rana Khallaf[2(✉)]

[1] Cairo University, Giza, Egypt
[2] Structural Engineering and Construciton Management,
Future University in Egypt, Cairo, Egypt
rana.khallaf@fue.edu.eg

Abstract. One of the main issues that impact the construction industry is the prevalence of wastes. These wastes have negative effects on the project including higher cost and need for larger inventory. One of the main tenets of lean is to eliminate these wastes in order to increase the value to the customer. This paper focuses on lean applications in construction projects, specifically on how to identify and classify wastes, and how to eliminate/reduce them. Firstly, literature review is conducted to collect the different classifications of wastes in construction. This is followed by a discussion of the eight types of wastes identified in context of construction projects and an application to a real-life case study, the Porto Sokhna Project in Egypt. This is performed to identify wastes that occurred and their impact as well as propose elimination/mitigation strategies. Project documents were analyzed and interviews were conducted to reach this. These wastes were then classified under eight main waste types. The effect of these wastes was then monitored and recorded in terms of cost and schedule. Finally, solutions were proposed to eliminate/mitigate the wastes.

Keywords: Lean construction · Project controls · Wastes · Case study

1 Introduction

Construction projects are known to produce high volumes of wastes as a result of poor planning and management of the construction process. These wastes not only refer to material, but also wastes in time, human resources, as well as rework. These wastes in turn can affect the cost, time, quality, and value of the construction process as well as final product. For example, wastes in construction include: excessive transport, unnecessary storage, idle equipment, excessive buffers between activities, long decision-making process, rework, and inefficient use of resources. These wastes usually lead to an increase in project cost, time, or both, and can also have an effect on the quality of the final product. Egypt's 2030 sustainable development vision proposes a multitude of megaprojects such as the construction of one million housing units for low-income citizens at a total cost of LE150 billion (Official Publication of the Government of Egypt 2015). Hence it is imperative to introduce lean to the construction sector in Egypt for the successful delivery of these megaprojects. This paper

I. El Dimeery et al. (Eds.): JIC Smart Cities 2019, SUCI, pp. 73–79, 2021.
https://doi.org/10.1007/978-3-030-64217-4_9

focuses on the identification of wastes in construction projects and proposes mitigation strategies for the Egyptian construction sector.

2   Lean Implementation and Elimination of Wastes

Lean as a concept focuses on maximizing value and minimizing waste. It was first applied in the manufacturing industry by Taiichi Ohno, who is considered the father of lean. Ohno (1988) proposed the first classification of wastes in production based on lean thinking, which are: inventory, motion, transportation, defects, over processing, overproduction, and waiting. Although lean started in manufacturing, its concepts are applicable to the construction area as well. Even though each construction is considered one-of-a-kind (Koskela et al. 2013) since it is a unique product created through a temporary endeavor, there are common wastes that can be observed. Over the years many researchers have attempted to classify wastes in the construction sector. Table 1 shows a select result of a literature review conducted to collect the different classifications of wastes from previous research. These wastes affect projects negatively and contribute to delays and cost over-runs. In order to countermand the effects of wastes, lean has been successfully implemented in the construction sector to deliver projects while reducing these wastes and improving project performance (Li et al. 2016). This research uses a combination of these classifications and applies it to a real-life case study.

Table 1. Classification of wastes in construction

Reference	Classification
Alarcon (1997)	(1) Controllable causes associated with flows (related to resources, equipment, and labor); (2) controllable causes associated with conversions (related to planning, procedures, and quality); and (3) controllable causes associated with management activities (related to supervision and decision-making)
Formoso et al. (1999)	(1) Overproduction; (2) substitution (of a material by a more expensive one with unnecessarily better performance); (3) waiting time; (4) transportation; (5) processing; (6) inventories; (7) movement; (8) production; and (9) others
Khanh and Kim (2014)	(1) Management/administration; (2) people; (3) execution; (4) material/machines; and (5) information/communication
Llatas (2011)	(1) Packaging (of materials to be transported on-site); (2) remains (left-over building materials); and (3) soil (left-over materials from excavation)
Nikakhtar et al. (2015)	(1) Construction site related (e.g. idle time, debris, and excess materials); (2) external factors related (e.g. design errors); and (3) construction processes related (from the operations or non value-added work)
Ramaswamy and Kalidindi (2009)	(1) Material (classified into excess inventory and scrap waste); (2) quality (e.g. rework); (3) labor; and (4) equipment. (3 and 4 are both classified into: waiting, idle, transportation, excess processing, and excess movement)
Watson (2014)	(1) Time; (2) cost; (3) quality; (4) safety; and (5) environment

Although lean has been introduced to the construction industry in Egypt, it is still considered in its infancy stage. The implementation of lean has been restricted to limited research and case studies and has not been applied to large-scale construction projects. Swefie (2013) conducted a survey and reported that 55% of the experts in the Egyptian construction sector were not aware of lean. In a similar survey, Abo-zaid and Othman (2018) reported that 30.4% were completely unaware of lean, which confirms that lean is slowly spreading in the market. They also reported that the highest ranked source of waste was from damaged materials during construction and improperly stored materials. Issa (2013) used the last planner system on a flour milling factory project in Egypt in order to reduce the effect of risk factors on project time. Reda and Khallaf (2019) surveyed experts in the Egyptian construction industry and reported that 91% were not aware of the Last Planner System (LPS), which is a collaborative planning tool used in lean practices. They also proposed a framework for LPS adoption and implementation in construction projects in Egypt.

3   Application on the Porto Sokhna Project in Egypt

3.1   Project Background

The Porto Sokhna Project was launched in 2007 with over 2.5 million square meters of residential units in a compound overlooking the Red Sea in Egypt. In this study, case study research was conducted using document analysis and structured interviews to collect data on the Porto Sokhna project. This was made possible because a number of the authors currently work on this project. Hence, the following methods were used for data collection:

- Project document analysis including contract documents, specifications, and project plans.
- Observation of the project through participation and meetings.
- Interviews with key personnel in the projects.
- Quantitative data analysis on project cost, schedule, and quality.

The following sections outline the wastes identified in the project and their impact on the project budget and schedule. Wastes were classified into eight categories: material, inventory, motion, time, cost, quality, labor inefficiency, and non-utilized equipment (a combination of the classification methods identified in Table 1).

3.2   Classification of Wastes

1) Material Wastes: Materials generally represent the highest percentage of expenses in construction. Hence, poor material management can cause large amounts of wastes during construction. In this case study, it was found that there are material wastes due to two main reasons: (i) bad storage, and (ii) usage of materials by the contractors. Using document analysis, the actual amount of work completed was compared to the material needed to finish the work and the percentage of waste identified was over 10% (the amount allowed by the contract).

2) Inventory: In order to ensure that all required materials are available for their time of use, contractors usually have an inventory either on or off-site. This is also done to reduce risk of supplier delay in transporting materials to the site. In this project, it was found that there are wastes stemming from excess inventory. For example, excess inventory was found for cement with two tons of the material beyond their expiration date. According to storage conditions, cement should not be stored over 60 days.

3) Motion: This type of waste focuses on the excess movement of labor, especially on-site during project operations. As the unnecessary motion of labor increases, this leads to time wasted between value-adding activities. In this project, it was observed that the presence of only one car hindered operations since engineers needed longer times to move from one project area to another across the large space. This led to poor project monitoring and performance of labor due to absence of continuous supervision.

4) Waste in time: In order to track project progress in terms of time/schedule, the Schedule Performance Index (SPI) was calculated and found to be 0.8. Since it was less than 1, this indicated that the project was behind schedule, with the main reason being variation orders requested by the owner/consultant.

5) Waste in Cost: The Cost Performance Index (CPI) was calculated to determine the project progress with respect to cost. A CPI of 0.7 was calculated, which meant that the project was over budget. This can be linked to the expired materials found and poor quality observed in some of the work.

6) Poor Quality: Poor quality leads to more defects and rework and can result from bad practices, material, or execution. In this project, a number of non-conformance reports were found with the consultant refusing to accept certain works from the contractor.

7) Labor Inefficiency: In order to identify the labor inefficiency, their performance rate was calculated and monitored. A drop in performance rate was observed, which led to an increase in the time to perform activities and the associated costs.

8) Non-utilized Equipment: Daily logs and project reports were used to determine the available equipment and their usage. For example, it was found that although there are four loaders, only two of those were in use. This means that the equipment was not utilized to the maximum possible.

3.3   Impact of Wastes on Project Budget and Schedule

Table 2 shows the wastes classified into the eight aforementioned types. Mitigation strategies are proposed for each type of waste as well.

Table 2. Wastes in the Porto Sokhna Project

Waste type	Waste in Porto Sokhna Project	Mitigation strategy
Material wastes	• Wastes in materials more than the allowable (contractual) 10% • Increase in cost • Delay in schedule from the additional time needed to purchase more materials and wait for their arrival	• Issue a fine to contractors that exceed the 10% limit on waste • Build an on-site storage space for inventory with a high ceiling to protect materials from adverse weather
Inventory	• Some materials had expired due to long periods of storage • Lack of protection from bad weather such as rain, which led to damage in material • Increase in cost from storing materials (which had expired in some cases)	• Decrease the amount of materials stored and apply the 'Just-in-Time' principle to procure material only before use and not in advance
Motion	• Poor monitoring and control of workers • Productivity decrease due to workers' effort in moving and maneuvering around the site on-foot • Increase in the time required for some activities	• Rent more cars for easy transportation of engineers in the site
Waste in time	• Increase in the expected duration to finish the project • Increase in cost to finish the project on-time	• Use Building Information Modeling (BIM) to create a 4D model to monitor and control the time schedule • Use 'pull' scheduling to eliminate wastes and use Kanban boards
Waste in Cost	• Increase in the expected cost to finish the project • Decrease in the quality in order to decrease the total cost • Increase in time in order to decrease the cost (instead of fast-tracking the project) • Disputes	• Apply BIM to create a 5D model to monitor and control cost • Apply value engineering principles to identify cost-reducing alternatives
Poor Quality	• Increase in time resulting from errors and rework • Increase in cost due to consumption of more materials for rework • Increase in defects which led to customer dissatisfaction • Increase in claims and variation orders	• Make a quality control plan and conduct random inspection tests • Impose a fine for any nonconformance issues

(*continued*)

Table 2. (*continued*)

Waste type	Waste in Porto Sokhna Project	Mitigation strategy
Labor inefficiency	• High temperatures led to lower productivity than expected	• Avoid times of high temperatures by having two shifts from 8 a.m. to 12 p.m. and another shift from 5 pm to 9 pm instead of 1 shift from 8 a.m. to 5p.m.
Non-utilized Equipment	• Increase in cost and time due to the increase in the number of non-utilized equipment (such as loaders)	• Create a resource optimization plan with a focus on loaders in this project specifically

After analyzing the wastes in this project, suitable mitigation/elimination strategies for the project are proposed as follows: (i) apply just-in-time to reduce inventory and eliminate resulting wastes; (ii) sign contracts with the suppliers to ensure their commitment to the deliveries' exact date and time (to minimize the cost of excessive inventory, reduce the need for transportation of the materials from storage to their place of use on-site, and lower waste levels); (iii) involve all stakeholders for the upcoming stages; (iv) hold weekly meetings (or daily meetings if required) to implement pull planning and create six-week look-aheads to create adaptive and not predictive schedule; (v) transform the CAD workflow to BIM workflow for future phases to create 4D and 5D models (adding cost and schedule) and achieve the collaboration and integration; (vi) improve the coordination between the engineers and foremen to improve work and with other departments, and train the personnel; (vii) create continuous flow, adopt standardized work, and apply value stream mapping; and (viii) mistake-proof the construction processes (Aziz and Hafez 2013; Khanh and Kim 2014; Nikakhtar et al. 2015). Other strategies include applying value engineering to propose alternatives (to eliminate wastes) and training the engineers and labor to educate them (on value-added, non-value added, and wastes in activities). Material Logistics Plan (MLP) is also proposed to be applied using BIM to better plan project resources. Combining the proposed lean principles with BIM can impact the project through: 1) increased quality; 2) reduced project time; 3) increased flexibility; 4) reduced defects; 5) decreased inventory (JIT); 6) enhanced visual management; 7) continuous improvement; and 8) collaboration in the design stage and throughout the project life cycle.

4   Conclusions

The construction industry in Egypt requires a cultural change in order to implement lean effectively. This can be achieved firstly through conducting awareness sessions to introduce lean and providing training courses for project participants. This paper aims to increase awareness and understanding on the amounts of wastes in construction projects in Egypt and the benefits that can be obtained from the application of lean in the construction sector. A case study of the Porto Sokhna Project was presented. Firstly, wastes and their impact on the project were identified. This was followed by proposing strategies to mitigate/eliminate these wastes using lean principles and techniques to be applied in the upcoming phases of the project. Some of these techniques include BIM, Material Logistics Planning, Just-in-Time, and Kanban Boards. In order to ensure the use of lean, the Egyptian government can impose regulations, which would be aimed at supporting the Sustainable Development Strategy for the 2030 Vision. This case study has proven lean to be a useful and practical method to improve construction projects by identifying and eliminating wastes.

References

Alarcon, L.: Tools for the Identification and Reduction of Waste in Construction Projects, 1st edn. CRC Press, London (1997)

Aziz, R., Hafez, S.: Applying lean thinking in construction and performance improvement. Alex. Eng. J. **52**(4), 679–695 (2013). https://doi.org/10.1016/j.aej.2013.04.008

Khanh, H.D., Kim, S.Y.: Identifying causes for waste factors in high-rise building projects: a survey in Vietnam. KSCE J. Civ. Eng. **18**(4), 865–874 (2014). https://doi.org/10.1007/s12205-014-1327-z.   https://pdfs.semanticscholar.org/a9c0/c73313a2a1dbe7df2463781a7f83e618901f.pdf

Koskela, L., Bølviken, T., Rooke, J.: Which are the wastes of construction? In: Proceedings of the 21th Annual Conference of the International Group for Lean Construction, Fortaleza, Brazil (2013)

Li, S., Wu, X., Zhou, W., Liu, X.: Renew. Sustain. Energy Rev. (2016). http://dx.doi.org/10.1016/j.rser.2016.12.112

Llatas, C.: A model for quantifying construction waste in projects according to the European waste list. Waste Manag. **31**(6), 1261–1276 (2011)

Murata, K., Tezel, A., Koskela, L., Tzortzopoulos, P.: Sources of waste on construction worksite: a comparison to the manufacturing industry. In: González, V.A. (ed.) Proceedings of the 26th Annual Conference of the International Group for Lean Construction (IGLC), Chennai, India, pp. 973–981 (2018). https://doi.org/10.24928/2018/0280

Nikakhtar, A., Hosseini, A.A., Wong, K.Y., Zavichi, A.: Application of lean construction principles to reduce construction process waste using computer simulation: a case study. Int. J. Serv. Oper. Manag. **20**(4), 461–480 (2015). https://doi.org/10.1504/IJSOM.2015.068528

Official Publication of the Government of Egypt: Sustainable Development Strategy: Egypt's Vision 2030. Egypt Economic Development Conference, Sharm El-Sheikh (2015)

Ohno, T.: Toyota Production System; Beyond Large Scale Production. Productivity Press, Portland (1988)

Ramaswamy, K., Kalidindi, S.: Waste in Indian building construction projects. In: Proceedings of 17th Annual Conference of the International Group of Lean Construction, pp. 3–14 (2009)

Reda, E., Khallaf, R.: A framework for last planner system implementation in Egypt. In: Proceedings of the Congrès International de Géotechnique - Ouvrages - Structures (CIGOS) Innovation for Sustainable Infrastructure in Hanoi, Vietnam (2019)

Swefie, M.: Improving project performance using lean construction in Egypt: a proposed framework. Masters Thesis, American University in Cairo (2013)

Watson, G.: Measurement of waste in concrete construction using lean construction methodologies. Dissertation Presented to the Faculty of Engineering and Surveying at the University of Southern Queensland (2014)

The Role of Geoinformatics in Renewable Energy Potential Estimation for Smart Cities - Emphasis on Solar and Wind Energy

Ahmed Agwa$^{(\boxtimes)}$, Moustafa Baraka, and Ahmed A. Sattar

German University in Cairo, Cairo, Egypt
ahmed.adel-mahmoud@guc.edu.eg

Abstract. Smart cities aim to enhance and sustain the quality of life for its residents. The focus in this research is to examine the potential location of constructing smart cities based on a major aspect of the availability and abundance of renewable energy that can be harvested to sustain such a city, with emphasis on wind and solar energy. Potential estimation for such renewable energy resources is investigated. Geoinformatics involves integrating and processing of geospatial data obtained from different earth measurement technologies, along with the capability of integrating attribute data (e.g. socio-economic). The ability of Geoinformatics to involve various data input in the process of decision-making has made Geoinformatics a viable approach for energy potential estimation. The role of Geoinformatics in estimating the potential of renewable energy is crucial and spatial analysis is utilized to build relations between different input parameters to reach for a potential production of energy at specific location.

Keywords: Smart City · Geoinformaics · Solar energy · On-shore wind energy · Potential estimation

1 Introduction

The term "city" is not a new term as it appeared since ancient times back to Sumer in 3500 to 3000 BC describing urban clusters governed by certain legal definitions. The term kept the same function implicitly till now but it evolved over years. Cities since the industrial revolution witnessed significant changes in their structure and witnessed increasing population [1]. According to United Nations Department of Economic and Social Affairs, 55% of the population of the world in 2018 are cities' residents with anticipation of increase to touch 68% by the year 2050 [2]. Subsequently, cities become hub areas of demands and consumption all over the world leading to arising calls for promoting and achieving environmental sustainability. Cities also become source of greenhouse gases (GHG) emissions [3]. Thus, since 1990's researches started to use the term "Smart City" as an indication of the ability of a city to enhance sustainably the quality of life for its residents and facing the new challenges related to the growth of population. One of the aspects that shapes the quality of life, also considered as one of challenges, is the energy sector especially the contribution of renewable energy sources

© The Author(s), under exclusive license to Springer Nature Switzerland AG 2021
I. El Dimeery et al. (Eds.): JIC Smart Cities 2019, SUCI, pp. 80–87, 2021.
https://doi.org/10.1007/978-3-030-64217-4_10

in fulfilling the energy demand as it is one of the indicators that measures how smart a certain city is [4]. In order to make use of renewable energy resources, planning including the investigation of the potential of these resources shall be implemented to assess the economic potential of exploiting these resources. A powerful tool that assists in investigating such potential by establishing spatial relations of different parameters is Geo-informatics.

Spatial and temporal datasets from different sources are crucial for Smart Cities planning. Planning for renewable energy, as one of the aspects of Smart Cities, is a good example as renewable energies such as solar and wind are intermittent temporally. Moreover, the availability and intensity of solar and wind energies also differ geospatially. Now, the question is "how to capture, store, mange and visualize such data?"... The answer is GEOINFORMATICS [5].

Geoinformatics necessitates concluding computational, analytical and statistical methodologies in order to better understand the geospatial data and solve many engineering challenges. This paper reviews the role of Geographic Information System (GIS), as an application of geoinformatics, in the estimation of solar and wind energies potential [5].

2 Methodology

The process of potential estimation of solar centralized systems and on-shore wind energies goes into different phases starting from estimating the theoretical potential and it ends up with estimating the economic potential. In other words, solar and on-shore wind energies potential estimation is sequential and cascading process that follows the following cascading down series of potentials (this method is widely known in literature as Top-down approach) [6, 7]:

1) Theoretical potential:
   The theoretical potential is the initial and broadest phase in the chain of potential estimation in which the total amount of the resource available within the area under consideration is introduced. For the solar energy, the governing parameter is the solar radiation, the higher the solar radiation is the higher the theoretical potential is. On the other hand, for the on-shore wind, the governing parameter is the wind speed at the turbine hub height which is translated into the corresponding wind power density which is the kinetic potential of wind energy and it is function in the third power of the wind speed average and air density. Areas with wind speed ranges from 7.5 m/s till 9.4 m/s or more and corresponding to wind power density from 500 W/m^2 till 978 W/m^2 or more are spots with promising wind potential [8].
2) Geographic potential:
   The geographic potential is the theoretical potential after identifying the areas where solar and on-shore wind energies are exploitable by excluding areas with geographic constraints out of the total area under consideration. The exclusion process entails defining the criteria based on which area funneling takes place. The exclusion criteria includes geographic constraints such as: urban built-up areas, water bodies, protected areas, highly elevated areas, sloped areas and distance to grid.

3) Technical potential:

The technical potential is the geographic potential after applying reductions due to losses that occurs because of converting the solar and wind energies into electricity. The losses take place as a result of technical limitations such as the efficiency of conversion and to-grid-transportation. For the solar energy, the ambient temperature plays a role in the power production process of photovoltaic cells and it heavily influences the efficiency of the cells. For the wind energy, the turbines power curves and spacing of turbines are the main points to be considered during this level of potential estimation.

4) Economic potential:

The technical potential is fine tuned in order to reach the level of the economic potential. Within this level of potential estimation; land price, initial cost, operation cost, and maintenance cost are considered. Additionally, the price of electricity and other socioeconomic factors are also considered (Fig. 1).

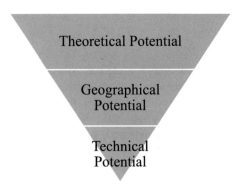

Fig. 1. Top-down methodology [7]

3 Data Acquisition

3.1 Wind Speed Data

Mentis et al. (2015) worked on estimating the onshore wind potential for the whole African continent based on annual and high resolution wind speed data obtained from VORTEX. The height above ground of VORTEX wind speed was 80 m. The weakness point of VORTEX wind speed data is its limitation of application in large scale studies because it is not ground validated. Mentis, et al also made use of satellite-based daily wind speed data at height of 10 m above the ground, from January 1983, from NASA Climatology Resource for Agroclimatology [8].

In EEA report (Europe's onshore and offshore wind energy potential), average wind speed data from the year 2000 till the year 2005 obtained from European Centre for Medium-Range Weather Forecasts (ECMWF) is utilized. The wind speed data is further modified to embrace the surface roughness that accounts for the influence of different land cover types over the wind speed. Corine Land Cover database (CLC) was used for this purpose [9].

Beata Sliz-Szkliniarz et al. (2011) estimated the wind energy potential of Kujawsko–Pomorskie Voivodeship. Wind speed data from the website of the National Climatic Data Centre formed the base of the potential estimation process. The wind speed dataset was concluded from the data obtained via 28 meteorological stations recording wind speed data along year 2005 till year 2009 in order to estimate the average daily wind speed data [10].

3.2 Solar Radiation Data

Angelis-Dimakis et al. (2010) divided the approaches by which solar irradiation (direct and diffuse radiation) is estimated into three approaches; first, obtaining field data from the widely spaced meteorological stations and in order to fill the gaps in between the stations extra measured data such as sunshine duration and cloud cover is required. Furthermore, interpolation techniques such as 3D inverse interpolation are applied. This approach also requires complementary aid (empirical model) to estimate the diffuse radiation. The second approach is utilizing satellite imagery obtained from geostationary satellites such as Meteosat. The last approach combines the first two approaches by utilizing satellite imagery data for areas with low meteorological stations density [7].

Korfiati et al. (20160 used freely available datasets including solar irradiation data for the purpose of estimating the global solar energy potential and photovoltaic cost. The irradiation data was acquired form NASA Surface meteorology and Solar Energy dataset (SSE – Release 6.0) [11].

Sun et al. (2013) considered Fujian Province, China to apply a GIS-based approach in the potential estimation of photovoltaic cells. The solar radiation map of Fujian was calculated through ArcGIS 9.3 specifically the solar radiation analyst module which require input data such as atmospheric effects, latitude, elevation, slope and aspect [12].

4 Data Processing

4.1 Wind Data

The onshore wind speed data captured by meteorological stations is recorded at certain height above the ground (anemometer height) which may differs from the hub height of turbines. Thus, extrapolation techniques are utilized in order to estimate the wind speed at the required hub height. Turbines hub height is one of the technical specifications offered by manufacturers for example the hub height of turbine Bonus 600 is 50 m and for turbines Vestas 82 and Nordex N80 are 80 m and 100 m respectively. The change in the vertical profile of the wind speed is determined by Log Law or Power Law Profile. The mathematical formula of Log Law is as the following [13]:

$$U_z/U_{z_r} = ln\left(\frac{z}{z_o}\right) \bigg/ ln\left(\frac{z_r}{z_o}\right) \tag{1}$$

where U_z is the wind speed recorded at height z, z is the anemometer height, U_{z_r} is the wind speed at the turbine hub height, z_r is the hub height (50, 80 or 100 m) and z_o is the roughness length. On the other hand, Power Law Profile mathematical formula is [13]:

$$U_z/U_{z_r} = (z/z_r)^{\alpha} \tag{2}$$

where α is the wind shear exponent which is correlated based on the surface roughness according to Counihan (1975). Justus (1978) also proposed a correlation based on the speed and height. Another correlation, proposed by NASA, relied on both of wind speed and surface roughness.

Wind speed which was calculated from extrapolation is then interpolated to fill the gaps in between meteorological stations. Interpolation of meteorological data such as wind speed is interpolated through deterministic or geostatistical methods. However, geostatistical methods namely kriging methods gives most accurate results compared to deterministic techniques [10, 14]. The interpolation process is validated by using cross-validation tool of ArcGIS Geostatistical Analyst. Once the wind speed interpolation surface is ready beside the air density, the theoretical potential is estimated as previously stated in Sect. 2.

The geographical potential is then estimated by overlaying layers of the exclusion parameters mentioned in Sect. 2 to identify the fraction of the area under consideration with constraints. Subsequently, the remaining area is suitable for wind turbine installations.

Statistical analysis of wind speed namely Probability Distribution is implemented to estimate the wind speed distribution over certain period with a view to assess the effect of the intermittency and change of the wind speed over the output power of wind turbines. Wind speed Probability Distribution is conducted through Rayleigh or Weibull approaches yet the second approach shows capability of embracing range of wind regimes. Moreover, the power curve of a wind turbine is applied for the sake of determining the turbine produced power as a function in the wind speed. The following mathematical expression gets the annual energy yield [8]:

$$E_{expected} = \mu.T \sum_{wind\ class} P(U) \cdot f(U) \tag{3}$$

where μ is availability factor that accounts for turbine outages, T is the hours of a year, $P(U)$ is the power curve and $f(U)$ is the probability based on wind classes from 0 to 25 m/s with 0.5 m/s interval. Further a reduction that accounts for the array and spacing of turbines is applied to get finally the technical potential.

4.2 Solar Data

The potential of the solar energy is initiated through estimating the solar radiation map of the area under study by utilizing the solar radiation analyst tool in ArcGIS. This tool takes into account the effect of many factors such as slope, aspect, the change in the sun angle and terrain shading by feeding the tool with Digital Elevation Model (DEM), which is mainly produced by the Shuttle Radar Topography Mission (SRTM), and the coefficient of atmospheric transmissivity (the fraction of solar radiation passes through the atmosphere). Once the solar radiation map is calculated, the theoretical potential is estimated by excluding areas with insufficient solar radiation. Furthermore, the process of defining the geographical potential almost follows the same exclusion criteria previously stated for wind geographical potential estimation [12, 15]. For estimating the annual technical potential of centralized PV system in Gigawatt hour, the following expression is used [16]:

$$Technical\ potential = \frac{Solar\ resource * efficincy * Available\ area}{Spacing\ factor} \quad (4)$$

where solar resource availability is in kilowatt hour/ m^2/year, PV module efficiency is percentage and the available area is in km^2.

Additionally, for estimating the annual technical potential of centralized Concentrated Solar Power (CSP) system in Gigawatt hour, the following expression is used [16]:

$$Technical\ potential = \frac{Solar\ resource * efficincy * Available\ area}{Spacing\ factor} \quad (5)$$

5   Results

The product of this potential estimation approach provides thematic maps of wind speed and wind power density classes at a certain turbine hub height within the initial phase of the methodology (theoretical potential) and for the solar energy, the solar radiation maps are presented namely Direct Normal Irradiance (DNI) map in case of investigating the potential of Concentrated Solar Plants (CSP) and Global Horizontal Irradiance (GHI) for investigating the potential of Photovoltaic (PV) cells installations. The next level of the methodology entails producing an exclusion map from which the exploitable area is calculated then producing the theoretical potential map and finally performing economical studies.

The maps produced out of this potential study direct planners and decision makers towards areas where wind and solar energies are of high importance. For example in case of Egypt, this approach shows some areas with superb wind potential such as Zafarana, Abu Darag, Ras Ghareb and Gebel Elzeit, and areas with good potential such as some spots at the eastern and western parts of the Nile [17]. Furthermore, the results of this approach entails numerical values of annual wind turbines and solar system energy output within the locations introduced out of this approach.

6 Conclusions

Applied geoinformatics represented in GIS in the context of this paper has been widely utilized in supporting the spatial analysis of different types of renewable energies. GIS spatial analysis support significantly the planning of solar and wind energies and the mapping of these resources in order to identify the locations with promising solar and wind potentials to planners, policy makers and developers to invest in such locations. Consequently, Constructing solar plants and wind farms within these areas leading to the penetration of renewable energy sources in the energy scheme supplying certain city therefore enriching the diversity of the city energy supply system and the contribution of renewable energies in fulfilling a fraction of the city energy demand. Thus, promoting the smartness of the city as According to ISO 37120 [4], the share of energy from renewables is a core indicator of cities smartness [18].

Potential estimation studies exploit many geospatial datasets which represent certain parameters such as land use, land cover, elevations, and infrastructure locations of highways, railways and power grids. The accuracy of the potential estimated relies heavily on these spatial datasets which are mainly obtained from satellite imagery and remote sensing so the spatial and temporal resolution of these datasets are of high importance.

References

1. Ramaprasad, A., Sánchez-Ortiz, A., Syn, T.: A unified definition of a smart city. In: International Conference on Electronic Government, pp. 13–24 (2017)
2. United Nations Department of Economic and Social Affairs. 2018 Revision of World Urbanization Prospects [Internet] (2018). Cited 3 Oct 2019. https://www.un.org/developm ent/desa/publications/2018-revision-of-world-urbanization-prospects.html
3. Albino, V., Berardi, U., Dangelico, R.M.: Smart cities–definitions, dimensions, and performance. In: Proceedings of the IFKAD, pp. 1723–1738 (2013)
4. ISO ISO. 37120 sustainable development of communities—indicators for city services and quality of life (2014)
5. University of Lowa. GEOINFORMATICS [Internet]. Cited 3 Oct 2019. https://informatics. uiowa.edu/study-opportunities/graduate-program/geoinformatics
6. Hoogwijk, M., Graus, W.: Global potential of renewable energy sources: A Literature assessment. EcoFys (2008)
7. Angelis-Dimakis, A., Biberacher, M., Dominguez, J., Fiorese, G., Gadocha, S., Gnansounou, E., Guariso, G., Kartalidis, A., Panichelli, L., Pinedo, I., Robba, M.: Methods and tools to evaluate the availability of renewable energy sources. Renew. Sustain. Energy Rev. [Internet] 15(2), 1182–1200 (2011). http://dx.doi.org/10.1016/j.rser.2010.09.049
8. Mentis, D., Hermann, S., Howells, M., Welsch, M., Siyal, S.H.: Assessing the technical wind energy potential in Africa a GIS-based approach. Renew. Energy [Internet] 83, 110–125 (2015). http://dx.doi.org/10.1016/j.renene.2015.03.072
9. Coppens, C., Gordijn, H., Piek, M., Ruyssenaars, P., Schrander, J., de Smet, P., Swart, R., Hoogwijk, M., Papalexandrou, M., de Visser, E., Horalek, J., Kurfürst, P., Jensen, P., Petersen, B.S., Harfoot, M., Milego, R., Clausen, N.-E., Giebel, G.: Europe's onshore and offshore wind energy potential (2009)

10. Sliz-Szkliniarz, B., Vogt, J.: GIS-based approach for the evaluation of wind energy potential: a case study for the Kujawsko-Pomorskie Voivodeship. Renew. Sustain. Energy Rev. [Internet] **15**(3):1696–1707 (2011). http://dx.doi.org/10.1016/j.rser.2010.11.045

11. Korfiati, A., Gkonos, C., Veronesi, F., Gaki, A., Grassi, S., Schenkel, R., Volkwein, S., Raubal, M., Hurni, L.: Estimation of the global solar energy potential and photovoltaic cost with the use of open data. Int. J. Sustain. Energy Plan. Manag. **9**, 17–29 (2016)

12. Sun, Y., Hof, A., Wang, R., Liu, J., Lin, Y.J., Yang, D.: GIS-based approach for potential analysis of solar PV generation at the regional scale: a case study of Fujian Province. Energy Policy [Internet] **58**(2013), 248–259 (2013). http://dx.doi.org/10.1016/j.enpol.2013.03.002

13. Manwell, J.F., McGowan, J.G., Rogers, A.L.: Wind Energy Explained: Theory, Design and Application. Wiley (2010)

14. Esri. Deterministic methods for spatial interpolation [Internet] (2008). Cited 7 Oct 2019. http://webhelp.esri.com/arcgisdesktop/9.3/index.cfm?TopicName=Deterministic_methods_for_spatial_interpolation#target. Text = A deterministic interpolation can either, known as an exact interpolator

15. Oloo, F., Olang, L., Strobl, J.: Spatial modelling of solar energy potential in Kenya. Int. J. Sustain. Energy Plan. Manag. **6**(February 2016), 17–30 (2015)

16. The International Renewable Energy Agency (IRENA). Estimating the Renewable Energy Potential in Africa. International Renewable Energy Agency (2014)

17. Hatata, A.Y., Mousa, M.G., Elmahdy, R.M.: Analysis of wind data and assessing wind energy potentiality for selected locations in Egypt. Int. J. Sci. Eng. Res. **6**(3), 604–609 (2015)

18. Resch, B., Sagl, G., Trnros, T., Bachmaier, A., Eggers, J.B., Herkel, S., Narmsara, S., Gündra, H.: GIS-based planning and modeling for renewable energy: challenges and future research avenues. ISPRS Int. J. Geo-Inf. **3**(2), 662–692 (2014)

Smart City Developments Using a BIM-Oriented Workflow

Daniel Sorial[(⊠)] and Mahmoud El Khafif

German University in Cairo, Cairo, Egypt
danielsafwatws@gmail.com, mahmoud.elkhafif@guc.edu.eg

Abstract. The term "smart city" is usually associated with eco-friendliness, resource efficiency and environmental sustainability. Also, Information and Communication Technology (ICT) is vastly used within these developments as a catalyst for solving social and business needs during the operation of smart cities, with user-friendly solutions, optimized infrastructure and more sustainable approaches.

In order to reach such criteria, adequate technologies need to be used from the very beginning of the smart city development. Building Information Modeling (BIM) is not a luxury anymore but a necessity as it facilitates collaboration and management of information and communication among teams involved in the construction process. BIM helps the construction process to start smart and stay smart as information gathered within the construction phase is later used to operate, maintain and manage these cities in an efficient way. This paper aims to create a suitable workflow that helps in achieving efficiency in the construction process using smart solutions as BIM. This workflow should not only help with the construction phase but also act as a stepping stone for the ultimate objective of creating a smart city.

Keywords: Smart · BIM · Building Information Modeling · Workflow · Development

1 Introduction

Smart city is a concept of urban development that allows you to safely integrate information and communication technologies and the Internet to manage the city's assets. In the last two decades, the concept of "smart city" has become more and more popular in scientific literature and international policies. The current scenario requires cities to find ways to manage new challenges. Cities worldwide have started to look for solutions which enable transportation linkages, mixed land uses, and high-quality urban services with long-term positive effects on the economy. The concept of the smart city is far from being limited to the application of technologies to cities. In fact, the use of the term is proliferating in many sectors with no agreed upon definitions. This has led to confusion among urban policy makers, hoping to institute policies that will make their cities "smart." (Albino et al. 2015).

Many definitions of smart cities exist. A range of conceptual variants is often obtained by replacing "smart" with alternative adjectives, for example, "intelligent" or

I. El Dimeery et al. (Eds.): JIC Smart Cities 2019, SUCI, pp. 88–96, 2021.
https://doi.org/10.1007/978-3-030-64217-4_11

"digital". The label "smart city" is a fuzzy concept and is used in ways that are not always consistent. There is neither a single template of framing a smart city, nor a one-size-fits-all definition of it (O'Grady and O'Hare 2012). First time this term was used was in the 1990s with major focus on the new Information and Communication Technologies (ICT) with the modern infrastructure taken into consideration. Meanwhile, California Institute for Smart Communities was among the first to really focus on how to integrate the word "smart" in the urban communities and start to develop these smart cities (Alawadhi et al. 2012).

Switching to the development part, smart cities usually start their development phases as normal construction sites then the addition of ICT and digital aspects comes later on after the full development of such cities. This being the usual case, is not really efficient as many data regarding the early life of these structures is lost after the development phase or not really used to its fullest because of the nature of the traditional construction methodologies. In order to fully embrace the idea of smart cities, full adoption of digital methods must happen from the early stages of the development of such cities. Building Information Modeling (BIM) can be of great benefit in such case as it allows us to digitalize all of the project development phases across all disciplines. BIM can also act as a data depository with information dating back to the earliest stages of the structure till operation and demolition (Mgbere et al. 2018). Such information is crucial to identify the best ways to integrate different elements of these cities with each other like traffic control, mobility sensors, power saving and energy analysis, environment friendly structures and low carbon prints and more.

Building Information Modeling can also help in improving the performance and quality of the structure, especially during the pre-construction phase. This allows for more careful evaluation of the construction which leads to a problem-free execution with value taken into consideration as this early evaluation allows for value engineering and study of design alternatives to optimize the overall quality of the building (Eastman et al. 2008). When this new methodology is compared to its usual counterpart, traditional ways usually are found to be much more time consuming with significant presence of human errors, design mistakes, clashes, inefficient processes and decisions that are based on insufficient data. All of this is usually present throughout the earliest stages of project developments till the final phases of execution. Effects of these problems persist till the operation phases of such buildings and can be seen in various forms as in inefficient designing, non-environmentally friendly structures with major carbon emissions and relatively large carbon footprint, time wastage in clashes, redesigning and reimplementation.

In this study, the main aim is to create a suitable workflow that helps in achieving efficiency in the construction process using smart solutions as BIM. This workflow should not only help with the construction phase but also act as a stepping stone for the ultimate objective of creating a smart city.

2 BIM Between Project Development and Analysis

When linking developments of smart cities and the use of BIM, a critical common point is shared between them which is productivity. Smart cities are always in need of efficient and optimized ways of management, operation and maintenance of the urban life. When combined with the amount of data analysis provided by the use of BIM within the construction and development phase as well as the operation and maintenance phase, this leads to a perfect balance between quality, cost and productivity with smart life in consideration (Talebi 2014).

In an increasingly competitive market and with pressure from clients and internal organizations to increase the return on investment (ROI) in projects, state of the art technological aids has emerged in order to minimize project costs, increase project control and amplify the construction industry's productivity (Dainty et al. 2006; Eastman et al. 2008). One of the most causes of decreased productivity, time delays and increased costs are changes, especially in design. Figure 1 shows cost of any design change during the project lifetime increases whilst the ability to impact functional capabilities decreases. And here comes the role of BIM as a method of data collection and decision making, the amount of data made available by 3D models and 4D simulations are more than adequate to produce a well-studied decision with all of consequences taken into consideration. In processes similar to line 4, a large part of tasks has to be carried out during the schematic design and design development. The usual design process has the highest workload within the construction documentation phase which at this point, any change in the design has a considerable extra cost and time.

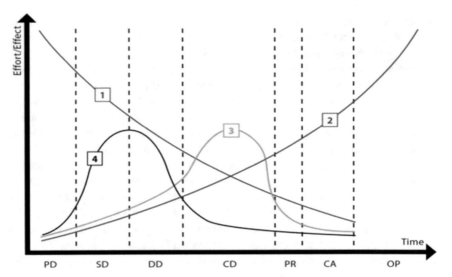

PD – Pre-design / **SD** – Schematic Design / **DD** – Design Development / **CD** – Construction Documentation / **PR** – Procurement / **CA** – Construction Administration / **OP** – Operation

–1– Ability to impact cost and functional capabilities   –3– Traditional design process
–2– Cost of design changes   –4– Preferred design process

Fig. 1. Added value, cost of changes and current compensation distribution for design services (CURT 2004)

Examining Fig. 1, it is clear that the use of BIM during the design phase has much more impact on the project as it is the phase with the highest capability of affecting the cost. This is one of many benefits of BIM implementation in a project.

When this methodology is applied to a construction project, not only the development part is of great benefit but also the analysis part. This can enhance the process of establishing a smart city by providing with all the benefits of project development alongside the analyses needed for creating more efficient structures. Such analyses methods can include energy analysis, design and structural analysis, 4D and time analysis, 5D and cost estimations and more. Each level aims to enhance the structure in way that compliments the concept of cost-quality-time (Talebi 2014).

3 Methodology

Implementation of BIM in a project life cycle needs thorough investigation of that project and proper analysis of the exact phase that the BIM methodologies will enter that cycle. For instance, applying BIM to a project in its early stages while still in concept or initial design is much easier than applying it mid-project. Of course, the earlier the implementation happens the better the results. Usual construction sites, even when taking smart cities construction sites in consideration are all following the traditional way of execution and process management, which lacks efficiency and produces lots of mistakes and human errors which leads to inevitable extra costs and time delays, these delays can cost up to 7% of the total project cost. While proper integration of BIM and modeling into a project life cycle can be complicated, this implementation process can be further explained using a flowchart showing both cases where BIM tools are either implemented or not and how much this affects the construction process.

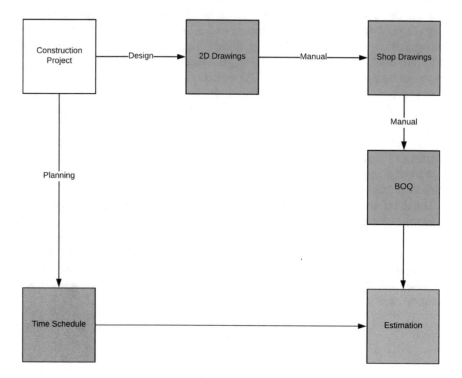

Fig. 2. Workflow of a non-BIM Project

First, Fig. 2 shows the typical construction project without the implementation of BIM tools and methods, where the integration between different fields is minimal to none, no automation is present, major appearance of human error and the absence of proper site monitoring and 4D analyses.

Following the flow chart, it is clear that most of the processes depended heavily on manpower and manual calculations. The absence of an automated system that can handle such jobs makes any modification much harder to achieve and direct implementations of such data with other fields of construction almost impossible without handling the metadata first. The regular result is a project which has not been optimized regarding design efficiency, quality and time and with the construction process is normally hardly of consideration in terms of constructability, smooth implementation process, avoidance of reworking, minimization of waste, etc. Second, Fig. 3 shows the typical construction project but with the help of BIM techniques and software which allows full integration of different fields and effective automation of time-consuming jobs.

The first thing obtained in the construction project are 2D drawings of the preliminary design. These drawings can be modeled and turned into a 3D model using different modeling programs available on the market. This 3D model automatically generated the required sections, shop drawings and bill of quantities of different elements with no human interference, therefore eliminating the human error factor. Also reviews and clash detections can be performed later on to enhance the model quality and the accuracy of the BOQ. These automatic products are linked to the model, any change in the model will automatically affect the sections and BOQs and vice versa. While modeling the 2D drawings, design checks can be done to the structure to ensure its safety and design efficiency while cutting down on extra safe elements. When the design deemed satisfactory, it can proceed to the budget check phase where cost and time estimations are the ultimate output. If the design was not found safe or economic, the whole structure can be redesigned using BIM integrated software and the optimized design can be updated straight to the 3D model.

If any aspect of the design whether it is the budget, time schedule, structural integrity or the design efficiency is deemed unsatisfactory, this specific aspect will be easily redesigned and integrated back to the system without the hassle of manually updating each step from the start. A time schedule is created using a scheduling program and then linked to the reviewed model using one of the 4D analysis programs to perform a full 4D analysis of the project. This can have a major impact on site monitoring efficiency and comprehension levels of the time schedule.

The developed workflow plan model aims to solve disputes, changes and claims before start of the project. It is also used to measure the impact of any change wanted in any aspect of the project; thus, a comprehensive analysis is performed and consequences of this change can be studied to avoid any unwanted issues or problems regarding the requested change. This not only saves time but also reduces cost for reworking and cause minimum claims within the project which leads to an undisturbed execution at site.

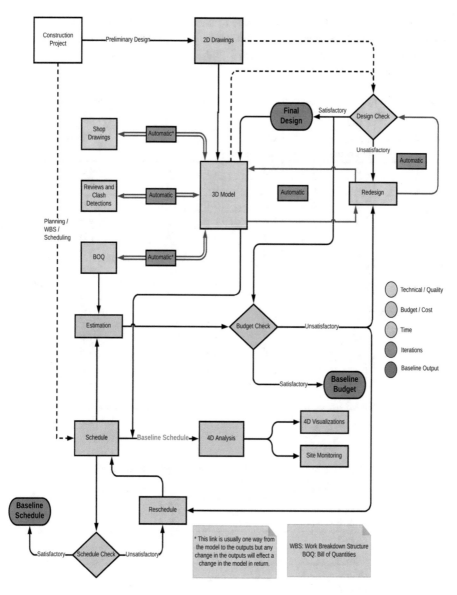

Fig. 3. Work flow of a project with BIM integration

4 Results

Comparing required durations of many site jobs, here are some technical site jobs with and without the help of BIM technologies as demonstrated. These data were collected from the site after surveying with 13 different engineers to estimate the time needed for certain jobs with the traditional methods using AutoCAD 2D, Excel and sometimes by hand and then, in comparison, estimate the same jobs but with the use of 3D programs like Revit. Skill levels may vary from one to another depending on the experience and whether if he/she had experienced this work before or not. All the durations are presented in man days: 8 h/Day (Tables 1 and 2).

Table 1. Comparison of durations needed for different technical jobs with traditional ways

Time needed for:	BOQ				Section detailing	Shop drawings	Total
	Concrete elements	Steel	Windows and doors	Bricks			
Eng. 1	7 d	7 d	1 d	7 d	4 d	4 d	30
Eng. 2	9 d	8 d	2 d	6 d	4 d	2 d	31
Eng. 3	7 d	6 d	1 d	5 d	6 d	3 d	28
Eng. 4	2 d	2 d	1 d	6 d	4 d	4 d	19
Eng. 5	4 d	3 d	1 d	4 d	3 d	7 d	22
Eng. 6	2 d	1 d	1 d	2 d	2 d	6 d	14
Eng. 7	2 d	1 d	1 d	3 d	4 d	4 d	15
Eng. 8	3 d	2 d	1 d	2 d	2 d	2 d	24
Eng. 9	5 d	2 d	2 d	4 d	3 d	3 d	19
Eng. 10	2 d	1 d	1 d	2 d	4 d	4 d	14
Eng. 11	4 d	4 d	1 d	1 d	3 d	4 d	17
Eng. 12	6 d	6 d	2 d	1 d	4 d	5 d	24
Eng. 13	4 d	5 d	1 d	2 d	2 d	2 d	16
Average time for each job	4.38 d	3.69 d	1.23 d	3.46 d	3.46 d	3.85 d	20.07 d

Table 2. Comparison of durations needed for different technical jobs with Revit

Time needed for:	Creating the model
Eng. 1	6 d
Eng. 2	5 d
Eng. 3	4 d
Eng. 4	6 d
Eng. 5	6 d
Eng. 6	6 d
Eng. 7	5 d
Eng. 8	4 d
Eng. 9	3 d
Eng. 10	4 d
Eng. 11	7 d
Eng. 12	2 d
Eng. 13	3 d
Average time for each job in man days	4.69 d

After comparing the results, an average total of 20-man days are needed to prepare the required drawings using the conventional methods compared to 4.69-man days needed with the help of 3D modeling. Thus, using 3D modeling techniques can speed up the process of calculating quantities and producing needed drawings by up to 4 times faster. All this saves valuable time and cost while decreases errors in such jobs. Moreover, the vast majority of design clashes that regularly disrupt the site processes can be eliminated before starting actual site implementation; this saves not only time and cost but also disputes and claims (Béquin 2018; Shami 2018). According to the national BIM library from the National Building Specification (NBS), 60% of those who used BIM reported improvements in efficiencies while only 22% reported that they would rather not adopt it. Also, most of client reviews are mixed which suggests that not all clients are seeing the benefits of using such options (NBS 2019). And finally, using BIM in the process of construction of the Bristol business school resulted in savings of over 2 million Euros (Marcubie 2017). Also according to a research done by the Danish government (BIM-Today 2017), using digital communication platforms in large scale construction sites can save up to 7% of the total project cost.

Switching our attention to Egypt, a case study done by (Essawy et al. 2017) mentioned that BIM is recommended to be used in the Egyptian construction industry, especially that its needed capabilities are readily available as it will contribute in achieving many benefits in the Egyptian construction industry. Another study gathered information from 42 Egyptian architects, engineers and contractors (Othman et al. 2018) showed that BIM has a very good diffusion within the Egyptian industry with 83%. Finally, a further study (Mohamed 2018) proved that BIM implementation has many benefits for the construction practices that positively influence the sustainability of the industry. As BIM adoption improves the management of the construction project information, enhances the quality of deliverables, offers better collaboration between the project team, facilitates and improves environmental building analysis, and contributes to materials waste reduction. All of these BIM implementations, in principal, apply certain parts and loops of the above workflow or similar approaches to successfully use BIM in the desired project.

5   Conclusion

When taking these results into consideration, this proves that using BIM and adopting such workflow within the lifecycle of a project helps with optimizing the construction process by cutting down labor hours while increasing the quality of the output work by nearly eliminating the human error aspect. This also allows minimizing claims and useless document circulations which usually waste time and efforts. Finally, the data gathered while constructing this structure will be the foundation stone of the smart city data repository which will effectively help in managing and operating such smart city to its maximum potential economically and environmentally. Applying such workflow in the early stages of a construction project is not only beneficial for the establishment of smart cities but also for any kind of construction process. This workflow can be adopted at any stage of structure development; however, the benefits and optimizations are much more prominent when implemented in the early stages such as the concept design phase.

According to the findings above, building information modeling is a beneficial approach and can help smart cities development greatly when implemented with a well-formed strategy, fluid workflow and calculated decisions.

References

Construction Users Roundtable (CURT). Collaboration, Integrated Information and the Project lifecycle in Building Design, Construction and Operation (2004). http://codebim.com/wp-content/uploads/2013/06/CurtCollaboration.pdf

Talebi, S.: Rethinking the project development process through use of BIM. Conference Paper January 2014 (2014)

BIM-Today: construction cost saving on digitally sharing information (2017). https://www.pbctoday.co.uk/news/bim-news/study-shows-construction-cost-saving-on-digitally-sharing-information/36006/

Marcubie: The use of BIM has resulted in cost savings of over £2 m for Bristol Business School. Upfront (2017)

Mohamed, A.: The Implementation of Building Information Modeling (BIM) Towards Sustainable Construction Industry in Egypt "The Pre-construction Phase". Master's thesis. AUC (2018)

Béquin, M.: Building Information Modelling's impact on claims in construction projects. PM World J. **VII**(V) (2018)

Shami, K.A.: Investigating the use Building Information Modeling (BIM) in managing construction claims. PM World J. **VII**(II) (2018)

NBS: The ninth NBS National BIM Report (2019)

O'Grady, M., O'Hare, G.: How smart is your city? Science **335**(3), 1581–1582 (2012)

Alawadhi, S., Aldama-Nalda, A., Chourabi, H., Gil-Garcia, J.R., Leung, S., Mellouli, S., Nam, T., Pardo, T.A., Scholl, H.J., Walker, S.: Building understanding of smart city initiatives. Lecture Notes in Computer Science, vol. 7443, pp. 40–53 (2012)

Albino, V., Berardi, U., Dangelico, R.M.: Smart cities: definitions, dimensions, performance, and initiatives. Article J. Urban Technol. (2015)

Dainty, A., Moore, D., Murray, M.: Communication in Construction. Taylor & Francis, Abingdon (2006)

Eastman, C., Teicholz, P., Sacks, R., Liston, K.: BIM Handbook: A Guide to Building Information Modeling for Owners, Managers, Designers, Engineers and Contractors. Wiley (2008)

Essawy, A., Elmikawi, M., Hosney, R.: Building Information Modeling (BIM) in the Egyptian construction industry. Al-Azhar Univ. Civil Eng. Res. Mag. (CERM) **39**(1) (2017)

Othman, M., Elbeltagi, E., Abdelshakour Hassan, M.: BIM implementation within the Egyptian AEC industry. In: 2nd International Conference Sustainable Construction and Project Management-Sustainable Infrastructure and Transportation for Future Cities (2018)

Mgbere, C., Knyshenko, V.A., Bakirova, A.: Building Information Modeling. A Management Tool for Smart City. Conference Paper, September 2018 (2018)

States DoT Roads and Bridges Network Inspection and Maintenance Practices

Amin K. Akhnoukh[1(✉)], Amr Abdel Hameed[2], and Abanoub Atteya[2]

[1] Construction Management Department, East Carolina University, Greenville, NC, USA
akhnoukhal7@ecu.edu
[2] Electrical Engineering Department, German University in Cairo, Cairo, Egypt

Abstract. The American Society of Civil Engineers (ASCE) infrastructure report card shows that the United States overall infrastructure score is D+, including a high percentage of structurally deficient and functionally obsolete roads and bridges. Due to the budget limitations at the Federal Highway Administration (FHWA), innovative approaches are being selected to increase the efficiency of infrastructure inspection, provide a smart decision technique for maintenance activities prioritization, and conduct maintenance using low-cost high-performance materials.

This paper provides a list of the state-of-the art practices followed by different State Departments of Transportation (DoTs), including the use of Automated Road Analyzers (ARAN) trucks for scanning, and Bridge Management Systems (BMSs) for maintenance, repair, and replacement of deficient structures. And finally, the use of economic high performance materials in construction of new projects and the maintenance activities of existing and aging inventory.

The expected outcomes of using the afore-mentioned techniques include early detection of structural deterioration, sufficient maintenance provided given the FHWA budget deficiency, and improved infrastructure conditions.

Keywords: Bridge management systems · Maintenance · Automated road analyzer · Infrastructure

1 Introduction

The American Society of Civil Engineers (ASCE) infrastructure report card shows that the United States overall infrastructure score is D+. The low rating of the US bridge network is attributed to the high percentage of structurally deficient and functionally obsolete bridges. Structural deficiency represents bridges with deteriorated structural members and lower load rating. While functionally obsolete bridge includes bridges that has insufficient geometrical design including extreme curvature, lanes with low-width, and approaches with high slopes. The Federal Highway Administration (FHWA) has an annual budget dedicated to bridge maintenance, repair, and replacement of bridges. However, the FHWA budget is insufficient, and increased budget deficit is directly impacting the conditions of the bridge network.

© The Author(s), under exclusive license to Springer Nature Switzerland AG 2021
I. El Dimeery et al. (Eds.): JIC Smart Cities 2019, SUCI, pp. 97–102, 2021.
https://doi.org/10.1007/978-3-030-64217-4_12

Due to the budget deficit the FHWA is currently researching different techniques for efficient and cost-effective roadway and bridges inspection and investigating smart decision techniques for maintenance activities, and perform maintenance using economic high performance materials.

This paper provides a list of the state-of-the-art practices followed by different State Departments of Transportation (DoTs) to maintain and possibly improve the conditions of the bridge networks including the use of automated road analyzer trucks (ARANs) for scanning, artificial intelligence (AI) and Internet of Things (IoT) for data collection and analysis. Finally bridge management systems application for prioritization of maintenance activities.

2   Automated Road Analyzers (ARAN Trucks)

The current road and bridge slabs conditions in the United States is affected by several deterioration factors including aging, high levels of loading due to increased truck weights and high average annual daily traffic, and chemical and environmental attacks as alkali-silica reactivity (ASR) [1] and chloride attacks resulting from de-icing salts [2]. Several research projects investigated the deterioration of highway and bridge networks and provided stochastic deterioration models to estimate potential future deterioration [3].

Roadway and bridge slabs deterioration includes rutting [4], honeycombing, and cracking [5]. Examples of roadway deterioration is shown in Fig. 1.

Fig. 1. (a) Rutting of asphalt and 1 (b) Honeycombing of hard pavement

Due to the size of roadway and bridge networks of the United States, several efficient testing methods for detection of annomalies are being researched by state DOTs including ground penetrating radars [6] and ultrasonic testing of bridge decks [7]. Advanced scanning techniques provided by automated road analyzer trucks (ARANs) are being used by few states for expedited and cost-efficient detedtion of defficient roadways and bridge slabs.

The ARAN application replaces the traditional method used to inventory highway and bridge assests. The ARAN truck, shown in Fig. 2, drives according to regular traffic speed and perform scans for asphalt. The quality and depth of the scan depends on the frequency of the used scanner and the speed of travel.

Fig. 2. ARAN truck for roadway and bridge deck scanning

The ARAN recording provides DOT personnel with the following data:

- High definition video capture to enable DOT personnel to view the highway and its surface conditions from their office
- Laser profilometery with high precision laser for measuring roughness and rutting of the highway. The ARAN laser can do longitudinal scans parallel to travel direction and transverse scans up to 4 m width
- Inventory data is produced by the ARAN to include the location of the distress, type of the distress, and its measures. Data output can be formatted and produced to be imported into a GIS or road management system for further analysis

To-date, different states are adopting the ARAN truck including Arkansas, Pennsylvania, Michigan, and Texas. The application of ARAN trucks expedite the data capturing, results in flawless data capturing, and doesn't require much human intervention in capturing the data and/or recording the results.

3   Bridge Management Systems

Bridge management systems (BMS) are currently used in the majority of the states to assist in prioritization of maintenance decisions for large bridge inventories. Currently, the BMS software commercially known as Pontis, as a product of the American Association of State Highway and Transportation Officials (AASHTO) taregts the following:

- Manage the complete lifecycle of bridges and other structures
- Store data relevant to the current condition of different bridge elements within the bridge inventory
- Provide possible bridge elements conditions in the future for resource allocation purposes

Pontis, as a BMS, incorporates a set of probabilistic models and detailed database to predict the maintenance and required repair activities. In addition, the BMS recommend optimum activities for bridge conditions improvement given schedule, budgets, and policies control. Despite of its robust capabilities, different state DOTs implement Pontis partially in their maintenance decisions. For instance, different DOTs utilize Pontis as a database to store the results of bridge inspections only. Other states utilize further capabilities of Pontis and enable the BMS program to perform optimized maintenance activities. Status of Pontis applications in the United States is shown in Fig. 3.

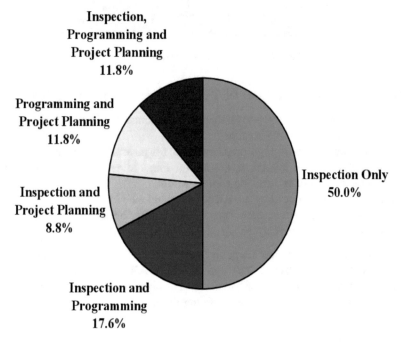

Fig. 3. Current status of State DOTs in Pontis BMS applications

4   Application of Economic Ultra-high Performance Materials in Construction Projects

The use of economic ultra-high performance materials in construction of new projects and in maintenance, repair, and replacement of existing projects. Economic ultra-high performance materials include the following:

- **Ultra-high performance concrete (UHPC):** developed in France in the 1990s as a special concrete with superior characteristics including high early strength, high final strength, high modulus of elasticity and rupture. The incorporation of UHPC mixes in current construction projects results in smaller structural members with lighter weight and longer spans [8–13]
- **Large prestress strands:** to enhance the capacity of precast/prestressed concrete members due to the large prestressing force applied to the prestressed members [14]
- **High grade welded wire reinforcement:** to enhance the shear performance of shallow bridge girders [15]
- **Incorporating micro and nano-sized supplementary cementitious materials (SCMs):** to produce concrete mixes with superior characteristics including mechanical characteristics, self-healing concrete characteristics, self-cleaning concrete members [16, 17], and long-term performance (durability)

5   Conclusions

The current status of roadway and bridge inventory within the United States requires continuous and optimized maintenance processes to maintain and improve the current inventory status. Due to budget limitations at the FHWA, state DOTs are currently researching and implementing new techniques for conducting roadway and bridge scans, and maintaining deficient members. Techniques includes the using of ARAN trucks, bridge management systems for decision making, and the utilization of high performance materials to increase the life span of construction projects.

6   Recommendation for Future Research

The development of new types of concrete requires the re-evaluation of design code charts and design curves to maintain a calibrated reliability indexes for new designs [18]. Additional research is required to explore the potential advantages of BIM applications in new construction projects and its potential benefits in reducing cost, time, and effort required to execute a given construction schedule [19, 20].

References

1. Akhnoukh, A.K., Kamel, L.Z., Barsoum, M.M.: Alkali-silica reaction mitigation and prevention measures for Arkansas local aggregates. World Acad. Sci. Eng. Technol. Int. J. Civil Environ. Eng. **10**(2), 95–99 (2016)
2. Shi, X., Akin, M., Pan, T., Fay, L., Liu, Y., Yang.Z.: Deicer impact on pavement materials: introduction and recent developments. Open Civil Eng. J. **3**, 16–27 (2009)
3. Morcous, G., Akhnoukh, A.: Stochastic modeling of infrastructure deterioration: an application to concrete bridge decks. In: Proceedings of the Joint International Conference on Computing and Decision Making in Civil and Vuilding Engineering, Montreal, Canada (2006)

4. Qi-sen, Z., Liang, C.Y., Lian, L.X.: Rutting in asphalt pavement under heavy load and high temperature. In: Proceedings of GeoHunan International Conference (2009)
5. Mackiewicz, P.: Fatigue cracking in road pavement. In: Proceedings of the IOP Conference Series, Materials Science and Engineering (2018)
6. Xia, T., Huston, D.: High speed ground penetrating radar for road pavement and bridge structural inspection and maintenance, Report on Project # SPR-RSCH017-728 (2016)
7. Rens, K.L: An ultrasonic approach for nondestructive testing of deteriorating infrastructure: use of direct sequence spread spectrum ultrasonic evaluation to detect embedded steel deterioration, A dissertation, Iowa State University (1994)
8. Soliman, N.A., Tagnit-Hamou, A.: Using glass sand as an alternative for quartz sand in UHPC. J. Constr. Build. Mater. **145**, 243–252 (2017)
9. Akhnoukh, A.K.: Development of high performance precast/prestressed bridge girders, A Dissertation, University of Nebraska-Lincoln (2008)
10. Graybeal, B., Tanesi, J.: Durability of ultra-high-performance concrete. Am. Soc. Civil Eng. J. Mater. Civil Eng. **19**(10), 848–854 (2007)
11. Akhnoukh, A.K.: Overview of nanotechnology applications in construction industry in the United States. Micro Nano-Syst. J. **5**(2), 147–153 (2013)
12. Akhnoukh, A.K.: The use of micro and nano-sized particles in increasing concrete durability. J. Part. Sci. Technol. **38**(5), 529–534 (2019)
13. Koh, K.T., Park, S.H., Ryu, G.S., An, G.H., Kim, B.S.: Effect of the type of silica fume and filler on mechanical properties of ultra-high-performance concrete. Key Eng. Mater. **774**, 349–354 (2018)
14. Akhnoukh, A.: Prestressed concrete girder bridges using large 0.7 inch strands. World Acad. Eng. Sci. Technol. Int. J. Civil Environ. Struct. Constr. Archit. Eng. **7**(9), 613–617 (2013)
15. Akhnoukh, A.K., Xie, H.: Welded wire reinforcement versus random steel fibers in precast/prestressed ultra-high performance concrete I-girders. Elsevier J. Constr. Build. Mater. **24**(11), 2200–2207 (2010)
16. Akhnoukh, A.K.: Implementation of nanotechnology in improving the environmental compliance of construction projects in the United States. J. Part. Sci. Technol. **36**(3), 357–361 (2018)
17. Elia, H., Ghosh, A., Akhnoukh, A.K., Nima, Z.A.: Using nano- and micro-titanium dioxide (TiO$_2$) in concrete to reduce air pollution. J. Nanomed. Nanotechnol. **9**, 1000505 (2018)
18. Morcous, G., Akhnoukh, A.K.: Reliability analysis of NU girders designed using AASHTO LRFD. In: Proceedings of the American Society of Civil Engineers Structures Congress, California (2007)
19. Meadati, P., Irizarry, J., Akhnoukh, A.: Building information modeling implementation – current and desired status. In: Proceedings of the International Workshop on Computing in Civil Engineering, Florida (2011)
20. Meadati, P., Liou, F., Irizarry, J., Akhnoukh, A.K.: Enhancing visual learning in construction education using BIM. Int. J. Polytech. Stud. **1**(2) (2012)

BIM and GIS Synergy for Smart Cities

Moustafa Baraka[✉]

Surveying and Geodesy, The German University in Cairo,
New Cairo City, Egypt
moustafa.baraka@guc.edu.eg

Abstract. An overview of Building Information Modelling (BIM) and Geographic Information Systems (GIS) is presented, where BIM and GIS are distinct and shared characterisitics are examined. Both BIM and GIS involve realistic modelling through digital data. Nevertheless, BIM and GIS models these data differently, with varying scope and scale for representation of world objects to accommodate various envisioned applications. Recent research on the use of BIM and GIS in smart cities recommend the integration of both, for the many technical, administrative, economic, and environmental benefits. The paper highlights the challenges for integrations, along with possible integration schema and applications of BIM and GIS in smart cities.

Keywords: BIM · GIS · Integration levels · Smart cities

1 Introduction

The research presented herein addresses the synergy in integrating BIM and GIS for smart cities applications. Geographic Information System (GIS) is a system designed to; capture, store, manipulate, analyze, manage, and present spatially georeferenced data, i.e. referencing data to geographic locations on the earth. Further, GIS links tabular attribute data with spatial data. GIS is capable of spatially analyzing, manage large datasets, and display information in a map and graphical representations (Dempsey 2001). Building Information Modelling (BIM) is a process for; creating and managing information on a construction project across the project lifecycle. BIM provides a digital description of building process and the built environment. The BIM model is developed based on information collected and updated from collaborating stakeholders within a project. BIM addresses key project phases of; planning, design, construction and operation. BIM enables stakeholders to enhance output of project phases all through the building lifecycle (National British Standards 2016).

Within the context of BIM and GIS, Ronsdorf (2016) defines a smart city as a smart place that' "uses connected, integrated systems and operations to become more efficient, sustainable and to achieve better outcomes for the people who live, work and visit, as well as the governments and businesses that operate within". BIM and GIS have proven tracks in addressing city needs in general. Kavanagh and Mulhall (2016) indicated that BIM and GIS would provide an operative means of integrating digital systems in the built environment of a smart city, for the sustainablitiy of the city and the well being of its inhabitants. BIM and GIS integration within smart city context, could be realized at the data level, the process level, or the application level to be shown in this research.

I. El Dimeery et al. (Eds.): JIC Smart Cities 2019, SUCI, pp. 103–111, 2021.
https://doi.org/10.1007/978-3-030-64217-4_13

Ronsdorf (2016) reports that GIS and BIM are a vital component in supporting smart cities. BIM and GIS support a smart city all through its lifecycle, be it at the design, building, and operating phases. BIM and. Both BIM and GIS would use open data for smart cities which facilitates their integration. While BIM provides a detailed geometric and semantic information of smart city, GIS provides geospatial analyses and modelling. As BIM and GIS share within their middle name the word "information", integrating BIM and GIS is intuitive, and would result in more benefits than if each were implemented separately. Further, combining authoring tools and analytical capabilites avialable within BIM and GIS would result in more combined benefits.

2   BIM and GIS Distinct and Shared Characterisitics

In order to achieve effective integration, BIM and GIS different distinct and shared common charateristics are examined in this section. Figure 1 shows the existing overlap between BIM and GIS, which would provide the basis for integration in many smart cities applications. Of interest to point out that the distinction in scale and coverage for BIM and GIS. While GIS scale range is from small to medium scale representation starting from worldwide to building coverage. BIM scale range is from medium to large representation, starting from building to its components. However, there is a clear overlap in scale and coverage for BIM and GIS addressing smart city extent to its smart buildings.

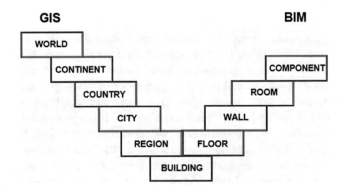

Fig. 1. Overlap between BIM and GIS. Source: Sani and Abdul Rahman (2018)

Figure 2 shows further details of the distinct and shared characteristics in terms of; tools and discipline. The figure identifies the tools and associated data available for GIS and BIM, along with the overlapping disciplines that are of interest to smart cities.

Further examination of BIM and GIS distinct and shared characteristics were presented by Karan et al. (2016). BIM scope is on 3D modelling and analysis of detailed geometric and attribute of man-made features within design to be built or existing built environment, along with modelling of; design, construction and maintenance. GIS scope

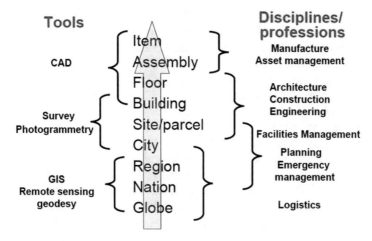

Fig. 2. BIM and GIS tools and disciplines. Source: Limp (2007)

is mainly on 2D geospatial modelling of planned and existing man-made objects within the physical natural environment, as well as topological and temporal analyses of natural and man-made objects interaction and impact.

Moreover, BIM focus on indoor environments, while GIS focus on existing natural and manmade objects in outdoor environment. Resulting in more benefits from the complementary and overlapping scopes of BIM and GIS. Table 1 gives a general comparison between BIM and GIS in terms of; scope, level of detail, spatial capability, coordinate system, popular file format, and popular software Zhang (2018).

Table 1. BIM and GIS general comparison. Source: Zhang (2018)

	BIM	GIS
Scope	Enhances interoperability in building industry. Models the building design to operation processes (planned or existing)	Represents the real world (natural and as-built environment), stores, manage, analyze and presents geospatial data
Level of detail	Detailed physical geometric/structure description	Limited physical structure description
Spatial capability	No geospatial analysis	Performs geospatial analysis
Coordinate system	Uses local coordinate system, and plane (flat) earth representation	Uses geographic coordinate system, spherical earth representation and map projections for regional representation
Popular file format	IFC, DWG	GML, XML, SHP
Popular software	Revit, FME	ArcGIS, QGIS

Current trends regarding 3D city modelling considers the joint use of BIM and GIS. While BIM provides a detailed geometric and semantically dense 3D information on the interiors components of a building or built environment within a city. GIS provides geospatial analyses and elaborate information of the georeferenced natural and man-made environment and surroundings within a city. BIM provide macro scale geometric analyses, focusing on components of a building. Where this analyses is complemented by the analyses provided by GIS at the macro scale, with further spatial analyses of city components (Zhang 2018).

Of importance in integrating BIM and GIS is to understand the prevailing coordinate systems representation in both. Kaden and Clemen (2017), identified in the coordinate systems prevalent in both BIM and GIS. Table 2 presents a comparison between BIM and GIS Coordinate Systems in terms of; approach, aim, representation and coordinate systems. Understanding BIM and transforming it into GIS coordinate systems world facilitate the georeferencing of BIM objects (e.g. buildings) within GIS.

Table 2. Comparison between BIM and GIS coordinate systems. Source: Kaden and Clemen (2017)

	BIM	GIS
Approach	Bottom-up approach: idea for a building, leading to a design model or plan with a coordinate system suitable for a real world representation	Top-down approach: real world objects, leading to a surveying with a coordinate system suitable for a 2D or 3D model
Aim	The correct representation of a planned building	The correct representation of real world objects
Representation	By relative placement of constructive elements (component-based, generative)	Representation by absolute positioning of topographic elements located on the earth surface
Coordinate SYSTEM	Local survey coordinate system (LSS), Project coordinate system (PCS)	Regional or global geographic coordinate reference system (CRS)

O'Neill et al. (2014) have pointed out some benefits where BIM capabilities would gain involving GIS capabilities. In which case a BIM-GIS environment would extend BIM building design tools to include GIS urban design tools, with the capacity to link geometric data of BIM to the physical data of GIS. Hence, allowing for access to 2D and 3D vector and raster data into a 3D city data modelling and visualization. Such an integration would help in analyzing existing and future conditions. Ultimately leading to developing baseline 3D model and database for smart cities, with ability to test development scenarios at various smart city sites, with multiple scenarios. Owing to the complementary characteristics of BIM and GIS at phases of engineering design, planning and construction, their integration in support of smart cities would be a "smart" approach.

3   BIM and GIS Integration

The trends in interoperability between BIM and GIS considers aspects of; ontology, systems, and standards and specifications (Limp 2007), where these trends lead to interoperable systems of both software and databases. Resulting a seamless; design, construction, operations, maintenance, management, with an increase in efficiency, and reduction in cost and time. Kavanagh and Mulhall (2016) mention concerns for the multi-scale data of BIM and GIS, with macro and meso data scale that is dominant in GIS, and micro scale data that is dominant in BIM. Along with considering the public sector objectives and governance in relation to city planning, in view of private sector objectives and governance in relation to building design within smart cities.

According Sani and Abdul Rahman (2018), BIM and GIS integration approaches can be addressed at three different levels, namely; data level, process level and application level. At the data level; data models and data structure are re-designed to conform to the targeted application, with a possibility to extend these models and structures to attain the objective of the intended application. At the process level; both BIM and GIS are included in a workflow to achieve this integration. The application level; considers new applications taking advantage of the different functionalities of BIM and GIS. Successful integration at the application level needs data interoperability. More elaboration are given next by Zhang (2018) on BIM and GIS three levels of integration):

- At the data level, the integration of BIM and GIS considers standard data models with open sources, namely; Industry Foundation Class (IFC) as a standard data model BIM, and City Markup Language (CityGML) as a standard data model for GIS. IFC is driven by Architecture, Engineering and Construction (AEC) industry, with several 3D data formats. IFC is the main open data schema used for information exchange of BIM objects within AEC. CityGML is the open standard data model and exchange format for 3D models of cities and landscapes Including also attributes, relations and components of a city, (Sani and Abdul Rahman 2018). Zhang (2018) stated that research is currently addressing the use of CityGML and IFC standards to achieve data exchange, with concerns for semantic conversion and georeferencing.
- Process Level Approach: integration considers to retain the format or structure of data, considering ontology differences of objects within BIM and GIS. A reference ontology would offer a comprehensive approach that provide the basis for integration. The use of both IFC and CityGML is recommended. A workflow is proposed by Kaden and Clemen (2017) considering examples of a BIM object (e.g. Mensa building on a university campus), natively defined within a BIM tool (e.g. Revit) and natively defined within a project coordinate system (PCS). Next, an intermediate local surveying coordinate system (LSS) is introduced to link BIM PCS (e.g. university campus) to a regional coordinate regional system (CRS). Finally, the CRS referenced BIM object (e.g. Mensa building along with university campus) are georeferenced to a national CRS. Figure 3 gives an outline for the workflow for the transformation of BIM objects from PCS to CRS coordinate system.

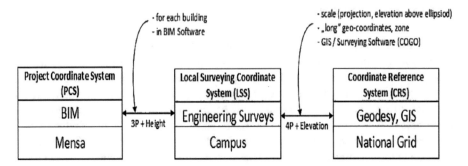

Fig. 3. Workflow for transformation of BIM objects between different coordinate systems. Source: Kaden and Clemen (2017)

–   Application Level Approach: integration considers data interoperability to achieve the intended use retaining all data sources. Integrating BIM and GIS information in applications for smart cities, such as; analyzing traffic noise, and construction site materials for cost optimization are cited. Further applications suitable for smart cities could be developed within this approach Zhang (2018). Abdul Basir et al. (2018) presented a conceptual overview of the interoperability of BIM and GIS. Figure 4 gives a proposed system architecture for BIM/GIS interoperability platform.

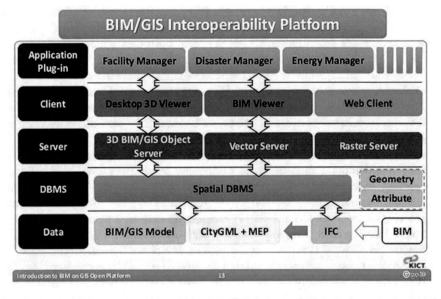

Fig. 4. A proposed system architecture for BIM/GIS interoperability platform. Source: Abdul Basir et al. (2018)

Incorporating GIS capabilities into BIM at the application level would benefit from GIS spatial analysis capabilities and tools in querying BIM objects. As GIS uses spatial query languages that are not readily available within BIM models, Borrmann (2010) developed a 3D spatial query language that would accommodate GIS-like spatial queries of BIM models. GIS spatial queries are applied to BIM models, providing spatial analyses of various building components. Examples of such spatial query dialog is presented in Fig. 5, where spatial queries are composed to provide analyses of location, relation, proximity, topology of various building elements (e.g. rooms, walls, slabs, columns, and utilities. The developed spatial query language addresses 3D building and can extend to address 3D city models. The spatial query language was implemented on top of object-relational Structured Query Language (SQL).

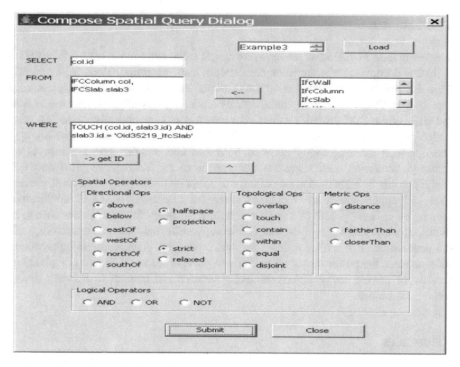

Fig. 5. Spatial query dialog interface for 3D building elements. Source: Borrmann (2010)

Karan et al. (2016) identified examples of integrated BIM-GIS applications in different stages of a construction project, namely; design, preconstruction, construction, and operation. Table 3 presents examples addressing the different phases of the life-cycle of a smart city Lifecycle.

Table 3. Examples of BIM-GIS applications at different phases of smart city lifecycle. Source: Karan et al. (2016)

Project stage	BIM-GIS application examples
Design	Digital modeling of building and landscape-level Components, Site selection and fire response management
Preconstruction	Identifying optimal number and location of tower cranes
Construction	Evaluation and visualization of construction Performance, Construction supply chain management
Operation	Detecting and mapping utility network Information, Facility management supply chain

4 Conclusion

Integrating BIM and GIS has proven syergy, with smart cities at the core. BIM and GIS distinct and common charateristics are catalysts for integration. Both BIM and GIS share the realistic representations through digital information, using many common data sources. Current trends regarding 3D city modelling considers the joint use of BIM and GIS. While BIM provides a detailed geometric and semantically dense 3D information on the interiors components of a building or built environment within a city. The trends in interoperability between BIM and GIS should considers aspects of; ontology, semantics, and standards and specifications. To achieve BIM and GIS integration can be addressed at one of three different levels, namely; data level, process level and application level. Examples given for integrated BIM-GIS applications support the different phases of a smart city through its lifecycle. Some issues need to be addressed when integrating BIM and GIS in support of smart cities. Considering scale and data issues, with further concern for data ownership. Also, considering reconciliation between the public sector objectives and governance in relation to city planning, in view of private sector objectives and governance in relation to building design within smart cities.

References

Abdul Basir, W., Majid, Z., Ujang, U., Chong, A.: Integration of GIS and BIM techniques in construction project management: a review. In: The International Archives of the Photogrammetry, Remote Sensing and Spatial Information Sciences, vol. XLII-4/W9 (2018)

Borrmann, A.: From GIS to BIM and back again—a spatial query language for 3D building models and 3D city models. In: Proceedings of the 5th International 3D GeoInfo Conference, Berlin, German, pp. 19–26 (2010)

Dempsey, C.: Geographic Information Systems (GIS) (2001). https://www.gislounge.com/. Accessed 14 June 2018

Kaden, R., Clemen, C.: BIM and GIS digital terrain models, topographic information models, city models, formats and standards. In: International Federation of Surveyors (FIG) Working Week 2017, Helsinki, Finland, 29 May–2 June (2017)

Karan, E., Irizarry, J., Haymaker, J.: BIM and GIS integration and interoperability based on semantic web technology. J. Comput. Civil Eng. **30**(3) (2016)

Kavanagh, J., Mulhall, T.: Smart cities/smart buildings – a tale of two Scales. In: International Federation of Surveyors (FIG) Working Week 2016: Recovery from Disaster, Christchurch, New Zealand, 2–6 May 2016

Limp, F.: Geospatial, 3D, Visualization and BIM Convergence at the intersection (2007). https://portal.opengeospatial.org/files/?artifact_id=20665. Accessed 12 Nov 2017

O'Neill, C., Gallas, T., Brandersky, R.: BIM and Smarter Cities: The Use of BIM for Planning (2014). https://bimforum.org/wp-content/uploads/2014/04/TGP_Boston-BIMForum-Final.pdf. Accessed 3 Dec 2017

National British Standards (NBS) (2016). https://www.thenbs.com/knowledge/what-is-building-information-modelling-bim. Accessed 4 July 2017

Ronsdorf, C.: 3D, BIM and IOT: scaling from buildings to smart cities. In: GIS Conference 2016, Geospatial World Forum 2016, Netherlands (2016)

Sani, M., Abdul Rahman, A.: GIS and BIM integration at data level: a review. In: The International Archives of the Photogrammetry, Remote Sensing and Spatial Information Sciences, vol. XLII-4/W9 (2018)

Zhang, Z.: BIM to GIS-based Building Model Conversion in Support of Urban Energy Simulation, M.Sc. Thesis, Lund University, Lund, Sweden (2018)

Tolerance Management, Failure, and Defects in Construction

Mohamed Kamel[1](✉), Omar Habib[2], Mohamed Farahat[3],
and Rana Khallaf[4]

[1] Jones Lang LaSalle, Cairo, Egypt
Mohamed.Mah.Kamel@gmail.com
[2] Orascom Construction, Cairo, Egypt
omar.habib95@outlook.com
[3] ElSewedy PSP, Cairo, Egypt
[4] Future University, Cairo, Egypt

Abstract. Tolerance in the construction industry has been regarded as a vague topic due to the paucity of information available on it. As a result, the effect of tolerance and its impact on a project has been neglected in many cases. This impact can sometimes be catastrophic to the testing and commissioning phase of the project and the overall project quality as well. Tolerance management as a science focuses on the acceptable margin of error or discrepancy. One of the main objectives of this paper is to raise awareness about tolerance in construction and how to approach tolerance failures and defects to reduce/eliminate waste. This paper focuses on tolerance management and addresses how it can be applied in the construction field. Literature review was conducted to collect data on tolerance management in various fields and report on its main principles. This paper also introduces two terms: tolerance failure and defects, and discusses the relationship between them. Furthermore, five categories of tolerance failures are introduced. Finally, preliminary solutions and mitigation strategies for tolerance failure categories are proposed.

Keywords: Tolerance management · Tolerance principles · Lean construction · Tolerance failure · Defects

1 Introduction

Tolerance is the ability to adapt to various external or internal condition indicators within acceptable limits based on actual performance. It defines variability in different aspects such as quality, time, geometry, behavior, cost, and other parameters. Tolerance management is considered the grey zone before a tolerance becomes a defect. It can be applied at each stage of a project starting from the brainstorming stage, design, executing, until project closing. Tolerance management needs to be monitored at an integrated level, not just at the discipline or even product level. It might also vary from one discipline to another. For example, in the construction industry, the civil department may execute works that would be acceptable according to its standards, while unacceptable for the mechanical works. An example is in a mega power plant project in

Beni Suef, Egypt, where pipeline systems were to be installed on supports over reinforced concrete (RC) foundations. The sequence of work was such that the reinforced concrete footings had to be constructed first (by the civil team), followed by the mechanical team installing the fabricated supports over the foundations. Afterwards, it was found that even though the RC foundation levels were acceptable and safe according to the civil team, they would pose a problem to the pipeline because their levels were not acceptable and had to be increased. Thus, the solution was to increase the level by inserting metal base plates in the RC foundation and then installing the pipeline system over the foundation. In view of the previously mentioned example, tolerances have a key role in the assembly and production process that requires tolerance management and/or visual management tools to be applied for facilitating the production (Da Rocha et al. 2018), because parts that are made independently are expected to function coherently with adjoining parts (Creveling 1997). Tolerance should be studied at specific stages at the individual level as well as in relation to other parts in the process to analyze the impact of changes on a more integrated level.

Consequently, to maintain tolerances within a project, it is preferable to visualize the value stream for the multiple processes involved and apply lean techniques to reach the optimum level of tolerance management (Milberg and Tommelein 2004), which would directly affect the product's quality, cost, and time to complete. Nowadays, projects have evolved to a larger scale with higher levels of complexity in design, construction, and also in the interaction between the various disciplines. Consequently, the use of acceptable margins is essential and translates to the use of design specifications in manufacturing, including tolerances, which have evolved to support design interchangeability and mass production. Lean construction tools are used to minimize waste and maximize the value (Jose et al. 2018). Consequently, this paper aims to link lean tools to their use in tolerance analysis in order to maintain control over a process (whether design or construction-related) from the early stages (Jose et al. 2018). This is conducted through four steps: (1) identify and analyze tolerance management key principles; (2) discuss the lean tools that could be helpful to maximize the value while minimizing the waste in effort and time while also managing the tolerance; (3) explain the tolerance failure model and tolerance defects; and finally (4) discuss tools that can be used to mitigate these defects to improve the quality of the process and outcomes.

2   Tolerance Management Principles

This section discusses the four main tolerance management principles. Firstly, tolerance is managed for products or outcomes that do not have a standard result/specification due to variances in the process. Such standard results are given for the product or the outcome based on the product requirements specified in the beginning. Consequently, clear communication links must be followed in order to reduce waste in time and effort and to meet the final product objectives. The output of such communication links is to clearly establish the product's specifications and acceptable margins of error. Moreover, the communication links aim to describe the relationship between different phases, inform the disciplines of their role in the product creation, and help them understand the impact of each part separately and also provide a means to monitor the entire process. Early involvement of disciplines is important and effective to ensure that they are involved from the early design phases and that they can co-work together (Kent and Becerik-Gerber et al. 2010). This helps create a vision for acceptable tolerances and early detection of deviation since tolerances are considered a communication of design, manufacturing, and inspection intent. Therefore, tolerance specifications should be complete prior to beginning the work to ensure that the intent is unambiguous (Milberg 2006).

Secondly, a datum is a geometrically ideal reference used to describe the nominal geometry of the feature and its tolerance (Henzold 1995). The datum is the principle that considers that each of the size, form, and features of the product has a specific datum, where every specified tolerance needs to be met independently (unless multiple tolerances exist for the product, in which case the most restrictive one applies) (Henzold 1995). Consequently, it is preferable to indicate all the tolerances, group them by functionality (tolerances for the same characteristics or for different ones), and analyze them. A clear list of tolerances can then be created for use throughout the project. To overcome such risks, it is recommended to improve the process by listing the datum/margins for major and minor risks. Those related to major risks would need to be forced into tolerance since any deviation from such margins will lead to failure in the whole process. However, for minor risks, it is preferable to widen their margins until they fit with the process function; notwithstanding that such widening needs to be monitored to avoid shifting of a minor tolerance into a major one and to keep the process under control (Milberg 2006).

Thirdly, tolerance analysis allows the designer to check the impact of the variation in the critical dimensions that results from the nominal geometry, feature tolerances, and also provides additional information to guide the modification of the geometry or feature tolerances if necessary (Milberg 2006). Accordingly, tolerance has two types. (i) a type that has a critical impact over the dimension so variation analysis would be the most appropriate tool to clarify the variation in the critical dimensions of the process; and (ii) a type of tolerances that focuses on the small changes by indicating their ratio to the whole process. The first type is the most critical one and has a severe impact on the outcome of the process (Milberg 2006).

Finally, it is important to reach the correct tolerance margins by choosing the right connections for design and the most appropriate interoperability between different disciplines. Tolerance analysis should be done to ensure that critical variables are acceptable. If they are not, the design should be modified until an acceptable solution is found (Gerth 1997), (Milberg 2006).

3 Tolerance Failure

Tolerance Failure refers to systems or products that do not conform to their specifications. However, they can still operate with a level of quality that is lower than expected. Even though a machine can still operate with a lower efficiency, there is no single failure point since it will continue to operate. For example, during the curing of a concrete column, in some instances it may not achieve the pre-specified strength, but it can still provide its main function, which is sustaining the loads on it. However, it will not provide its sub-function, which is resisting an earthquake. So this column would be considered in the tolerance failure zone (Fig. 1).

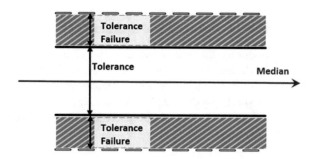

Fig. 1. Tolerance failure range.

Another issue that must be considered is tolerance accumulation. This can extend to work structuring and assembly sequence. Even though the units of a system may each conform to the specifications separately, the entire system combined may not conform due to tolerance accumulation (Milberg 2007). For example, an extension can be made to the case of the previously mentioned concrete column. If several columns happen to be in the tolerance failure zone, they may be acceptable, however, combined they may not be able to sustain the slab loads (their main function) due to the accumulation of tolerance failures. Several classifications of tolerance failure modes were found in literature (Milberg 2006 and Talebi et al. 2016). This paper proposed five main root causes of tolerance failures based on the literature review conducted.

4   Root Causes of Tolerance Failure

4.1   Inefficiency of Standards/Lack of Codes and Standards Integration

Codes and standards should be developed to cope with the development of the construction stage taking into consideration the different construction methods and the different sequences of work. Also, codes and standards should take into consideration the accumulation of normal tolerances throughout the stages of construction. Many reasons lead to inefficiency of codes and standards, such as:

4.1.1. The Construction Industry is Very Complex so Codes and Standards Cannot Define All of the Construction Tolerances.

4.1.2. The construction industry is well known for its fragmentation such that it is composed of various disciplines and trades. Thus, the formation of tolerance in codes and standards is based on different concepts and each discipline has its own definition and threshold of tolerance, which leads to tolerance failure due to dissonance between the different disciplines.

4.1.3. Some standards are very ambiguous to be understandable by all of the project team.

4.1.4. Specifications and tolerances should be compatible with the process capability.

4.2 Incomplete/Insufficient Drawings and Specifications

Drawings and specifications are the main communication channel between the designer and contractor from the design stage to the construction stage. Thus, any missing or insufficient data may lead to tolerance failures during construction. Incomplete data about the materials or project specifications with multiple interpretations may lead to tolerance failures as well due to the use of different materials or construction methods based on wrong assumptions.

4.3 Lack of Standardization

Standardization is developing certain procedures to achieve specific outcomes without variation between them. All work during the stages of construction must go through the standardization process to ensure they effectively fit together and to make tolerance management achievable. For example, there are various types of standards used between countries and sometimes within the same country such that the designer creates a design with certain intent of tolerance to accommodate these variations. After that, the consultant checks the construction works according to other tolerance standards that could be different from the designer's standards. This fragmentation in construction sometimes leads to errors, wastes, and even failures.

4.4 Lack of Awareness

Tolerance management is an important topic that is not widely discussed in engineering communities. To effectively apply tolerance management, all project participants and teams need to be made aware of it and its consequences. Also, lack of tolerance awareness among workers leads to poor workmanship, which makes tolerance management difficult to achieve. Tolerance concepts should also be introduced to engineering disciples at the university level to educate students.

4.5 Lack of Tolerance-Based Softwares

Currently available software takes into consideration many important factors in construction such as coordination, modeling, and energy analysis. However, no software has been made to identify or monitor tolerance. Creating such technology will lead to the analysis of a big dataset and track the accumulation of tolerances in the components.

5 Defects

This section delineates the difference between tolerances (in general), tolerance failure, and defects. The concept of defects, their types, and their effect is also discussed. In general, defects are work that does not conform to the specifications of the contract. Defects can be classified into two categories; patent and latent. Patent defects are those that can be detected through inspection and supervision; they are mostly known to be apparent and can be seen visually or by further investigation ("Patent defects in construction" 2018). On the other hand, latent defects are those that cannot be detected easily visually or even by inspection. They are hidden variations that can manifest at any time without previous signs ("Latent defects in construction", n.d.). Once they appear, they turn to be patent rather than latent defects. Moreover, defects under any contract are considered to be the responsibility of the contractor to rectify and amend within a reasonable time.

In the case of patent defects, the engineer/consultant normally instructs the contractor to rectify defects and damages as soon as possible and the defects notification period helps protect the employer's rights. In the case of latent defects, they are difficult to identify and assess because the time of occurrence of the defect can be unknown. For example, a foundation base may not be properly reinforced against earthquakes although the concrete used and dimensions of the foundation are properly designed, executed, and inspected. Such defects stay latent until an actual earthquake occurs. When the earthquake hits the foundation, only then will this defect appear. Consequently, it can then be decided whose fault it is and who is responsible for resolving the current situation (based on the contract, defects notification period, and the risk sharing between project participants). Currently, there is an ongoing issue regarding the allocation of variations in construction under the umbrella of tolerance or defects. To clear the misunderstanding, a clear definition of tolerance and defects needs to be explicitly stated and followed.

For example, to construct any concrete element, several resources and conditions are needed including materials, mixing equipment, skilled labor, proper pouring conditions, curing process, and inspection. Materials forming concrete have a great impact on its quality and any differences in these materials would affect it. If components with lower grades were used (that would not affect the structural integrity of the element or prevent the owner from using the functionalities of the building), then it may be within the tolerance boundaries and is up to the inspection team (and based on the specifications) to accept or reject such variations. Otherwise, it would be deemed a defect that cannot be accepted. The same applies to all processes and personnel required to make any other elements. Additionally, not all processes accept the same amount of tolerance as others. Low and high tolerance levels exist and are adjusted to fit the criteria set by experts to avoid defects. For example, if a window arm were taller by a centimeter or even less it would not fit into the opening in a wall and cannot be used. On the other hand, if a concrete beam is wider by a small margin (which occurs sometimes due to inaccuracies in design/construction), it can be overlooked as a tolerance margin and may not affect the functionality or visual elements. Therefore, tolerance and defects are treated differently depending on the elements themselves.

6 Tolerance Mitigation Strategies

In order to avoid the vast range of tolerances in any activity, certain strategies must be applied. These strategies aim to decrease the tolerance gap and control certain parameters that affect tolerances and eliminate discrepancies between them. According to Milberg (2006), these mitigation strategies can be classified into three types: 1) tolerance allocation strategy, 2) design change strategy, and 3) process change strategy, which are discussed below in detail.

6.1 Tolerance Allocation Strategies

Tolerance allocation strategies can be used when some of the tolerance requirements are governed by assembly concerns in a tolerance loop and some of the available process capabilities are tighter than the corresponding tolerance requirements (Milberg 2006). According to Milberg (2006), tolerance allocation is based on two parts: (i) assigning tolerances to individual parts based on available limits/standards for the combination of these parts; and (ii) there are unlimited ways to assemble the parts together while maintaining the combined tolerances. For instance, a typical floor may consist of a reinforced concrete (RC) slab, mortar, stone, epoxy, or any other paint. The thickness of each layer has a tolerance level as per the design requirements. We can assume that the floor thickness should not exceed 70 cm for height and elevation purposes as it would negatively affect the function of the building. This 70 cm is the critical dimension for which all elements' tolerances should obey (part (i) of tolerance allocation). This means that the RC slab can be 50 cm, the mortar 4 cm, the stone 3 cm, and different layers of epoxy paint can be 13 cm to reach the 70 cm limit. Hence, we can control the tolerance level of each element separately and allocate tolerance as deemed fit and based on the available resources. This is tolerance allocation. Also, based on that, there are an infinite number of possibilities to allocate tolerance, which would still not affect the critical dimension (part (ii) of tolerance allocation). Moreover, tolerance allocation can serve cost reduction purposes. If the thickness of the RC can be decreased to a minimum (save all tolerance for other elements), the difference in cost of the RC can be saved (since RC is more expensive) and the tolerance of the mortar or epoxy can be increased since they are relatively cheaper. This would be done without affecting the structural integrity of the final product. However, this strategy is not sufficient by itself and possibly requires additional support (from the other strategies).

6.2 Design Change Strategies

In this strategy, some elements of the design are changed to suit tolerance criteria either by removing/adjusting parts of the design that do not conform to tolerance levels or by adding some measures so that the non-conformed tolerance criteria are no longer required. This can be applied to the individual parts of the design or the assembly process as well. Design change strategies focus on changing the critical dimension, nominal geometry, connection design, and datum priority (Milberg 2006).

6.3   Process Change Strategies

Process change strategies involve process or inspection modifications focused on controlling the variation of individual components that enter the assembly (Milberg 2006). This strategy depends on assessing and evaluating the process before it is approved for construction. Hence, a process would not be accepted/performed in the first place in order to avoid rejection of tolerance requirements in the future.

7   Conclusions

This paper aimed to address tolerance and defects in order to distinguish between them and discuss their application in construction. Hence, it is important to elucidate on tolerance and delineate accepted margins and references in each project to avoid failures/defects. Literature review was conducted to collect data on each topic and present the principles, root causes, and strategies that can be applied. Moreover, the authors discussed failures that would hinder both the tolerances and processes from progressing, and proposed mitigation tools to overcome such failures and defects. From the discussion, tolerance management was shown to have a significant impact on the workflow and quality of the process and project. In addition, tolerance management not only affects activities and projects but can also manifest at larger scales and cause inherent damages to cities. For example, tolerance failures in smart cities would reduce their resilience and can cause disruptions in the networks and even lead to digital failures. This underlines the importance of increasing awareness on tolerance and defects to minimize/avoid wastes and failures thus increasing value to the customer.

References

Construction tolerances, https://www.designingbuildings.co.uk/constructiontolerance. Last accessed 2019/10/25

Creveling, C.: Tolerance Design: A Handbook for Developing Optimal Specifications. Pretence Hall, Upper Saddle River (1997)

Da Rocha, C., Tezel, A., Talebi, S., Koskela, L.: Product modularity, tolerance management, and visual management: potential Synergies. In: 26th Annual Conference of the International Group for Lean Construction (2018)

Gerth, R.: Tolerance analysis: a tutorial of current practice. In: Zahang H.C. (ed.) Advanced Tolerancing Techniques. Wiely-Interscience, New York (1997)

Henzold, G.: Handbook of Geometrical Tolerancing Design, Manufacturing, and Inspection. John wiley, Hoboken (1995)

Jose, J., Prasanna, P., Prakash, F.: Lean design strategy of waste minimization in construction industries. Int. J. Appl. Eng. Res. 13, 64593–4598 (2018). ISSN 0973–4562

Kent, D., Becerik-Gerber, B. : Understanding construction industry experience and attitudes toward integrated project delivery. J. Constr. Eng. Manage. 815–825 (2010)

Latent Defects in Construction. https://sitemate.com/resources/articles/quality/latent-defects-in-construction/. Accessed 07 Oct 2019

Milberg, C. : Application of Tolerance Management to Civil Systems. Ph.D. Dissertation, University of California, Berkeley, pp. 289–296 (2006)

Milberg, C. : Tolerance Considerations in Work Structuring. Proceedings IGLC-15 East Lansing, Michigan, USA, pp. 233–243 (2007)

Milberg, C., Tommelein, I.: Tolerance mapping – partition wall case revisited. In: Proceedings of the 12th Annual Conference of the International Group for Lean Construction (2004)

Patent Defects in Construction. https://www.designingbuildings.co.uk/wiki/Patent_defects_in_construction./. Accessed 30 Sep 2019

Talebi, S., Koskela, L., Shelbourn, M., Tzortzopoulos, P.: Critical review of tolerance management in construction. In: Proceedings 24th Annual Conference of the International Group for Lean Construction, Boston, USA, pp. 63–72 (2016). sect. 4

Application of Modified Invasive Weed Algorithm for Condition-Based Budget Allocation of Water Distribution Networks

Nehal Elshaboury[1(✉)], Eslam Mohammed Abdelkader[2], and Mohamed Marzouk[2]

[1] Housing and Building National Research Center, Giza, Egypt
nehal_ahmed_2014@hotmail.com
[2] Cairo University, Giza, Egypt

Abstract. Water Distribution Network is one of the critical components in water supply systems. Most water systems around the world are subjected to severe aging and deterioration which could lead to disastrous failures or sudden shutdowns. Optimizing maintenance and repair works of these systems triggers the need for a rigorous budget allocation model. In view of this situation, this paper presents a model for optimizing maintenance and replacement of the water networks using a set of metaheuristic algorithms. It introduces a modified invasive weed optimization algorithm to amplify the search mechanism of the classical invasive weed optimization algorithm by enhancing both exploration and exploitation of the exhaustive search space. The proposed optimization algorithm is validated through comparisons with the particle swarm optimization algorithm, shuffled frog leaping algorithm, and artificial bee colony algorithm. The capabilities of the developed model are exemplified through its application in a case study in Shaker Al-Bahery, Qalyubia governorate, Egypt. The results reveal that the proposed method exhibited superior results when compared to the aforementioned algorithms, which eventually leads to the establishment of more efficient decision-making models.

Keywords: Water Distribution Network · Budget allocation · Metaheuristic algorithms · Modified invasive weed optimization algorithm · Search mechanism

1 Introduction

Water systems are deemed key infrastructures that are used in taking, treating, and supplying water to consumers (Kamiński et al. 2017). Water Distribution Networks (WDNs) would be the primary focus of this research since they are considered to be as one of the most significant elements in providing healthy drinking water. Besides, their associated investments constitute the largest weight (i.e. 80%) in maintenance budgets (Kleiner and Rajani 2000). WDNs are complex systems that consist of a huge number of connected pipelines and accessories (i.e. hydrants and valves). The problem of aging and deterioration is a natural tendency of WDN infrastructures. The deferred maintenance and rehabilitation of such assets are associated with overwhelming replacement

© The Author(s), under exclusive license to Springer Nature Switzerland AG 2021
I. El Dimeery et al. (Eds.): JIC Smart Cities 2019, SUCI, pp. 121–131, 2021.
https://doi.org/10.1007/978-3-030-64217-4_15

costs and thus demand judicious spending and efficient planning (Alegre and Coelho 2012). The lack of standards, guidelines, or best practices highlights the necessity of devising a strategy to prioritize and schedule the maintenance and renewal plans and to spend the allocated limited budget efficiently and effectively (Shahata 2013).

The optimization of WDN has been a green research topic for a long time. It initially focused on a single indispensable objective of cost minimization because the economic objective is the main motivation for water utilities. Currently, the optimization problem involves multiple opposing objectives along the whole life cycle of WDN. The problem generally comprises cost minimization and benefit maximization, no matter how many objectives there are and how to quantify them. It is popular to adopt two objective functions because a) it is the simplest multiple objective optimizations, and b) excessive objectives will result in complex searching for solutions (Zhou 2018). In this research, the problem allows the minimization of a cost function as well as the maximization of a condition function while considering budget constraints. One of the most popular ways to deal with multi-objective optimization is to incorporate the multiple criteria into a single objective function by using weighting factors. It has been reported that these transformation and simplification approaches tend to simplify the problem, reduce the objective dimensions, and reduce the computational load. The single objective optimization equips decision-makers with only one optimal solution.

There has been an extensive endeavor to address the optimization of water infrastructure assets. Dridi et al. (2008) presented a strategy to minimize the total cost of pipe renewal while considering the hydraulic criterion. The number of pipes to be replaced as well as the optimal time for renewal was identified using three different optimization techniques: IGA (Island Genetic Algorithm), NPGA-2 (Niched Pareto Genetic Algorithm 2), and NSGA-II (Non-dominated Sorting Genetic Algorithm-II). Mohamed (2009) developed short and long term budget allocation plans for water systems based on their breakage rates and condition assessment. The results showed that various rehabilitation alternatives are almost close for both models. Tantawy (2012) developed a framework to optimally allocate available budget for assets using GA. The problem was formulated in either of two ways: a) maximization of the average condition within certain repair budget, and b) minimization of the cost of repair to achieve desired condition index.

Osama (2014) employed GA to develop an optimization model for water, sewer and road networks by accounting for the criticality assessment, level of service, condition assessment, and life cycle cost. El-Masoudi (2016) presented a repair prioritization and fund allocation model for water pipelines by deploying a GA optimization technique. The model attempted to minimize the total life cycle cost of repairs for all pipelines considering acceptable overall network condition, planning horizon, and limited yearly budgets. Ismaeel (2016) developed a budget allocation model utilizing GA and greedy heuristics to allocate the budget among the water network components. The model aimed at maximizing the condition of the total network while satisfying the allowable annual budget. Shin et al. (2016) showed the use of GA technique to find a near-optimal rehabilitation schedule of water pipelines. The model aimed at minimizing the whole life cycle cost during the study period while considering restricted budget constraints.

Zangenehmadar (2016) presented a model to optimally allocate the budget for water pipelines by utilizing GA technique. It accounted for the cost, breakage rate and constrained budget of the project. El-Ghandour and Elbeltagi (2018) optimized the design and rehabilitation of water networks using the five evolutionary algorithms; which are particle swarm optimization (PSO), modified shuffled frog leaping (SFL), ant colony optimization (ACO), memetic algorithm (MA), and GA. Zhou (2018) developed an optimal rehabilitation model for water pipelines. The objectives of the model included life cycle cost minimization, hydraulic reliability maximization, and burst number minimization. The objectives were subjected to financial and hydraulic performance constraints. The optimization problem was solved using a modified NSGA-II technique.

2 Applications of Meta-heuristics

Meta-heuristics are bio-inspired optimization algorithms that are usually applied to solve complex and large exhaustive search space problems. They are characterized by their capabilities to overcome the shortcomings of inferior accuracy, local minima and premature convergence of hill-climbing derivative-based algorithms. Banharnsakun (2017) presented an image segmentation method that involved artificial bee colony algorithm and Otsu function for pavement surface distress detection. The developed method attained better segmentation results when compared to the differential evolution algorithm, artificial bee colony algorithm, genetic algorithm, and Ostu function. Marzouk and Mohammed Abdelkader (2019) introduced a hybrid fuzzy multi-objective non-dominated sorting genetic algorithm II to select the most sustainable materials subject to time, cost and environmental constraints. Marzouk and Mohammed Abdelkader (2017) introduced a building information modeling-based method for minimizing construction emissions. The construction activities were investigated as per project time; project life cycle cost; project environmental impact; and primary energy consumed by them.

Aziz et al. (2017) utilized two swarm optimization algorithms, namely whale optimization (WO) algorithm and moth-flame optimization (MFO) algorithm for multi-level thresholding. The objective was to search for the optimum threshold values that maximized the Otsu's function. They highlighted that for a higher number of thresholding problems, the MFO algorithm achieved higher fitness function values, peak signal to noise ratio and structural similarity index when compared to the WO algorithm. On the other hand, the MFO algorithm required longer computational time to search for the optimal values. Mohammed Abdelkader et al. (2019a) presented an optimization-based methodology for the computation of optimum thresholds of ground penetrating radar. They compared the genetic algorithm, particle swarm optimization algorithm and shuffled frog leaping algorithm based on some performance indicators such as mean average performance, hypervolume indicator and inverted generational distance such that they inferred that the shuffled frog leaping algorithm provided superior performance when compared to other meta-heuristics.

Mohammed Abdelkader et al. (2019b) presented a computerized hybrid Bayesian-based method for modeling the deterioration of concrete bridge decks. Genetic algorithm was utilized to compute the missing conditional probabilities of the variable transition probability matrices in the stochastic optimization model. Mohammed Abdelkader et al. (2019c) introduced a self-adaptive multi-objective invasive weed optimization-based method for the detection of crack images in reinforced concrete bridges. The proposed method outperformed multi-objective genetic algorithm-based method, multi-objective particle swarm-based method and multi-objective harmony search-based method as per average values of mean-squared error, peak signal to noise ratio and structural similarity index.

3   Model Development

The mechanism of the budget allocation model of a WDN is illustrated in Fig. 1. The model involves a) performing the optimization algorithms, and b) computing the optimal solutions.

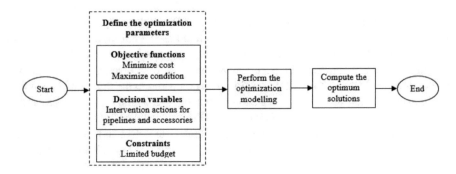

Fig. 1. Components of the proposed framework

The optimization model incorporates two objective functions as illustrated in this section. The first evaluated function is to minimize the total cost required to attain the improvements suggested to the network as per Eq. 1.

$$LCC = \frac{1}{(1+r)^t} \sum_{j=1}^{K} \sum_{i=1}^{N} C_{ij} \tag{1}$$

Where; C_{ij} represents the cost of repair strategy i applied to network component j at time t, r refers to the discount rate, and T refers to the number of years. In this research, the discount rate is taken at 7%.

The second objective of this model is to maximize the structural condition of the overall network at the end of the study period, as per Eq. 2.

$$R_{NM} = 1 - \sum_{i=1}^{m} \Pi_{j=1}^{n} Q_j \qquad (2)$$

Where; R_{NM} refers to the mechanical reliability of the network, m is the number of minimum cut sets in the network, Q_j is the probability of failure of the component (segment), and n is the number of components (segments) in the corresponding minimum cut set.

In this research, the fitness function is a combination of the total cost and the structural network condition. The aggregation of these two parameters is based on user-defined weights. Equal importance for the two parameters will establish 50% weights for each. These two functions are aggregated into a single function, as per Eq. 3.

$$Overall\ performance = (0.5 \times LCC) + (0.5 \times R_{NM}) \qquad (3)$$

The decision variables refer to the available intervention actions for pipelines and accessories. There are four possible repair strategies (i.e. full replacement, major repair, minor repair or no action) for pipelines and accessories. For pipelines, minor repairs are intended to restore a small section of the pipeline using compression coupling while major repairs involve replacing larger sections of the pipeline using telescopic coupling. Accessories that are subjected to minor defects should be repaired on the spot if operating conditions allow. If not, the accessories should be dismantled from the site in order to perform a full repair. Otherwise, in case of severe damage, the accessories should be replaced with new ones in order to prevent severe failures. El-Masoudi (2016) suggested the estimated repair costs associated with the four repair scenarios (i.e. no action, minor repair, major repair, and a full replacement) to be 0%, 20%, 50%, and 100% of the total replacement cost, respectively. The degree of improvement associated with any decision variable if applied (except for doing nothing) is demonstrated in Table 1 (El-Masoudi 2016).

Table 1. Predicted improvement after repair (El-Masoudi 2016)

Condition	Replacement	Major	Minor
Excellent	–	–	–
Good	Excellent	–	–
Moderate	Excellent	Good	Good
Poor	Excellent	Good	Moderate
Critical	Excellent	Good	Moderate

The costs of repair and rehabilitation are subjected to annual budget limitations as per Eq. 4. In this research, the appropriate annual budget for maintenance and renewal is assumed to be 10% of the total project value.

$$\sum_{j=1}^{K} \sum_{i=1}^{N} C_{ij} < Budget \qquad (4)$$

4 Meta-Heuristic Optimization Algorithms

In this research, a modified invasive weed optimization (MIWO) algorithm is introduced to reach the most optimal budget allocation plan. The proposed optimization algorithm is validated through comparison with the PSO, shuffled frog leaping (SFL), and artificial bee colony (ABC) algorithms. The basic concepts of the PSO, SFL and ABC in addition to the needed references are illustrated as in the below sub-sections.

4.1 Modified Invasive Weed Optimization Algorithm

The Invasive Weed Optimization (IWO) algorithm was proposed by Mehrabian and Lucas in 2006, which was originated from the colonial behavior of weeds. IWO is based on simulating the invasive behaviour of weed in colonizing and finding the most suitable place for growth and reproduction. Weeds are robust and undesirable plants that grow spontaneously and they can have a harmful effect on both farms and gardens. More details about the classical invasive weed optimization algorithm can be found in Mohammed Abdelkader et al. (2019c). MIWO algorithm is primarily focusing on three areas of improvement in order to amplify the search paradigm of the IWO algorithm. These enhancements are aiming at improving the exploration and exploitation of the IWO algorithm.

The first area of improvement is introducing a fitness function-based spatial dispersion search mechanism to search for the global optimum solutions. The second area involves an exponential-based search function instead of quadratic search function. This offers a faster mechanism to search the different sub-regions of the search space. The third area is utilizing the entropy equation as a measure of dispersion and variability instead of the standard deviation. The maximum entropy requires less computational effort to search the space than the standard deviation.

4.2 Particle Swarm Optimization Algorithm

The particle Swarm Optimization algorithm was introduced by Kennedy and Eberhart in 1995. It was inspired by the movement of birds and fish in search of their food. The initial particles are created and assigned initial velocities. Each particle in the swarm is characterized by its current position, best position, and velocity. The particle best position is archived and automatically updated whenever a new better position is reached. The global best position among the stored particle best positions is stored and updated once a better position is found. The new positions of the particles are updated by considering randomized values towards some directions. These changes are calculated using the velocity. The velocity of any particle in a swarm relies on particle best position, global best position, and the random function. This approach is repeated until a stopping criterion is reached. Examples of termination criteria are provided as follows: achieving a satisfactory solution, reaching to a maximum number of iterations, and finding constant fitness for a certain number of iterations. The global solution is achieved by the current position of the best particle in the last iteration.

4.3 Shuffled Frog Leaping Algorithm

The shuffled frog leaping algorithm was developed by Eusuff and Lansey in 2003. It imitates the behavior of frogs when searching for the location with the maximum amount of food (Eusuff et al. 2006). The SFL algorithm consists of a set of frogs that is partitioned into memeplexes. The individual frogs in different memeplexes hold ideas that can be influenced by other frogs' ideas, and experience a process of memetic evolution. Memetic evolution enhances the performance of the individual frog towards a goal. The algorithm performs simultaneously a local search in each memeplex. After a certain number of memetic evolution steps, the memeplexes are forced to mix and new memeplexes are formed through a shuffling process (Liong and Atiquzzaman 2004). The shuffling process accelerates the searching procedure and ensures the global exploration of the solutions. The local search and the shuffling processes continue until defined convergence criteria are satisfied (Eusuff and Lansey 2003).

4.4 Artificial Bee Colony Algorithm

Artificial bee colony (ABC) algorithm which was developed by Karaboga in 2005 simulates the behavior of honey bees searching for food sources close to their hive. In the ABC algorithm, there are three types of bees: employed bees, onlooker bees, and scout bees. The first half of the colony consists of the employed artificial bees and the second half includes the onlookers. The employed bees exploit the food sources and share the information of these food sources to the onlooker bees. Then, the onlooker bees decide whether or not to exploit the food sources based on the nectar amount of food sources. The employed bee whose food source has been exhausted by the bees becomes a scout. The scout bees search for new food sources through random searching (Zhu and Kwong 2010).

5 Model Implementation

This research proposes an optimization model for managing the replacement and rehabilitation of water network components. The capabilities of the developed model are exemplified through its application in a case study in Shaker Al-Bahery, Qalyubia Governorate, Egypt. The network layout is illustrated in Fig. 2. The project's objective is to provide the area with drinking water from Atlas pump station, which is at a distance of about 800 meters and is related to El-Salam water networks in Cairo. The case study project covers a land area of approximately 336,100 square meters. The network consists of 173 segments with a total length of 10.3 km. The network serves a population of approximately 19,859 people for the year 2018. The total project cost was estimated to be 3.96 Million Egyptian Pounds. The budget allocation model of the network is determined in major stages, as described below.

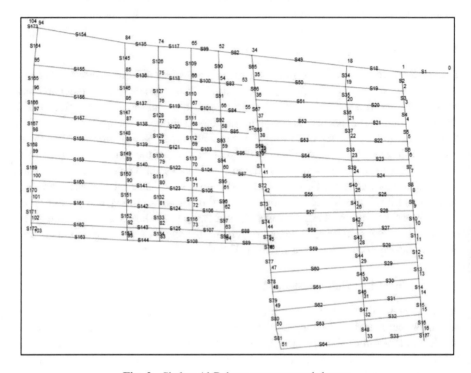

Fig. 2. Shaker Al-Bahery water network layout

As demonstrated earlier in the model development section, the first objective function is to minimize the total cost associated with the intervention actions while the second objective function is to maximize the overall network condition at the end of the study period. After constructing the two objective functions, they are aggregated and evaluated based on a fitness function. Four decision variables are considered (do nothing, minor repair, major repair, and replace). The main constraints are the allowable planning horizon (3-years plan) and budget constraints. After identifying the inputs, the next step comprises MIWO, PSO, SFL and ABC algorithms implementation as described in the below sub-sections.

The optimization algorithms are performed using Matlab. The problem setup comprises clarifying the input fitness function, the number of decision variables, and lower bound and upper bound of variables. In PSO, the personal learning coefficient ($c1$) and global learning coefficient ($c2$) have values of 2 and the inertia weight (w) is taken as 1. In SFL, the memeplex size is chosen to be 100 while the number of memeplexes is selected to be 5. In this research, the population size is chosen to be 250 and the maximum number of iterations is assumed to be 200 for all meta-heuristics in order to provide a fair comparison between them. The convergence curves of the meta-heuristic-based budget allocation models are shown in Fig. 3. It is worth noting that MIWO-based budget allocation outperformed other meta-heuristic budget allocation models such that the overall performances of ABC, PSO, SFL and MIWO are 4.3391, 1.1736, 6.4709 and 0.9475, respectively. SFL algorithm achieved the least performance, which exemplifies its under-performing capacity in dealing with similar budget allocation models. Moreover, it can be inferred the proposed method attained the best performance because of its superior exploration and exploitation capacity.

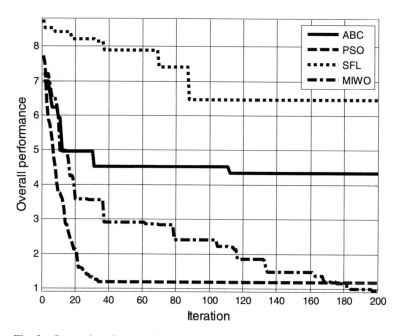

Fig. 3. Comparison between the meta-heuristic-based budget allocation models

6 Conclusion

This research presented a methodology to help water municipalities for efficient budget allocation and better scheduling of needed intervention strategies. The model provided decision-makers with the most optimum repair and replacement strategy as well as the optimum moment for its application based on the cost of intervention actions and their associated degree of condition improvement. In this paper, a modified invasive weed optimization algorithm was introduced to enhance the search mechanism of the classical invasive weed optimization algorithm. This algorithm was tested on a water distribution network in Shaker Al-Bahery, Egypt. The performance of the proposed method was validated through comparison with particle swarm optimization algorithm, shuffled frog leaping algorithm, and artificial bee colony algorithm. The results demonstrated that the proposed modified invasive weed optimization-based budget allocation model exemplified superior performance compared to the aforementioned algorithms.

References

Alegre, H., Coelho, S.T.: Infrastructure asset management of urban water systems (2012). https://www.intechopen.com/books/water-supply-system-analysis-selected-topics/infrastructure-asset-management-of-urban-water-systems. Accessed 04 June 2018

El Aziz, M.A., Ewees, A.A., Hassanien, A.E.: Whale optimization algorithm and moth-flame optimization for multilevel thresholding image segmentation. Expert Syst. Appl. **83**, 242–256 (2017)

Banharnsakun, A.: Hybrid ABC-ANN for pavement surface distress detection and classification. Int. J. Mach. Learn. Cybern. **8**(2), 699–710 (2017)

Dridi, L., Parizeau, M., Mailhot, A., Villeneuve, J.: Using evolutionary optimization techniques for scheduling water pipe renewal considering a short planning horizon. Comput.-Aided Civ. Infrastruct. Eng. **23**(8), 625–635 (2008)

El-Ghandour, H., Elbeltagi, E.: Comparison of five evolutionary algorithms for optimization of water distribution networks. J. Comput. Civ. Eng. **32**(1), 1–10 (2018)

El-Masoudi, I.: Condition assessment and optimal repair strategies of water networks using genetic algorithms. MSc Thesis, Mansoura University, Mansoura, Egypt (2016)

Eusuff, M., Lansey, K.: Optimization of water distribution network design using the shuffled frog leaping algorithm. J. Water Resour. Plan Manag. **129**(3), 210–225 (2003)

Eusuff, M., Lansey, K., Pasha, F.: Shuffled frog-leaping algorithm: a memetic meta-heuristic for discrete optimization. Eng. Optimiz. **38**(2), 129–154 (2006)

Ismaeel, M.: Performance-based budget allocation model for water networks. MSc Thesis, Concordia University, Montreal, Canada (2016)

Kamiński, K., Kamiński, W., Mizerski, T.: Application of artificial neural networks to the technical condition assessment of water supply systems. Ecol. Chem. Eng. S **24**(1), 31–40 (2017)

Kleiner, Y., Rajani, B.: Considering time-dependent factors in the statistical prediction of water main breaks. In: Proceedings of American Water Works Association Infrastructure Conference. American Water Works Association, Denver (2000)

Liong, S., Atiquzzaman, M.: Optimal design of water distribution network using shuffled complex evolution. J. Instrum. Eng. **44**(1), 93–107 (2004)

Marzouk, M., Mohammed Abdelkader, E.: A hybrid fuzzy-optimization method for modeling construction emissions. Decis. Sci. Lett. **9**, 1–20 (2019)

Marzouk, M., Mohammed Abdelkader, E.: Minimizing construction emissions using building information modeling and decision-making techniques. Int. J. 3-D Inf. Model. **6**(2), 1–22 (2017)

Mohammed Abdelkader, E., Marzouk, M., Zayed, T.: An optimization-based methodology for the definition of amplitude thresholds of the ground penetrating. Soft Comput. **23**, 1–24 (2019a)

Mohammed Abdelkader, E., Zayed, T., Marzouk, M.: A computerized hybrid Bayesian-based approach for modeling the deterioration of concrete bridge decks. Struct. Infrastruct. Eng. **25**(19), 1178–1199 (2019b)

Mohammed Abdelkader, E., Moselhi, O., Marzouk, M., Zayed, T.: A multi-objective invasive weed optimization method for segmentation of distress images. Intell. Autom. Soft Comput. **26**, 1–20 (2019c)

Mohamed, E.: Fund allocation and rehabilitation planning for water main projects. MSc Thesis, Concordia University, Montreal, Canada (2009)

Osama, A.: Fuzzy-based methodology for integrated infrastructure asset management. MSc Thesis, Cairo University, Cairo, Egypt (2014)

Shahata, K.: Decision-support framework for integrated asset management of major municipal infrastructure. Ph.D. Thesis, Concordia University, Montreal, Canada (2013)

Shin, H., Joo, C., Koo, J.: Optimal rehabilitation model for water pipeline systems with genetic algorithm. Proc. Eng. **154**, 384–390 (2016)

Tantawy, M.: Framework for infrastructure asset management. Ph.D. Thesis, Mansoura University, Mansoura, Egypt (2012)

Zangenehmadar, Z.: Asset management tools for sustainable water distribution networks. Ph.D. Thesis, Concordia University, Montreal, Canada (2016)

Zhou, Y.: Deterioration and optimal rehabilitation modeling for urban water distribution systems. Ph.D. Thesis, Delft University of Technology, Delft, The Netherlands (2018)

Zhu, G., Kwong, S.: Gbest-guided artificial bee colony algorithm for numerical function optimization. Appl. Math. Comput. **217**(7), 3166–3173 (2010)

GIS-BIM Data Integration Towards a Smart Campus

Yousif Ward$^{(\boxtimes)}$, Salem Morsy, and Adel El-Shazly

Cairo University, Giza, Egypt
youssif@outlook.com

Abstract. Smart cities are the key information for many purposes such as improvement of the educational environment, emergency response, and facilities management. Smart cities require the development of a digital system that can manage, visualize, share and exchange the attribute data and spatial data in a user-friendly environment. Building Information Modeling (BIM) and Geographic Information System (GIS) are widely used as the modeling sources of smart cities. In this paper, GIS and BIM data were integrated to implement applications that help in the creation of a smart campus for the Faculty of Engineering, Cairo University, Egypt. Campus data of various objects such as buildings, roads, trees, etc. were acquired using satellite remote sensing images, terrestrial images, and traditional surveying techniques (e.g., GPS and total station), as well as field visits. The acquired data were integrated and presented in two formats; (i) 2D interactive map of the campus on the web that represents information of various objects with geographic coordinates in real-time and (ii) smart building cellphone application that uses quick response (QR) codes to display building attributes from stored geodatabase upon scanning. The presented applications have proved that the smart campus implementation can be achievable and meet management requirements at universities level.

Keywords: GIS · BIM · Smart campus · QR codes · Interactive map

1 Introduction

Smart cities are evolving worldwide, because their comprehensive digital environment that improves the efficiency, security and management of urban systems. The concept of smart cities is based on the use of geospatial data concerning the urban natural or built environment as well as urban services. The successful implementation of a smart city project requires the development of a digital system that can manage and visualize the geospatial data in a user-friendly environment [1].

Currently, Building Information Modeling (BIM) and Geographical Information Systems (GIS) are broadly used as the modelling sources of smart cities. BIM looks on the individual structures, while GIS focused at positions of these structures and relation with surrounding objects. Using BIM and GIS provide valuable information for planning, maintenance, emergency response, and facilities management [1–3]. Therefore, BIM and GIS data integration is essential in smart cities in order to achieve the aforementioned applications.

I. El Dimeery et al. (Eds.): JIC Smart Cities 2019, SUCI, pp. 132–139, 2021.
https://doi.org/10.1007/978-3-030-64217-4_16

Barret and Finch [4] showed the benefit of BIM to the management of a plant during three phases, namely design, construction and operation. El-Gamily [5] created a geodatabase for different elements of infrastructure facilities and services on the building bases for a selected number of schools. Then, a GIS-based interactive application was developed to manage different facilities at the school level. Deng et al. [6] developed a platform for combining traffic noise evaluation in indoor and outdoor environments based on BIM and 3D GIS integration.

A comprehensive review of the existing researches on the integrated application of BIM and GIS was conducted by [3]. The study focused on application objects, application phases, integrated patterns, and platforms used. The objects of interest included buildings, infrastructures and urban districts, while the application phases consisted of planning and design (P&D), construction, operation and maintenance (O&M) and demolition. The integrated patterns included three scenarios; extract data from BIM into GIS, extract data from GIS into BIM, or extract data from both BIM and GIS into another system. Finally, the platforms used for the integrated application of BIM and GIS, were studied. Theses platforms could be software systems or web platforms. Results showed that integration of BIM and GIS was mainly used for building as an application object in P&D and O&M phases. Extracting data from BIM into GIS was the mainstream way of BIM and GIS integration where ArcGIS was the most commonly used platform [3].

With the construction industry moving towards integrating BIM as standard, we integrated BIM and GIS data to create applications that help in the creation of a smarter campuses or cities. This integration improves the educational environment and the comprehensive management level of the universities [7, 8]. In the paper, smart campus has been addressed by GIS and BIM integration in order to i) create an interactive map of Cairo University campus on the web with geographic coordinates in real-time and ii) smart building cellphone application using quick response (QR) codes that display building attributes from the geodatabase upon scanning.

2   Smart Campus Implementation

With the concepts of smart campus attracting the industry, methods of managing spatial and attribute information has become critical using GIS and BIM. GIS offers advanced and user-friendly capabilities for smart campus projects [1, 7, 8]. BIM can be used to create, manage and share the digital information of vertical structures such as buildings [3], while GIS can be used to store, manage, query, analyze and visualize campus geographic data [8].

2.1   Study Area and Data Acquisition

The campus of Cairo University in Giza, Egypt was chosen to be studied in this paper as an educational environment. The campus includes a variety of land objects such as buildings, vegetation areas, asphalt, trees, obelisks, pools, etc. The whole campus was utilized to produce a 2D interactive map, while the Faculty of Engineering campus only was utilized in the smart building cellphone application.

A worldview-3 (WV-3) satellite image, downloaded from ArcGIS online database, was used in digitizing different objects in order to create a campus digital map. The image was collected on April 19th, 2017 with resolution of 0.31 m (see Fig. 1). Then, the image was georeferenced and rectified using arbitrary ground control points within the campus. Leica Viva GS15 smart antenna was utilized to measure those arbitrary points from known ground control points using differential static method. Leica Geo Office software was utilized in points processing.

Fig. 1. Worldview-3 image of Cairo University campus with Faculty of Engineering campus in red frames.

Terrestrial images were captured for different building facades of Faculty of Engineering campus using a smart phone to be used in 3D model construction. In addition, 3D points were measured on different facades using laser total station Sokkia Set 3X. A combination of terrestrial images and 3D points measurements, in addition to the ground control points, were used to create 3D model for buildings. Autodesk software packages were used in model creation including Recap, AutoCAD and 3DS Max.

Field visits were conducted in order to fed the database with information which known as "attribute data". With the focus on buildings, the attributes data mainly include building number, building name, building area, number of floors, number of lecture rooms, etc.

2.2   Overall Workflow

Figure 2 shows the overall workflow towards smart campus construction by offering two applications; 2D interactive campus map and smart building cellphone application.

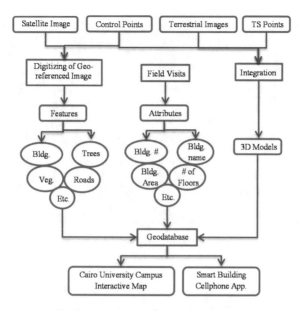

Fig. 2. Smart campus workflow implementation.

3 Smart Campus Applications

In order to make the data obtained throughout this research available in a user-friendly format, we stored the data in a cloud database that can be accessible through the internet. Thus, the central database on the web can be utilized by applications where the end-user can expand the possibilities for different usage scenarios. Our applications neither relied on third-party software nor require a certain level of knowledge by the user. The database was stored using Structured Query Language (SQL) and the interface was implemented using Hypertext Preprocessor (PHP) to manage the database. We have written database management interface on the web that allows us to modify the geodatabase in the cloud and on-demand.

3.1 2D Interactive Web Map of Cairo University Campus

The WV-3 satellite image was digitized using ArcGIS software to extract objects of Cairo University campus. These objects include buildings, vegetation areas, asphalt, trees, obelisks, and pools. Then, a website has been developed containing a GIS interactive map. The web map offers three display modes as shown in Fig. 3; Digitizing Overlay which overlays the digitized map on the satellite image, Digitizing Only which displays the digitized map only, and Satellite which displays the WV-3 image.

While digitizing many campus objects, we focused on buildings within the Faculty of Engineering campus. The user is able to display building attributes, stored in the cloud database, by clicking on a building without using any third-party software. The interactive map is available at: https://graduation.yward.net/map/. For instance, Building #2 (Administration Building) has all information added in its attribute table and can be used as a good example of how the map functions as shown in Fig. 4.

Fig. 3. 2D interactive web map of Cairo University with Digitizing Only mode displayed.

Fig. 4. The interactive map displaying information of the administration building.

3.2   Smart Buildings Cellphone Application

The second application of smart campus presented in this paper is a cellphone application that uses quick response (QR) codes to display building attributes upon scanning as shown Fig. 5. QR codes tags were printed and pinned near each building entrance. In this time, almost every smartphone comes with a QR reader pre-installed; which falls in line with our direction to make users do not utilize any software or a third-party application. Thus, they are able to access the geodatabase data we have established. Upon scanning the code using a smartphone, building information are displayed. The application also displays photo(s) and a 3D model of the building, if available (see Fig. 6). The information requests are handled by a PHP code on the web that identifies the building using a unique identification number and then queries the database to get the relevant information to the building once scanning its QR code.

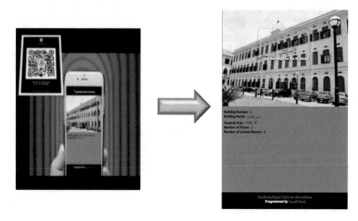

Fig. 5. Smart buildings cellphone application.

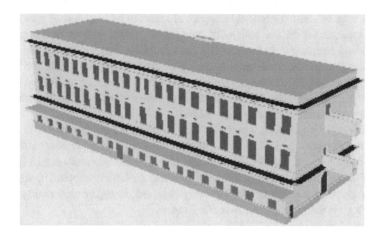

Fig. 6. 3D model of the administration building.

4 Discussion

This paper covers a few developed GIS-BIM solutions to achieve a smarter campus in user-friendly practical applications. The applications discussed in the previous section are fully functioning examples of how the cloud database and map can be utilized.

The interactive Cairo University campus map offers three display modes as aforementioned and it is connected to a central geodatabase on the internet. Thus, this map's applications are countless, for example, it can be used by the public (students) to state the number of available lecture rooms in a certain building, which faculty offices reside in the said building, whether a building has Wi-Fi or not, which lectures are being held in the said building and so on. For internal facilities management, the map can be used to track attributes related to the quality assurance or asset management. For instance, the map can display building maintenance requirements such as the number of defective or broken projectors, chairs, tables, ACs, etc. It can be also utilized to achieve previous use cases by allowing students to report broken/malfunctioning items in a certain building. Thus, the students can see the map with the public information while the faculty staff can see the extended map with internal/confidential information.

Smart building QR identification codes application is very practical to access the information in the cloud geodatabase. By simply scanning a QR code next to a building's entrance, the user is able to know all building's information and explore building's 3D model. In addition, the application allows using a dedicated form to report issues about the building. This application complements the interactive map with how accessible it is and how straight forward the experience is.

5 Conclusions and Future Works

The paper shows how GIS and BIM could help in the implementation of a smart campus. Two applications have been presented in context of a smart campus. First, 2D interactive map of Cairo University campus, displaying information of various objects has been represented on the web with geographic coordinates in real-time. Second, smart building cellphone application has been implemented to display building attributes from stored geodatabase upon scanning its unique QR code. With the realization of these applications, smart campus can provide a comprehensive intelligent perception information service platform to meet the management requirements at universities level that promotes teaching and scientific research.

The cellphone application can be extend to include students attendance tracking using Global Positioning System and QR combination. The application automatically creates a location-based radius around a certain class. Then, students scan a QR code to mark their attendance. Another cellphone application is currently being implemented to navigate a place of interest within the campus based on augmented reality and positioning using Global Navigation Satellite System.

References

1. Shahrour, I.: Use of GIS in Smart City Projects. GIM International. https://www.gim-international.com/content/article/use-of-gis-in-smart-city-projects. Accessed 11 Feb 2019
2. TWNKLS a PTC company: Can Augmented Reality bridge the gaps between BIM and GIS? https://twnkls.com/blogs/can-augmented-reality-bridge-the-gaps-between-bim-and-gis/. Accessed 22 Mar 2019
3. Ma, Z., Ren, Y.: Integrated application of BIM and GIS: an overview. Proc. Eng. **196**, 1072–1079 (2017)
4. Barrett, P., Finch, E.: Facilities Management: The Dynamics of Excellence, 3rd edn. Wiley, Hoboken (2014)

5. El-Gamily, H.I., Al-Rasheed, K.: Deploying an interactive GIS system for facility and asset management: case study-ministry of education, Kuwait. J. Geogr. Inf. Syst. **7**(02), 191 (2015)
6. Deng, Y., Cheng, J.C., Anumba, C.: A framework for 3D traffic noise mapping using data from BIM and GIS integration. Struct. Infrastruct. Eng. **12**(10), 1267–1280 (2016)
7. Bi, T., Yang, X., Ren, M.: The design and implementation of smart campus system. JCP **12** (6), 527–533 (2017)
8. Jiang, Y.: Design and development of smart campus system based on BIM and GIS. In: 6th International Conference on Energy, Environment and Sustainable Development (ICEESD), pp. 52–54. Atlantis Press, Thailand (2017)

Risk Management Methodology for Green Building Construction Projects Using Fuzzy-Based Multi-criteria Decision-Making

Reem Ahmed[1]([⊠]), Ayman Nassar[1], and Mohamed Afifi[2]

[1] German University in Cairo, Cairo, Egypt
reem.aboulrous@gmail.com
[2] American University in Cairo, Cairo, Egypt

Abstract. Construction projects usually encounter a significant number of unforeseen events referred to as project risks. Those risks typically incur a noteworthy impact on the overall project duration, cost as well as quality. Therefore, a systematic means of construction project risk planning, analysis and management is required to ensure those negative impacts are brought to a minimum, especially in the case of green building construction given that it is a relatively new field of application in Egypt. However, a great deficiency in proper record keeping is experienced in Egyptian construction projects in general, which increases the complexity of the construction and diminishes possibilities of innovation. Accordingly, the current risk management procedure is mostly reactive, qualitative and involves a great deal of uncertainties associated with the subjective data collection methods. To overcome those limitations, this paper presents a predictive risk management framework that identifies, clusters and evaluates the most important risks in the Egyptian green construction industry. A novel Fuzzy-based Multi-Criteria Decision-Making methodology is utilized as part of this study to capture the uncertainties in the industry experts' opinions, as well as consider a degree of indeterminacy to their decisions. The application of the aforementioned methodology results in the quantification of the priority level of each of the identified project risks that in turn suggests a risk management strategy to each individual risk. The risk planning procedure proposed in this study was applied to Egyptian construction projects and resulted in a significant declination in money losses and time-related delays, and accordingly, a reduction in the level of disputes between project participants and stakeholders.

Keywords: Construction · Risk management · Risk analysis · MCDM · AHP · Fuzzy logic

1 Introduction

Green Building Construction is a term that refers to buildings constructed on an environmentally-friendly or an energy-saving basis, from planning, construction, renovation to disassembly. Numerous advantages are associated with green buildings and sustainable housing, most importantly the higher profitability achieved through the

I. El Dimeery et al. (Eds.): JIC Smart Cities 2019, SUCI, pp. 140–150, 2021.
https://doi.org/10.1007/978-3-030-64217-4_17

reduction of energy consumption and accordingly the increase experienced in the life cycle of the assets by using more natural concepts and materials. However, municipalities and cities all over the world are facing considerable challenges of making urban life safer, healthier and more energy-efficient. One of the most evident challenges faced are the risks encountered throughout the green building construction process. Therefore, managing and dealing with such risks become inevitable in order to implement a successful green building construction project aiming at the contribution towards a safe, sustainable and smart city.

Risk management is a project management practice that is considered of great importance as Royer (2000) stated that: "Experience has shown that risk management must be of critical concern to project managers as unmanaged or unmitigated risks are one of the primary causes of project failure". Risk management is thus in direct relation to the successful project accomplishment as it helps ensure a project is completed on time, within budget and with the sought-after performance (Turner 1999).

The most widely-accepted risk management process is basically constructed from four iterative phases, namely: Risk Identification, Risk Estimation, Risk Response Planning, Execution and Monitoring.

When dealing with risks, the potential for improvement should also be taken into account, for example, to undertake the project with fewer resources or to make use of an unexpected window of opportunity. Risks are at the very core of the business: risks and opportunities are linked; there are no opportunities without risks related to them. Thus, risks raise the value of a project; usually, higher risks bring greater opportunities (Miller 2001).

One of the most important decisions in a project is the response to risks: how to deal with each risk type. This is directly linked to this research; as this study will examine how risks are identified and analyzed in project networks and which party takes responsibility for risk management. Before the decisions of risk allocations are ready to be made, the attitude that project actors have towards the risk has to be determined. Before a project starts, every player's strategy, as well as the ability to bear and manage risks, has to be known before risks are assigned to them (Klemetti 2006).

In this context, this research would first review the practices followed in the literature for risk analysis, response and allocation purposes, identify and analyze the most commonly experienced risks in construction projects especially in areas similar to the Egyptian dynamic environment and hence develop a risk management framework that aims to decrease dispute levels in construction projects between the various parties.

2   Background

A notable amount of efforts has been conducted in the area of risk management within the context of green building construction projects and is continuously gaining momentum over the recent years (Yang and Zou 2014).

A study by Qin et al. in 2016 has identified and assessed the most commonly experienced risks in the Chinese construction industry and has resulted in the conclusion that the most important and critical risks were: Complicated approval procedures due to government bureaucracy, Lack of design experience on green buildings and Poor maintenance in green buildings. Zou and Couani identified 40 risks that are most experienced in green building construction and evaluated their importance value by means of a survey questionnaire given out to experts from the Australian construction industry. This survey resulted in a ranked list of green construction risks, with Higher investment, Additional costs in skills development and Lack of expertise as the top three risks identified (Zou and Couani 2012).

On the other hand, Yang et al. conducted a stakeholder-associated risk model to examine the risks in green building construction projects based on Social Network Analysis. They found that diverse stakeholders recognized ethical/reputational risks more widely and that technological risks were not important as perceived (Yang et al. 2016).

Despite the relatively diverse literature available on the area of risk management in green construction projects, the applicability of the integration between Fuzzy Logic and Multi-Criteria DecisionMaking techniques remain questionable. Multi-Criteria Decision-Making (MCDM) is a very important tool in the assessment and evaluation of various industry problems. MCDM is the process of evaluating all the alternative options and choosing one of them or prioritizing them based on their performance against preset criteria and attributes. Over the years, numerous methods have been developed to solve multiple attribute decision-making problems. One of the most significant tools in MCDM, and the one considered mostly used among all disciplines is the Analytic Hierarchy Process (AHP) (Ahmed et al. 2019).

The AHP is a powerful decision-making method developed by Thomas Saaty in 1971. It mainly aims at ranking alternatives in order to prioritize them. Another very important MCDM is the Multi-Attribute Utility Theory (MAUT) which was first developed as an attempt to apply objective measurement to decision making. The basic hypothesis incurred in this method specifies that within any decision problem there is a real valued function or utility (U), defined by the set of feasible alternatives that the decision-maker seeks to maximize (Olson 1996). Each alternative results in an outcome, which may have a value on a number of different dimensions. MAUT seeks to measure these values, one dimension at a time, followed by an aggregation of these values across the dimensions through a weighting procedure. MCDM methods are typically evaluated using Classic Logic which permits only Boolean values of decision either true (1) or false (0). However, it has been greatly argued in previous researches that human decisions are usually made based on imprecise and non-numerical information, therefore, Fuzzy Logic was introduced by Zadeh (1965) to capture the vagueness and imprecise information provided by decisionmakers. The Fuzzy Logic represents the degree of truth of a decision made on a scale from 0–1 representing the extreme boundaries of a decision being completely false or completely true. Applying Fuzzy Logic to MCDM has proved to be of great benefit and reliability in evaluating human responses rather than Classic Logic-based methods (Ahmed et al. 2019).

Therefore, this study was conducted to fill this gap and apply two methods into assessing green construction project risks that are Fuzzy Analytic Hierarchy Process and Fuzzy Multi Attribute Utility Theory which is considered a novelty of this study. In the following sections, the framework developed for identifying and evaluating the project risks as well as the results of the risks' prioritization are elaborately discussed.

3   Methodology

In the following figure, a flow chart of the methodology adopted as part of this study is demonstrated to get a clearer picture of the procedure that has been followed starting from the literature review up until the calculation of each risk's priority score and the ranking of the corresponding risk mitigation strategies (Fig. 1).

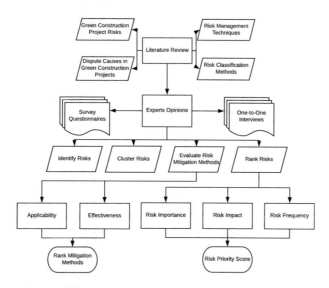

Fig. 1. Overview of research methodology

3.1   Risk Identification

Based on the literature search and the opinions of the experts interviewed, the following risks were identified and grouped into six different clusters categorizing risks of relevant natures. The clusters developed are namely: Design Risks, External Risks, Organizational Risks, Project Management Risks, Legal Risks and Construction Risks. Those six clusters are further divided into a total of 42 risks that are most evident in Egyptian construction projects that are relevant to green buildings. A detailed illustration of the identified risks as well as the clusters they are part of is given in the following two tables, Table 1 and 2.

Table 1. Clusters of risks identified (part 1)

Design	External	Organizational
– Change of design requirements by owners – Lack of qualified professionals with proper design expertise – Energy saving uncertainty – Unclear objectives and requirements by the owner – Defective design and inaccurate quantity estimation	– Force Majeure/Acts of God – War threats and political instability – Corruption and bribes – Criminal acts – Inflation – Currency fluctuation – Unexpected inclement weather – Owner's sudden bankruptcy – Pollution produced as part of construction process	– Owner's shortage of funds – High target for Green Mark Rating – Contractors' incompetence – Lack or departure of qualified staff – Labor strikes and disputes – Conflicts due to differences in culture – Unfairness in tendering – Delayed payment to contractors

Table 2. Clusters of risks identified (part 2)

Project Management	Legal	Construction
– Lack of scope of work definition – Unreasonably imposed tight schedule – Poor quality of work – Delay of material supply by suppliers – Quality problems of supplier material – Lack of management staff – Poor communication among stakeholders	– Changes in green building laws and regulations – Delays in obtaining site access and right of way – Difficulty in claiming insurance compensation – Unclear contract conditions – Import/Export restrictions	– Unfamiliarity with green technologies and construction processes – Workers' safety and health issues – Accidents during construction – Low productivity of labor and equipment – Unpredicted technical problems in construction – Subcontractors' poor performance – Shortage in manpower supply and availability – Unforeseen site conditions

3.2 Risk Importance

The framework developed as part of this study helps identify the main risks evident and experienced in Egyptian construction projects and assess those risks according to several factors. In this section, the developed framework is discussed along with the methods used for calculation purposes of the various indicators for risk evaluation.

The primary step in the framework presented in this study is to assess the importance index (II) of the risks identified in the literature search and the experts' survey.

An importance index is generated for each identified risk based on the risks' significance in the construction project and its effect on causing disputes if not resolved and dealt with. This indicator is evaluated based on the perceptions of the surveyed experts from their point of view and experience in previous construction projects. Each expert is given a value representing the weight given to his answers to the survey, based on their experience, knowledge and background as shown in the equation below, Eq. 1.

$$\text{II} = \sum_{i=1}^{n} ai \times ei \tag{1}$$

where II: the Importance Index for each risk factor, n: Number of experts interviewed, a: the Level of Importance of the risk factor as perceived by Expert "i", and e: is the Weight given for Expert "i" based on his background and experience as compared by other experts included in this study.

The Importance Index for each risk factor is evaluated on a 5-point scale by means of a pair-wise comparison between the identified risk factors. And the final weights for the risk factors are calculated by means of a Fuzzy Analytic Hierarchy Process (F-AHP) method developed by Aydin and Kahraman (2010) as an improvement of the original AHP method developed by Saaty (1982). The scale used for the F-AHP method is presented in Table 3.

Table 3. Importance index pair-wise comparison scale

Rating scale		Linguistic description
	(0,0,0)	Just Equal
1	(0,1,3)	Equally Important
3	(1,3,5)	Moderately More Important
5	(3,5,7)	Strongly More Important
7	(5,7,9)	Very Strongly More Important
9	(7,9,9)	Extremely More Important

After the weights are calculated, the consistency ratio is then determined in order to verify the assigned weights. Weights are assigned to components only if the Consistency Ratio (CR) has a value less than 0.1.

3.3 Risk Impact

The second step in the risk assessment process adopted in this study is the evaluation of the impact value (IV) of each of the identified risks in the case of their occurrence in a construction project. This is done by means of a 5-point Likert scale that measures the effect of the risks on the project performance metrics and success factors on the basis of four different criteria, namely: Cost, Schedule, Safety and Quality as shown in Table 4.

Table 4. Risk impact rating scale

Scale	Cost	Schedule	Safety	Quality
Catastrophic	>20%	>15%	Fatality	>10%
Major	15–20%	10–15%	Severe Injury	8–10%
Moderate	10–15%	5–10%	Medical Treatment	5–8%
Minor	5–10%	1–5%	First Aid	2–5%
Negligible	<5%	<1%	No Injury	<2%

The method used in this fragment of the study is a Fuzzy Multi-Attribute Utility Theory (F-MAUT) method that is developed by Kahraman and Kaya (2012). The F-MAUT method used assesses the weights of each of the four criteria first using the traditional AHP method (Saaty 1982), and then gives fuzzy rating to each of the alternatives (risks) with respect to each of the four criteria/attributes.

3.4   Risk Frequency

The following step in the risk assessment framework developed is to assess the frequency value (FV) of each of the identified risks based on their probability of occurrence as per the experts judgements. The same F-MAUT method and the exact course of action of the Risk Impact Assessment is followed in this part as well. However, the measuring scale is different due to the different nature of the evaluation purpose. The rating scale for the frequency of each of the risks is given in Table 5.

Table 5. Risk frequency rating scale

Scale	Frequency	Probability
Extremely Frequent	1 in 2 Projects	>=50%
Frequent	1 in 4 Project	>=25%
Common	1 in 10 Projects	>=10%
Unlikely	1 in 20 Projects	>=5%
Extremely Rare	1 in more than 20 Projects	<5%

3.5   Risk Priority Score

The final step in the risk analysis process is the calculation of a priority index for each of the identified risk factors based on their importance, impact and frequency levels as shown in Eq. 2. The Risk Priority Score (RPS) is calculation by multiplying the previously attained indicators to get the RPS value for each risk.

$$\text{RPS} = \text{II} \times \text{IV} \times \text{FV} \times 10^6 \qquad (2)$$

where RPS is the Risk Priority Score, II is the Importance Index, IV is the Impact Value and FV is the Frequency Value 10^6.

3.6   Risk Mitigation Methods

In this context, the possible solutions or response actions to be taken to mitigate or minimize the effects of the identified risks have been identified and were proposed to experts to rank them as per the previously stated Fuzzy-AHP methodology. The seven identified risk mitigation strategies are: Efficient communication with owner to grasp their goal of the Green Mark Standard, Outlining clear roles and responsibilities for all stakeholders, Planning and allowing for contingency funds, Using past successful green building projects as reference, Constant design evaluation and verifications, Implementing training programs to upgrade labour knowledge and skills of new technologies and finally ensuring that the contract language is clear and precise and outlines each stakeholder's liability and limit of power and influence on the project. The risk mitigation strategies identified are evaluated by the experts involved in this study as per two main criteria: their "effectiveness" in solving problems and challenges related to green building construction projects, as well as their "applicability" with respect to each of the identified green building projects risks.

4   Model Implementation Results

One hundred and twenty two key personnel were invited to participate in the surveying process including project managers, contractors and consultants involved in projects related to the green building industry were requested to fill the survey questionnaire developed in this research. A total response rate of 41.3% was achieved in the due course of data collection that was concentrated within Cairo, Egypt.

4.1   Risk Importance Weights

Each of the forty-two identified risks was clustered with risks of a common nature, and the total number of clusters was six where each cluster was weighed against the other with respect to their relative importance. The importance weights are calculated using the F-AHP method where local and global weights are obtained. The global weights are further used in a later section to calculate the overall priority of each of the risk factors. The weights derived as per the analysis of the expert responses ranked the clusters as follows: External Risks, followed by Construction Risks, Legal Risks, Organizational Risks, then Design Risks and Project Management Risks. A sample of the weights developed for the Project Management cluster and its constituents is given in Table 6.

Table 6. Importance weights for the Project Management cluster

Cluster	Risk factors	Local weights	Global weights
Project Management 11%	– Lack of scope of work definition	16%	0.0176
	– Unreasonably imposed tight schedule	23%	0.0253
	– Poor quality of work	18%	0.0198
	– Delay of material supply by suppliers	14%	0.0154
	– Quality problems of supplier material	10%	0.0110
	– Lack of management staff	7%	0.0077
	– Poor communication among stakeholders	12%	0.0132

4.2 Risk Impact Results

The impact of each of the identified risks is evaluated based on the F-MAUT discussed in the methodology. Upon calculating the Impact Values (IV), it was evident that the most serious risk factor is the risk of Force Majeure/Acts of God followed by the Changes in Green Building Laws and Regulations, while the risk with the least effect on the green building construction project with respect to the four criteria of evaluation is the risk of Delay of Material Supply by Suppliers. The next figure illustrates the results developed as part of the Impact Value calculation process for the External Risks cluster (Fig. 2).

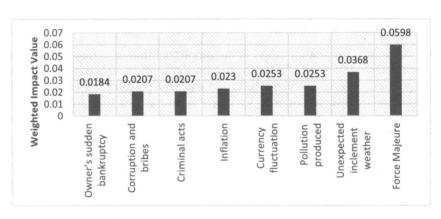

Fig. 2. Impact value results for the external risks' cluster

4.3   Risk Frequency Results

Consequently, the frequency of occurrence of each of the identified risks is evaluated based on the F-MAUT discussed in the methodology. Upon calculating the Frequency Values (FV), it was evident that the most frequent risk factor in the Egyptian green construction industry is the risk of Lack of qualified professionals with proper green building design expertise, while the risk with the least probability in the green construction projects is the risk of Lack of management staff.

4.4   Risk Priority Results

Considering that the importance, impact and frequency criteria for each of the identified risks are equally weighted, the priority index was calculated for each identified risk as per the stated equation in the methodology section. The results of the Risk Priority Score for the Legal Risks cluster is provided in the following table (Table 7).

Table 7. RPS results for the Legal Risks cluster

Cluster	Risk factors	RPS
Legal Risks	Unclear contract conditions	3.29868
	Difficulty in claiming insurance compensation	6.6759
	Delays in obtaining site access and right of way	9.848916
	Import/Export restrictions	15.833664
	Changes in green building laws and regulations	18.8496

4.5   Risk Mitigation Methods Ranking

The ranking provided by the experts interviewed has graded the Efficient communication with the owner regarding their expected or sought-after Green Mark Standard as the most important method, and the most likely to mitigate many risks from the ones identified. This was followed by conducting training programs for labor and workers, as well as outlining clear roles and responsibilities for all stakeholders involved in a project.

5   Conclusion

Green building construction projects have attained a rapid development over recent years due to its positive efficacy of saving energy and resources consumptions. The overall purpose of this study was to shed the light on practices of risk identification, assessment and management in the green construction industry of various countries as a means to minimize disputes between parties and apply them in Egyptian construction projects through a survey aiming at key personnel in the construction industry. The survey questionnaire was developed to identify the most experienced risks in the Egyptian construction industry and assess the perceptions of the construction industry experts of risk frequency and impact evaluation as well as risk importance index calculation. The analysis of the results ranked the risks based on different factors by means of methods integrating the multi-criteria decision making techniques with fuzzy logic which has not been considered before in green construction project risk assessment and analysis. A Fuzzy AHP method is used for weighing the importance of the different risk factors identified, while a Fuzzy MAUT method is used for assessing the risk frequency and impact values. This framework is deemed to be a very useful tool for project managers and decision makers in controlling and mitigating the project risks as it presents an efficient prioritization tool for the project risks as well as a suggestion for risk response procedures that can help in risk avoidance and prevention.

References

Ahmed, R., Nassar, A.: The effect of risk allocation on minimizing disputes in construction projects in Egypt. Int. J. Eng. Tech. Res. **5**(3), 1–6 (2016)

Ahmed, R., Afifi, M., Nassar, A.: Using multi-criteria decision making (MCDM) methods in Egyptian construction projects. In: Canadian Society for Civil Engineering Annual Conference, Laval, QC, Canada (2019)

Aydin, S., Kahraman, C.: Multiattribute supplier selection using fuzzy analytic hierarchy process. Int. J. Comput. Intell. Syst. **3**(5), 553–565 (2010)

Hameed, A., Woo, S.: Risk importance and allocation in the Pakistan Construction Industry: a contractors' perspective. KSCE J. Civ. Eng. **11**(2), 73–80 (2007)

Issa, U.H., Farag, M.A., Abdelhafez, L.M., Ahmed, S.A.: A risk allocation model for construction projects in Yemen. Civ. Environ. Res. **7**(3), 78–88 (2015)

Kahraman, C., Kaya, İ.: A fuzzy multiple attribute utility model for intelligent building assessment. J. Civ. Eng. Manag. **18**(6), 811–820 (2012)

Klemetti, A.: Risk Management in Construction Project Networks. Helsinki University of Technology Laboratory of Industrial Management, Helsinki, Finland (2006)

Miller, A.: A systematic approach to risk management for construction. Struct. Survey **19**(5-2001), 245–252 (2001)

Olson, D.: Decision Aids for Selection Problems. Springer (1996)

Qin, X., Mo, Y., Jing, L.: Risk perceptions of the life-cycle of green buildings in China. J. Cleaner Prod. **126**, 148–158 (2016)

Royer, P.S.: Risk management: the undiscovered dimension of project. Project Manag. J. **31**, 6–13 (2000)

Saaty, T.L.: The analytic hierarchy process: a new approach to deal with fuzziness in architecture. Archit. Sci. Rev. **25**(3), 64–69 (1982)

Turner, J.R.: The Handbook of Project-Based Management: Improving the Processes for Achieving Strategic Objectives, 2nd edn. McGraw-Hill, London (1999). 529 p.

Yang, R., Zou, P.: Stakeholder-associated risks and their interactions in complex green building projects: a social network model. Build. Environ. **73**, 208–222 (2014)

Yang, R., Zou, P., Wang, J.: Modelling stakeholder-associated risk networks in green building projects. Int. J. Proj. Manag. **34**, 66–81 (2016)

Zadeh, L.: Fuzzy Sets. Inf. Control **8**(3), 338–353 (1965)

Zietsman, J., Rilett, L.R., Kim, S.J.: Transportation corridor decision-making with multi-attribute utility theory. Int. J. Manag. Decis. Making **7**(2–3), 254–266 (2006)

Zou, P., Couani, P.: Managing risks in green building supply chain. Archit. Eng. Des. Manag. **8**, 143–158 (2012)

Development of Wind and Flood Vulnerability Index for Residential Buildings

Carol Massarra[1(✉)], Carol Friedland[2], and Amin Akhnoukh[1]

[1] East Carolina University, Greenville, NC, USA
massarrac19@ecu.edu
[2] Louisiana State University, Baton Rouge, LA, USA

Abstract. Building vulnerability assessment is an important technique for managing disaster, performing hazard mitigation, and managing disaster reduction practices. In the engineering field, vulnerability is mainly assessed based on the quantitative approach, which is a simulation-based technique that provides information regarding the potential loss or damage rather than accounting for indicators influencing building vulnerability. Identifying indicators that significantly contribute to building vulnerability is a key element to develop vulnerability index, which is a tool for understanding performance of building subjected to hurricane hazards and helping decision-makers to prioritize evacuation. This study qualitatively assesses residential building vulnerability through the development of Wind and Flood Building Vulnerability Index (WFBVI). WFBVI is categorized as a three-level (low-moderate-high) index using Analytical Hierarchy Process (AHP). A dataset of single-family homes damaged by 2005 Hurricane Katrina is used to demonstrate implementation of the WFBVI. The application leads to valuable results on how hurricane building vulnerability can be reflected by quantifiable wind and flood index across spatial scales. Overall, the results provide engineers with insights on the actual performance of residential structures in areas subject to severe wind and flood hurricane hazards and provide qualitative information for developing effective strategies to mitigate future risk and improve decision-making processes.

Keywords: Hurricane hazards · Vulnerability index · Indicators

1 Introduction

In the United States, major hurricanes have caused damage to human, society, structures and infrastructure more than any other natural hazards [1]. On a case by case basis, hurricanes are very different, which make them unique. Approaches for assessing building vulnerability are either quantitative (e.g. vulnerability curves), semi-quantitative (e.g. vulnerability metrics) or qualitative (e.g. vulnerability indicators) approaches [2, 3]. However with the multi-hazard nature of hurricane assessing vulnerability become more difficult and challengeable [4, 5]. From an engineering perspective, efforts to understand and reduce the negative consequences of hurricanes are linked to the quantitative approach [e.g. 6–8], where curves that link the intensity of hazard to the expected damage or loss are developed. With multi-hazard nature of the

I. El Dimeery et al. (Eds.): JIC Smart Cities 2019, SUCI, pp. 151–159, 2021.
https://doi.org/10.1007/978-3-030-64217-4_18

hurricane, vulnerability assessment poses a wide range of challenges. These challenges originated from non-equality of vulnerability for elements exposed to a hazard [9], the complexity of vulnerability concept [2, 10] and the variation of vulnerability across spatial and temporal scales [4, 5, 11]. The use of a qualitative approach [3], where vulnerability is assessed through index that ranges from 0 to 1 [12] is considered a very effective approach to simplify vulnerability assessment.

In the social science field, the qualitative approach is well defined and broadly applied to consider the vulnerability of an object at risk. In the engineering field, less attention has been given to the qualitative approach. This may be strongly related to the fact that in the engineering field, vulnerability is considered to be hazard-specific and mainly focused on a quantitative approach that is based on simulation techniques. However, the quantitative approach provides information regarding the potential loss or damage rather than accounting for the different factors influencing the vulnerability of elements at risk. This drawback is related to the high degree of generalization [13] and the requirement of a significant amount of input data, and computation capabilities required to develop the vulnerability curves. Modeling physical vulnerability using index can overcome this problem, as the focus of this method is to identify and better understand the principal factors that contribute to the configuration of vulnerability [2]. Many people are living in coastal areas, which bring issues related to loss of lives and destroyed of structures during hurricane events. The solution to these issues may lie in the structure of a smart city. There are many ways in which smart cities can help government and local officials prepare for natural disasters. This paper develops building vulnerability index for assessing the physical vulnerability of residential buildings subjected to simulations wind and flood hazards. The approach aims to qualitatively assess residential building vulnerability through the development of Wind and Flood Building Vulnerability Index (WFBVI). WFBVI is categorized as three-level index (low-moderate-high) using Analytical Hierarchy Process (AHP). A dataset of single-family homes damaged by 2005 Hurricane Katrina is used to demonstrate implementation of the WFBVI. Understanding building performance during hurricane events through the development of building vulnerability index provides an approach for a smarter, more accurate and equitable disaster recovery.

2   Methodology

The methodology depends on qualitatively assessing building damage by determining a specific building vulnerability index that characterizes the building vulnerability subjected to simultaneous wind and flood hazards. The index is generated using a set of weighted indicators by applying the Analytical Hierarchy Process (AHP) method [14].

2.1   Wind and Flood Characteristic Vulnerability Indicators

Documentation of past events data and assessment of damage aid the identification of indicators that influence the ability or inability of the residential buildings to function well during natural hazard events [3]. The selection of the most relevant indicators affecting the behavior of a structure is determined by the physical mechanism and type of impacting process, as well as the characteristics of the building [13]. Since vulnerability to natural hazards is not uniform [15] within the study area, a variety of indicators are identified, and information relating to each indicator is collected. Building and hazard attributes are used to describe the vulnerability indicators to be used in this study. These attributes describe hazard intestines and characteristics of buildings and their surroundings influencing building physical vulnerability. These indicators were selected following and examination of vulnerability studies of wind [e.g. 6, 16–20], flood [e.g. 21–27] and tsunami [e.g. 15, 28–32]. Table 1 shows the characteristic vulnerability indicators to be used in the development of the vulnerability index.

Table 1. Characteristic vulnerability indicators used to develop the vulnerability index and related hazards.

Indictor (*I*)	Wind	Flood
Building attributes		
Number of stories	X	X
Foundation type		X
Hazard Attributes		
Wind speed	X	
Inundation depth		X
Wave height		X
Water speed		X

For each *I* identified in Table 1 and based on the field observed data, the levels of the vulnerability indicators are identified. For example, levels of the number of stories obtained from field survey are one-and two-story.

2.2   Vulnerability Indicator Weighting Matrix

The Analytical Hierarchy Process (AHP) is one of the multiple criteria decision-making method that was developed by Saaty [14]. The method provides measures of judgement consistency and derives priorities among criteria and simplifies preference ratings among decision criteria using pair wise comparison. Generally, the comparison is performed using expert opinion. However, for this study, an alternative approach based on Random Forest (RF) classification method substitutes the expert opinion. From RF Model, measure of indictor importance M, which is based on how much the RF model prediction accuracy decreases when the indicator I is excluded from the analysis will be obtained and normalized. Each I identified in Table 1 is listed in a weighting matrix in such a way that one indicator I_j in the row of the matrix is compared with other indicator I_k in the column of the matrix, where $k > j$. The matrix is developed using a rating system with the following descriptors: 1- Equal Important, 3 - Slightly Important, 5 - Strongly Important, 7 - Very Strong Important, and 9 - Extreme Important [14]. Each I is given a score based on a criterion that is based on the and how much the difference between the normalized measure of importance $(M_{I_j} - M_{I_k})$ of the two indicators contributes to building collapse Table 2. The matrix is then normalized and the values in each row is average to get the corresponding indicator weight. The procedure is used for weighting the indicators and their levels.

Table 2. Criterion used to score the indicators

Criteria	Score	Description
$(M_{I_j} - M_{I_k}) > 0\% \leq 5\%$	1	Two indicators contribute equally to the collapse
$(M_{I_j} - M_{I_k}) > 5\% \ and \ \leq 10\%$	3	An indicator is slightly favored over another
$(M_{I_j} - M_{I_k}) > 10\% \ and \ \leq 20\%$	5	An indicator is strongly favored over another
$(M_{I_j} - M_{I_k}) > 20\% \ and \ \leq 30\%$	7	An indicator is very strongly favored over another
$(M_{I_j} - M_{I_k}) > 30\%$	9	An indicator is extremely favored over another

2.3   Consistency Analysis

A consistency analysis is performed to check the consistency of the calculated weights. Consistency index (*CI*) is given as $CI = \frac{\lambda_{max} - n}{n-1}$, where λ_{max} is the highest weight in the matrix, and n is the number of indicators. The consistency ratio *CR* is given as $CR = \frac{CI}{RI}$, where *RI* is a random inconsistency index [14]. *CR* of 0.1 or below is considered acceptable.

2.4   WFBVI Calculation

WFBVI for each building is calculated using the normalized weights of the characteristic indicators (W_i) and the normalized weights of their levels (w_n) and is given as

$$WFBVI = \sum_{i,n=[1]}^{m} W_i w_n \qquad (1)$$

Table 3 lists the *WFBVI* criterion and description.

Table 3. *WFBVI* criterion and description.

Criteria	Description
WFBVI $\geq 0 \ and \ \leq 0.2$	Low vulnerability
WFBVI $> 0.2 \ and \ \leq 0.3$	Moderate vulnerability
WFBVI > 0.3	High vulnerability

3   Case Study

A dataset containing observations describing hazard intensities and building attributes for $N = 866$ single-family homes in the three counties of coastal Mississippi (Hancock, Harrison, Jackson) that border the Gulf of Mexico is used for the application of the methodology. These homes ranged in damage from no damage/very minor damage to collapse [33]. The characteristic vulnerability indicators are wind speed, surge depth, wave height, water velocity, foundation type and number of stories. Observed hazard indicators were computed on the SL16 mesh, which was developed and validated for the devastating Gulf hurricanes of 2005 and 2008 (Dietrich et al. 2012). Observed building indicators were collected from field survey. Table 4 describes the observed hazard and building attribute indictors used in this study.

Table 4. Observed hazard and building attribute indictors

Symbol	Indicator	Range/Level
$U_{3,max}$	Maximum 3-second gust wind speed	[47.63–67.99] m/s
$H_{S,max}$	Maximum significant wave height	[0–3.20] m
D_{max}	Maximum water depth above local ground level	[0–7.94] m
U_{max}	Maximum water speed	[0–2.80] m/s
X_{FT}	Foundation type	Slab, Elevated
X_{NS}	Number of stories	One-Story Two-Story

3.1   Vulnerability Indicator Weighting Matrix

Based on Random Forest (RF) classification method, the characteristic vulnerability indicators were ranked and normalized. The ranked and normalized measure of importance are given is Table 5. Maximum water depth above local ground level and foundation type were ranked as most important hazard and building attributes, respectively for predicting building collapse.

Table 5. Normalized measure of importance

Indicator	Mean decrease accuracy	M %
D_{max}	120.0445	29%
$U_{3,max}$	118.4744	29%
U_{max}	87.62792	21%
$H_{S,max}$	68.14943	16%
X_{FT}	11.76954	3%
X_{NS}	7.89812	2%

Based on (AHP), the ranked indicators defined in Table 5 were pair wised and a comparison normalized matrix was developed. AHP results in a matrix with CR = 0.036361. Table 6 defines the normalized matrix with weights (λ) for each characteristic vulnerability indicators.

Table 6. Normalized pairwise matrix and weights (λ) for characteristic vulnerability indicators

	X_{NS}	X_{FT}	$H_{S,max}$	U_{max}	$U_{3,max}$	D_{max}	λ
X_{NS}	0.04	0.04	0.02	0.03	0.04	0.04	0.03
X_{FT}	0.04	0.04	0.02	0.03	0.04	0.04	0.03
$H_{S,max}$	0.11	0.11	0.06	0.08	0.06	0.06	0.08
U_{max}	0.18	0.18	0.06	0.08	0.08	0.08	0.11
$U_{3,max}$	0.32	0.32	0.42	0.40	0.39	0.39	0.37
D_{max}	0.32	0.32	0.42	0.40	0.39	0.39	0.37

Observed values of the characteristic vulnerability indicators were pair wised and a comparison normalized matrix was developed. Table 7 defines the weights for each level of the observed values of the vulnerability indicators.

Table 7. Weight λ and CR of the observed values of the vulnerability indicators

Indicator	Level	λ	CR
D_{max}	0–0.3	0.05	0.07
	0.3 to 0.6	0.22	
	>0.6	0.73	
$U_{3,max}$	96–100.5	0.11	0.02
	100.5–104.5	0.26	
	>104.5	0.63	
U_{max}	0–0.5	0.11	0.02
	0.5–1.5	0.26	
	>1.5	0.63	
$H_{S,max}$	0–0.3	0.06	0.07
	0.3–0.6	0.22	
	>0.6	0.72	
X_{FT}	Slab	0.88	0
	Elevated	0.13	
X_{NS}	One-story	0.88	0
	Two-Story	0.13	

Based on Eq. 1, WFBVI was developed for each building in the study area. An example of computed WFBVI is given in Table 8.

Table 8. Wind and Flood Building Vulnerability Index (WFBVI)

Indictor	Value	W	w
D_{max}	5.21 (m)	0.37	0.72
$U_{3,max}$	51.99 (m/s)	0.37	0.11
$H_{S,max}$	1.04 (m)	0.11	0.72
$U_{3,max}$	0.61 (m/s)	0.08	0.26
X_{FT}	Slab	0.03	0.03
X_{NS}	One-Story	0.03	0.03
WFBVI	0.45	Very high vulnerability	

4 Discussion

In this study, wind and flood building vulnerability index WFBVI at individual building scale was developed to estimate vulnerability of residential building collapse. The purpose of the paper is to qualitatively assess vulnerability of residential buildings to better understand performance of building subjected to hurricane hazards and help decision-makers to prioritize evacuation and prepare for future events. Current methodological approach considers a set of hazard intensities and building characteristics resulting in a more advanced differentiation of vulnerability based on a three-level index (low-moderate-high). The contributions and findings of this paper are:

- The Wind and Flood Building Vulnerability Index (WFBVI) was obtained through statistical classification using Random Forest variable importance procedure.
- The use of variable importance proved to be important for the identification and prioritization of indicators and their values, and minimize bias resulting from subjective expert judgment.
- Building vulnerability was qualitatively assessed and a set of indicators that contribute to building collapse was identified.

5 Conclusions

The developed methodology is useful for researchers and model developers who have access to datasets describing single family home damage and performance during extreme events (e.g. hurricane, flood) to aid the development of multi-hazard building vulnerability index. The application led to valuable results on how hurricane building vulnerability can be reflected using index rather than curves. The results provide engineers with insights on the actual performance of residential structures subject to multiple hurricane hazards. Additionally, the results provide decision makers with qualitative information for developing effective strategies to mitigate future risk and improve decision-making processes. The developed index provides an approach to inform and improve design practices and to identify potential improvements in building design and construction practices. The approach may serve as a critical tool for improving building code design and construction requirements in order to move toward smart cities and more accurate and equitable disaster recovery. Future work will extend the data to include more building attributes (e.g., foundation materials) and hazard attributes (e.g., frequency of hazard).

References

1. Pielke Jr., R.A., Gratz, J., Landsea, C.W., Collins, D., Saunders, M.A., Musulin, R.: Normalized hurricane damage in the United States: 1900–2005. Nat. Hazards Rev. **9**(1), 29–42 (2008)
2. Birkmann, J.: Measuring Vulnerability to Natural Hazards: Towards Disaster Resilient Societies. TERI Press, New Delhi (2006)
3. Kappes, M., Papathoma-Köhle, M., Keiler, M.: Assessing physical vulnerability for multi-hazards using an indicator-based methodology. Appl. Geogr. **32**(2), 577–590 (2012)
4. Kappes, M., Keiler, M., von Elverfeldt, K., Glade, T.: Challenges of analyzing multi-hazard risk: a review. Nat. Hazards **64**(2), 1925–1958 (2012)
5. Marzocchi, W., Garcia-Aristizabal, A., Gasparini, P., Mastellone, M.L., Di Ruocco, A.: Basic principles of multi-risk assessment: a case study in Italy. Nat. Hazards **62**(2), 551–573 (2012)
6. Pinelli, J.-P., Simiu, E., Gurley, K., Subramanian, C., Zhang, L., Cope, A., Filliben, J.J., Hamid, S.: Hurricane damage prediction model for residential structures. J. Struct. Eng. **130** (11), 1685–1691 (2004)
7. Khanduri, A., Morrow, G.: Vulnerability of buildings to windstorms and insurance loss estimation. J. Wind Eng. Ind. Aerodyn. **91**(4), 455–467 (2003)
8. Powell, M., Soukup, G., Cocke, S., Gulati, S., Morisseau-Leroy, N., Hamid, S., Dorst, N., Axe, L.: State of Florida hurricane loss projection model: atmospheric science component. J. Wind Eng. Ind. Aerodyn. **93**(8), 651–674 (2005)
9. Wu, S.-Y., Yarnal, B., Fisher, A.: Vulnerability of coastal communities to sealevel rise: a case study of Cape May county, New Jersey, USA. Climate Res. **22**(3), 255–270 (2002)
10. Barroca, B., Bernardara, P., Mouchel, J.-M., Hubert, G.: Indicators for identification of urban flooding vulnerability. Nat. Hazards Earth Syst. Sci. **6**(4), 553–561 (2006)
11. Li, Y., Xiong, W., Hu, W., Berry, P., Ju, H., Lin, E., Wang, W., Li, K., Pan, J.: Integrated assessment of China's agricultural vulnerability to climate change: a multi-indicator approach. Clim. Change **128**(3–4), 355–366 (2014)
12. Downing, T., Aerts, J., Soussan, J., Barthelemy, O., Bharwani, S., Ionescu, C., Hinkel, J., Klein, R., Mata, L., Martin, N.: Integrating social vulnerability into water management. In: International Conference on Adaptive & Integrated Water Management (2005)
13. Godfrey, A., Ciurean, R., Van Westen, C., Kingma, N., Glade, T.: Assessing vulnerability of buildings to Hydro-meteorological hazards using an expert based approach–an application in Nehoiu Valley, Romania. Int. J. Disaster Risk Reduct. **13**, 229–241 (2015)
14. Saaty, T.L.: How to make a decision: the analytic hierarchy process. Eur. J. Oper. Res. **48**(1), 9–26 (1990)
15. Papathoma, M., Dominey-Howes, D., Zong, Y., Smith, D.: Assessing Tsunami vulnerability, an example from Herakleio, Crete. Nat. Hazards Earth Syst. Sci. **3**(5), 377–389 (2003)
16. Jain, V.K., Guin, J., He, H.: Statistical analysis of 2004 and 2005 hurricane claims data. In: 11th Americas Conference on Wind Engineering, San Juan, Puerto Rico (2009)

17. Vickery, P.J., Skerlj, P.F., Lin, J., Twisdale Jr, L.A., Young, M.A., Lavelle, F.M.: HAZUS-MH hurricane model methodology. II: damage and loss estimation. Nat. Hazards Rev. **7**(2), 94–103 (2006)

18. Huang, Z., Rosowsky, D.V., Sparks, R.: Long-term hurricane risk assessment and expected damage to residential structures. Reliab. Eng. Syst. Saf. **74**(3), 239–249 (2001)

19. Unanwa, C., McDonald, J., Mehta, K., Smith, D.: The development of wind damage bands for buildings. J. Wind Eng. Ind. Aerodyn. **84**(1), 119–149 (2000)

20. Schraft, A., Durand, E., Hausmann, P.: Storms over Europe: Losses and Scenarios. Swiss Reinsurance Co., Zurich (1993)

21. van de Lindt, J.W., Taggart, M.: Fragility analysis methodology for performance-based analysis of wood-frame buildings for flood. Nat. Hazards Rev. **10**(3), 113–123 (2009)

22. Scawthorn, C., Flores, P., Blais, N., Seligson, H., Tate, E., Chang, S., Mifflin, E., Thomas, W., Murphy, J., Jones, C.: HAZUS-MH flood loss estimation methodology. II. Damage and loss assessment. Nat. Hazards Rev. **7**(2), 72–81 (2006)

23. Messner, F., Meyer, V.: Flood damage, vulnerability and risk perception–challenges for flood damage research. Springer (2006)

24. Thieken, A.H., Müller, M., Kreibich, H., Merz, B.: Flood damage and influencing factors: new insights from the August 2002 flood in Germany. Water Resour. Res. **41**(12), 1–17 (2005)

25. Zhai, G., Fukuzono, T., Ikeda, S.: Modeling flood damage: case of Tokai Flood 2000. J. Am. Water Recourses Assoc. **41**, 77–92 (2005)

26. Dutta, D., Herath, S., Musiake, K.: A mathematical model for flood loss estimation. J. Hydrol. **277**(1), 24–49 (2003)

27. Tang, J.C., Vongvisessomjai, S., Sahasakmontri, K.: Estimation of flood damage cost for Bangkok. Water Resour. Manag. **6**(1), 47–56 (1992)

28. Shuto, N., Matsutomi, H.: Field survey of the 1993 Hokkaido Nansei-Oki earthquake tsunami. Pure Appl. Geophys. **144**(3–4), 649–663 (1995)

29. Tsuji, Y., Matsutomi, H., Imamura, F., Takeo, M., Kawata, Y., Matsuyama, M., Takahashi, T., Sunarjo, Harjadi, P.: Damage to coastal villages due to the 1992 Flores Island earthquake tsunami. Pure Appl. Geophys. **144**(3–4), 481–524 (1995)

30. Maramai, A., Tinti, S.: The 3 June 1994 Java tsunami: a post-event survey of the coastal effects. Nat. Hazards **15**(1), 31–49 (1997)

31. Reese, S., Cousins, W., Power, W., Palmer, N., Tejakusuma, I., Nugrahadi, S.: Tsunami vulnerability of buildings and people in South Java–field observations after the July 2006 Java tsunami. Nat. Hazards Earth Syst. Sci. **7**(5), 573–589 (2007)

32. Papathoma, M., Dominey-Howes, D.: Tsunami vulnerability assessment and its implications for coastal hazard analysis and disaster management planning, Gulf of Corinth, Greece. Nat. Hazards Earth Syst. Sci. **3**(6), 733–747 (2003)

33. Massarra, C.C., Friedland, C.J., Marx, B.D., Dietrich, J.C.: Predictive multi-hazard hurricane data-based fragility model for residential homes. Coast. Eng. **151**, 10–21 (2019)

Smart Cities - Policy and Regulatory Frameworks

Mahmoud El Khafif[✉] and Nora Salem[✉]

The German University in Cairo, Cairo, Egypt
{mahmoud.elkhafif, Nora.salem}@guc.edu.eg

Abstract. The paper aims at highlighting the policies and regulatory frameworks of smart cities. For that purpose, the paper will first outline the international framework, including Goal 11 of the United Nations Sustainable Development Goals (SDGs), which aims at making cities inclusive, safe, resilient and sustainable. The latter necessitates the development of smart cities, since highly sophisticated ICT-based applications and services will not only contribute to greater energy efficiency but also improve the safety and wellbeing of inhabitants through for instance automated streetlights and better transportation solutions. Considering that more than 55% of the world's population lives in urban areas, the switch to digital technologies allowing for smarter and more inclusive cities, where waste, resource consumption and environmental impacts are significantly reduced, is imperative to reach the SDGs. The paper will then investigate the existence of an internationally recognized definition for smart cities and different international assessment mechanisms. Hereinafter, the paper will look at the implementation of various initiatives at the national level of different countries, among them Egypt. The paper will close with addressing the critical question of dealing with highly sensitive data in an ethically responsible manner in the scope of smart cities. Against that background, the European Union's legal framework for data protection in the virtual space will be assessed against a global citizen-driven initiative on promoting digital rights in smart cities.

Keywords: Smart cities · Regulatory · Assessment · International framework · National initiatives

1 Introduction

With a considerably growing number of smart city developments, a global policy and regulatory framework for smart cities that provides a definition for smart cities, identifies Key Performance Indicators as well as unified standards for compliance and comparison becomes increasingly vital. For that purpose, this paper aims at investigating existent frameworks at the international level (2), highlighting different national smart city initiatives at the regional level (3) and outlining data protection regulations and initiatives (4).

Against this background, the paper will first outline the broader umbrella of smart cities, namely Goal 11 of the United Nations Sustainable Development Goals (SDGs), which aims at making cities inclusive, safe, resilient and sustainable (Sect. 2.1). The

© The Author(s), under exclusive license to Springer Nature Switzerland AG 2021
I. El Dimeery et al. (Eds.): JIC Smart Cities 2019, SUCI, pp. 160–169, 2021.
https://doi.org/10.1007/978-3-030-64217-4_19

latter necessitates the development of smart cities, since highly sophisticated ICT-based applications and services will not only contribute to greater energy efficiency but also improve the safety and well-being of inhabitants through for instance automated streetlights and better transportation solutions. Considering that more than 55% of the world's population lives in urban areas, the switch to digital technologies allowing for smarter and more inclusive cities, where waste, resource consumption and environmental impacts are significantly reduced, is imperative to reach the SDGs. The paper will then investigate the existence of an internationally recognized definition for smart cities (Sect. 2.2) and different international assessment mechanisms (Sect. 2.3).

Hereinafter, the paper will look at current national initiatives towards developing smart cities at the regional level, including South-East Asia –Singapore, Seoul, Shanghai– (Sect. 3.1); North America –New York, Boston, Montreal– (Sect. 3.2); the MENA Region –Dubai, Egypt– (Sect. 3.3); and Europe –London, Helsinki, Barcelona– (Sect. 3.4).

Section 4 addresses the critical question of dealing with highly sensitive data in an ethically responsible manner in the scope of smart cities. For that purpose the paper outlines on the one hand a regional intergovernmental legal framework for data protection in the virtual space and on the other hand a global citizen-driven initiative on promoting digital rights in smart cities.

2 International Framework

2.1 United Nations Sustainable Development Goals

In 2015, Heads of all 193 United Nations (U.N.) Member States adopted the Sustainable Development Agenda, which includes a total of 17 Sustainable Development Goals (SDGs) with numerous indicators to building a more sustainable, safer, more prosperous planet for all humanity by 2030. The goals cover a range of key areas including poverty; hunger; health; quality education; gender equality; clean water and sanitation; affordable and clean energy; decent work and economic growth; industry, innovation and infrastructure; reduced inequalities; sustainable cities and communities; responsible consumption and production; climate action; life below water; life on land; as well as peace, justice and strong institutions.

In order to achieve the ambitious agenda, the UN Development Programme (UNDP) provides support to countries, monitors the agenda's progress, collaborates with partners, fosters cooperation and facilitates funding [1].

Goal 11 aims at making cities inclusive, safe, resilient and sustainable. To this end, common urban challenges such as congestion, minimal funds for the provision of basic services and energy, a lack of adequate housing, declining infrastructure and transportation systems as well as rising pollution continue to exist. These challenges will further amplify considering that the 4.2 billion people (55% of the world's population) living in urban cities today are estimated to reach 6.5 billion people by 2050, while cities occupy merely 3% of the Earth's land. [2] 90% of the rapid urbanization will take place in the developing world.

The challenges can be addressed through the inclusion of highly sophisticated Information and Communication Technology (ICT) - based applications and services which can contribute to greater energy efficiency, resource consumption, improved safety - through for instance automated streetlights or better transportation solutions - and thus to increased well-being of inhabitants and economic opportunity. Developing smart cities is subsequently imperative to Goal 11, but moreover also to the remaining 16 SDGs given that all goals are interdependent, meaning achieving one goal contributes to the achievement all other SDGs.

2.2 Definition

While a definition of what constitutes a smart city would be of great value, the debate around an internationally commonly agreed definition is still subject to controversy. [3] This is due to multifactorial reasons, inter alia the resistance of national governments to agree and consequently submit themselves under one exclusive paradigm. However, the most commonly referred to definitions include:

A smart city is "when investments in human and social capital and traditional (transportation) and modern (ICT) infrastructure fuel sustainable economic growth and a high quality of life, with a wise management of natural resources, through participatory government." [4].

"A smart sustainable city is an innovative city that uses information and communication technologies (ICTs) and other means to improve quality of life, efficiency of urban operation and services, and competitiveness, while ensuring that it meets the needs of present and future generations with respect to economic, social and environmental aspects." [5].

"A new concept and a new model, which applies the new generation of information technologies, such as the internet of things, cloud computing, big data and space/geographical information integration, to facilitate the planning, construction, management and smart services of cities. [...] Developing Smart Cities can benefit synchronized development, industrialization, information, urbanization and agricultural modernization and sustainability of cities development. The main target for developing Smart Cities is to pursue: convenience of the public services; delicacy of city management; livability of living environment; smartness of infrastructures; and long-term effectiveness of network security." [6].

These definitions entail at its core common elements, including of usage of ICT, for the purpose of improving infrastructure to ensure economic, social and environmental sustainability.

2.3 Assessment Mechanisms

Against this background, multiple intergovernmental organizations, industry organizations and academic institutions have adopted different systems to measure a city's smartness. Those include but are by no means limited to the (1) International Organization for Standardization (ISO); (2) the "United 4 Sustainable Smart Cities Initiative" under the umbrella of the UN Economic Commission for Europe, International Telecommunication Union and 15 UN bodies; and (3) IMD's Smart City Index 2019.

2.3.1   International Organization for Standardization (ISO)

The ISO has developed international standards, principles and requirements for performance metrics related to smart community infrastructures in 2015 (ISO/TS 37151). The technical specifications outlined in ISO/TS 37151 assess the smartness of community infrastructure –such as energy, water, transportation, waste and ICTs– along the lines of using enhanced technological performance that is designed, operated, and maintained to contribute to sustainable development and resilience of the community. [7] The technical specifications aim to measure the performance of smart community infrastructures for the purpose of enabling community managers, infrastructure operators, development agencies and investors to plan cities as well as to evaluate and compare the performance consistently.

2.3.2   United 4 Sustainable Smart Cities Initiative

The UN Economic Commission for Europe, UN Habitat, International Telecommunication Union and other stakeholders created the "United 4 Sustainable Smart Cities Initiative", which developed a methodology for key performance indicators (KPIs) for smart sustainable cities in order to establish the criteria to evaluate ICT's contributions in making cities smarter and more sustainable, and to provide cities with the means for self-assessments. [8] This intricate assessment mechanism outlines 3 main dimensions with numerous sub-dimensions: (1) Economy with 42 sub-dimensions; (2) Environment with 17 sub-dimensions; (3) and Society and Culture with 29 sub-dimensions. [9] The sub-dimensions range from access to Wifi in public areas, water and electricity supply ICT monitoring, e-Government, public transportation and bicycle networks, pedestrian infrastructure, air pollution, solid waste treatment, renewable energy consumption, student ICT access, gender income equality, resilience plans, emergency service response time, to cultural infrastructure.

2.3.3   IMD Smart City Index 2019

The IMD's World Competitiveness Center in collaboration with Singapore University of Technology and Design issued its first edition of the IMD Smart City Index 2019. [10] It compares 102 cities worldwide along the usage of technology in 5 key areas, each of which entails various indicators: (1) health and security; (2) mobility; (3) activities; (4) work and school opportunities; and (5) governance. The indicators range from online reporting of city maintenance problems, free public Wifi, website monitoring air pollution, arranging medical appointments online, car-sharing Apps, online ticket sales for public transport, online purchasing of tickets for cultural events online access to job listings, teaching IT skills in schools, to online voting for governmental processes [11].

3   Initiatives of Smart Cities Around the Globe and Their Specific Approach

The ranking of cities around the globe with regards of their smart developments differ with the rating agencies and their assessment criteria. Thus, various rankings exist.

The following examples of cities from different parts of the world that have embarked to become a smart city is therefore only a mall selection among many. The choice is mainly based on differentiating criteria and objectives and does not necessarily represent the technical advance.

3.1   South-East Asia

- **Singapore**

The IMD Smart City Index ranks Singapore as the smartest city in the world.
Singapore rating show a strong performance in infrastructure, public safety, lifelong learning opportunities, availability of green spaces, and online services through governmental portals as well as overall and efficient adoption of technology for the benefit of its citizen. A cashless payment system has been introduced which is now widely used.

Singapore's government has launched the Smart Nation Initiative by end of 2014; the 2017 budget included 2.4bn$ to promote this initiative, mainly through purchases from their start-ups, i.e. by actively supporting the own economy rather than granting subsidies [12, 13].

Another component of Singapore's initiative is the installation of a large network of wireless sensors on 110,000 lamp posts which are designed to collect for urban planning, maintenance and security [14].

- **Seoul**

The South Korean government with its "Global Digital Seoul 2020" plan focusses on offering efficient and comprehensive services to its citizens and intends to employ leadership through the beneficiaries. In 2013, the Mobile Seoul website has been launched that offers 60 real-time services in 11 categories which include transportation/mobility, culture and e-government as well as information on employment, real estate and others. Other websites and apps offer emergency alerts and services for elderly and disabled citizens [15].

- **Shanghai**

The approach in Shanghai differs from the previous ones. It is mainly technology-driven and uses Big Data to enable and enhance an intelligent urban management system and deliver digital public services. The city is equipped with a full-coverage fiber-optics network. Governmental departments work with data sharing and via cloud. Big Data is used for public services and health data analytics [15].

3.2   North America

- **New York**

The development of smart city components started in NYC in 2007 and the city plans forward into 2040, using data as the base to improve the handling of energy, climate change, air quality, etc. The One NYC initiative objectives are diversity, inclusivity, equity, growth, resiliency and sustainability for its citizens as well as the delivery of governmental services in a spirit of collaboration and interdependency [15].

- **Boston**

The Boston approach considers the citizens' ability to define what 'smart' means to them and their urban environment to create. Citizens shall feel being part of the team that creates a modern community rather than being recipients of services that have been designed for them for the purpose of efficiency and cost savings. Government aims at more positively interact with its citizens [16].

- **Montreal**

Montreal puts democratic principles at the forefront of its approach to the smart city development by introducing sharing of accessible and transparent data as well as public spaces to attract the participation in decision making of urban issues. Forums for public discussions have been established on the city's policy decisions [16].

3.3   MENA Region

- **Dubai**

The "Smart Dubai 2021" initiative aims at creating the happiest city and has been transforming Dubai by e-government with over 100 smart initiatives and 1,000 smart services by more than 20 governmental departments and their private associates in less than 3 years to make Dubai a world-leading city through technological advance and a personalized approach to its residents and visitors.

Terms like 'efficient', 'seamless', safe' and personalized' characterize Dubai's approach that looks at the collaboration between all public and private stakeholders to streamline all its resources for maximum efficiency for residents' benefits and business by customer, financial, resource and infrastructure impact. Smart Dubai defines 6 strategic objectives as smart living, smart economy, smart governance, smart mobility, smart environment and smart people that are measured by KPIs.

- **Egypt**

In Egypt, the establishment of smart cities is being discussed in connection with the government's Sustainable Development Strategy 2030 and its initiative to implement new urban centers, among which the New Administrative Capital, aiming at inclusivity and sustainability. The formulation of an "Egyptian Smart City Code" is under consideration.

"In order to be smart, cities need to be sustainable and inclusive" was the key message of a conference in March 2019, organized by the UN-Habitat Egypt Office and the Egyptian Ministry of Housing, Utilities & Urban Communities during which ideas on aspects like smart urban infrastructure, smart economy, smart energy, smart mobility and smart urban governance were discussed [15].

3.4 Europe

- **London**

The renewed "Smart London 2.0" initiative and its "Vision 2020" concentrates on citizens' services, especially for the vulnerable population. For an inclusive digital service, the city has established a fast network of free Wifi access and offers educational services for digital skills.

- **Helsinki**

Becoming the world's most functional city to serve its citizens stands at the forefront of Helsinki's smart city vision. For this, a policy of open government and transparency has been adopted to identify the needs of the population and have the citizens participate in the decision making process. "Kalasatama", Helsinki's smart innovation district, has become a model for citizens' participation and attracts visitors as well [15].

Together with 5 other cities a common Six City Strategy has been formulated.

- **Barcelona**

Barcelona strongly uses the Internet of Things (IoT) in its approach to become a smart city. The application of IoT-based services has led to significant savings in resources like water, has helped in job creation, supported a healthier environment and improved the city's income situation, e.g. through parking revenues. Barcelona targets to improve and simplify its IoT portal to enable more citizens to participate through better connectivity.

4   Regional and Global Data Protection Regulations and Initiatives

An issue of present and future discussions will be where to draw the line between the State's access to sensitive personal data and individual privacy rights, namely data protection rights, within the scope of using ICTs when developing smart cities.

The European Union's (EU) General Data Protection Regulation (GDPR) is a recently adopted regulatory framework for data protection from misuse by public and private actors. [17] The GDPR regulates the processing by an individual, a company or an organization of personal data relating to individuals in the EU and attributes considerable financial sanction in case of violation. The right to data protection stems from the universally and regionally acknowledged human right to privacy from interference by public authorities as outlined for instance in Article 8 European Convention on Human Rights (1953), Article 17 of the International Covenant on Civil, Political Rights (1976), or Article 12 of the Universal Declaration of Human Rights (1948). With the technological advancement, many States and regional and universal human rights mechanisms recognized the necessity of expanding the space of protection beyond the physical space at home to the virtual space [18].

The far reaching data protection regulations entailed in the GDPR are a positive step on the path of dealing with highly sensitive data in an ethically responsible manner. However, these EU regulations merely protect individuals residing in the EU, thus other individuals remain without protection. Similar data protection regulations at the global level continues to be hoped for, while it is assumed that such endeavors will result in broad resistance as these obligations might be presumed to slow down Smart City projects [3].

However, beyond the scope of enforceable conventions and regulations, citizens and local government-driven initiatives are coming into existence. The three cities of Amsterdam, Barcelona and New York have launched in 2018 the initiative "Cities Coalition for Digital Rights" aiming at protecting digital rights of residents and visitors. [19, 20] This initiative is based on 5 shared principles to create policies, tools and resources in order to promote and safeguard digital rights. The 5 principles serve as a base for policy discussions in coordination with the UN Human Settlements Programme (UN Habitat) and include (1) universal and equal access to the Internet, and digital literacy; (2) privacy, data protection and security; (3) transparency, accountability, and non-discrimination of data, content and algorithm; (4) participatory democracy, diversity and inclusion; and (5) open and ethical digital service standards. [19, 20] This digital rights initiative is based on the idea that individuals are entitled to human rights when accessing/using the internet and digital technologies/services the same way people enjoy human rights offline. The digital society should therefore protect these human rights to safeguard its participants from the risks of pervasive technologies in real and virtual spaces as well in their interactions with private and public institutions and administration [16, 19].

5   Conclusion

When assessing the selected smart city initiative (Sect. 3), it can be concluded that most of the cities mentioned –which are as said only a representative selection without prejudice to other smart city developments– have developed smart city components in line with the core elements of the definition of a smart city (Sect. 2.2).

However, the approaches which define the objectives of the developments differ, mainly between the cultural regions. Some developments –mainly those in South East Asia– target foremost technical efficiency and security through the application of technology, data creation and data management. Others –mainly the cities from the Western hemisphere– clearly put the service to their residents at the center of the smart development. Citizen rights and protection of privacy and individual freedoms form part of their objectives.

As for the measuring and assessment systems (Sect. 2.3), a unified and generally acknowledged approach would be helpful to guide cities in their process of becoming smart. A financial incentive system with easier funding –possibly through organizations like the World Bank, the International Monetary Fund or the regional/national development agencies– would benefit from such unified assessment system.

Thus, one recommendation is a unified global assessment systems spearheaded by global policy makers, such as UN Habitat; as well as greater cooperation between industry and policy makers, in order to achieve a common standardization and its practical implementation.

While new smart technologies offer a wide range of opportunities for communities with more efficient use of material and non-material resources, more sustainability, more comfort, safety and security, it should be noted that technology represents a powerful tool with possible side or negative effects, e.g. through misuse of its functions. Therefore, a responsible use of smart technologies, an appropriate regulatory framework and the protection of civil rights and freedoms need to be part of every approach to smart cities. This will, in turn, promote a wide acceptance and active participation through residents which is the essential base of successful implementation and further progress.

References

1. UNDP Homepage. https://www.undp.org/content/undp/en/home/sustainable-development-goals.html. Accessed 13 Oct 2019
2. UNDP Homepage. https://www.undp.org/content/undp/en/home/sustainable-development-goals/goal-11-sustainable-cities-and-communities.html. Accessed 8 Oct 2019
3. Weber, M., Podnar Zarko, I.: A regulatory view on smart city services. Sensors **19**, 415 (2019)
4. Caragliu, A., Del Bo, C., Nijkamp, P.: Smart cities in Europe. J. Urban Technol. **18**, 65–82 (2011)
5. International Telecommunications Union (ITU): Smart Sustainable Cities - An Analysis of Definitions, Geneva, Switzerland, p. 1 (2014)
6. International Organization for Standardization (ISO): Smart Cities, Preliminary Report, Geneva, Switzerland, p. 2 (2014)
7. International Organization for Standardization (ISO): Smart Community Infrastructure – Principles and Requirements for Performance Metrics – Technical Specification ISO/TS 37151 (2015). https://www.sis.se/api/document/preview/918844/. Accessed 5 Oct 2019
8. United 4 Smart Sustainable Cities. https://www.unece.org/urban-development-housing-and-land-management/united-4-smart-sustainable-cities-u4ssc.html. Accessed 14 Nov 2019
9. United 4 Smart Sustainable Cities: Collection Methodology for Key Performance Indicators for Smart Cities (2017). https://www.unece.org/fileadmin/DAM/hlm/documents/Publications/U4SSC-CollectionMethodologyforKPIfoSSC-2017.pdf. ISBN 978-92-61-25231-1. Accessed 14 Nov 2019
10. IMD Homepage (2019). https://www.imd.org/research-knowledge/reports/imd-smart-city-index-2019. Accessed 14 Nov 2019
11. IMD Smart City Index 2019, p. 11 (2019)
12. Singapore, Prime Minister's Office: "PM Lee Hsien Loong at launch of Smart Nation, 24 November 2016". www.pmo.gov.sg (2016)
13. Bloomberg.com: "Singapore Is Seeking Business With Startups to Spur Them", 22 August 2017 (2017). Accessed 12 Oct 2019
14. Channel News Asia: "Commentary: Singapore's Smart Nation vision blurry without a success story" (2015)
15. Smart Cities Library™: Global Resources and Promising Practices for Building and Refining Accessible and Inclusive Smart Cities from Darren Bates LLC (2017)
16. UN Information Center Cairo (2019)
17. European Union Regulation 2016/679 (2016)

18. ECtHR, Guide on Article 8 of the European Convention on Human Rights, 31 August 2019 (2019)
19. Smart Cities Library™: "Amsterdam, New York and Barcelona Coalition for Digital Rights", 14 November 2018 (2018)
20. Cities for Digital Rights Homepage. https://citiesfordigitalrights.org. Accessed 12 Oct 2019

Energy-Efficient Layered IoT Smart Home System

Bassant Abdelhamid[1], Awab Al-Habal[2], and Ahmed Khattab[2(✉)]

[1] ECE Department, Ain Shams University, Cairo, Egypt
babdelhamid@ieee.org
[2] EECE Department, Cairo University, Giza, Egypt
akhattab@ieee.org

Abstract. The Internet of Things (IoT) is one of the most prominent technologies that has recently penetrated in many application domains including smart homes. One of the main challenges in IoT applications is to design low-power systems such that they have prolonged lifetimes or can be powered via the ambient power harvested from the environments they are operating in. In this chapter, we propose an energy-efficient IoT-based smart home system design. The proposed design is based on a layered IoT architecture and aims to optimize the used hardware and software to reduce the energy consumed by the sensor nodes to prolong their lifetime. Experimental results show the effectiveness of the proposed system to monitor and control the ambient conditions inside the smart home by efficiently controlling its lighting and cooling devices using an energy-efficient design.

Keywords: Internet of Things (IoT) · Smart home · Layered architecture · Energy efficient

1 Introduction

In the past few years, the Internet of Things (IoT) gained significant interest in many fields due to its applicability and the advantages it brings to a wide range of applications such as precision agriculture, smart grids, and smart homes. Smart home is one of the application domains that has received vast research and industrial interest, resulting in a wide spectrum of smart home IoT and wireless sensor network (WSN) solutions [1–15]. The smart home literature can be classified into three categories according to their objectives which are to develop either: (1) platforms that just aim to monitor the smart home appliances to provide information about the home's electricity usage as in [5–7], (2) platforms that monitor residents with special needs to track and analyze their behavior as in [2, 8], or (3) platforms to monitor and control the on/off operation of the appliances to reduce the home's power consumption as in [1, 9–15]. In the latter set of platforms, the design goals were to facilitate the remote access and control of the smart home using cellular or WiFi-enabled smart devices connected to the Internet, and/or to minimize the power consumption of the smart home. However, the power consumption of the IoT system itself was never taken into consideration during the system design. Thus, such systems are either powered through the mains electricity source or has a

I. El Dimeery et al. (Eds.): JIC Smart Cities 2019, SUCI, pp. 170–177, 2021.
https://doi.org/10.1007/978-3-030-64217-4_20

very limited lifetime if powered using batteries. To the best of our knowledge, no paper focused on the lifetime of the nodes themselves in IoT-based smart home systems except [15]. The reported power consumption of the IoT system proposed in [15] is 9.8 days. This motivates us to design an IoT system that has a prolonged lifetime such that it can either operate for a longer duration or can be powered using the ambient energy freely available in the system's environment such as the Radio Frequency (RF) energy.

The main contributions of this chapter are the design and implementation of a layered IoT architecture aiming to reduce the overall power consumption, and the optimization of the energy consumption of the sensor node by: (1) Using a low-power controller and properly operating it in lower power modes whenever possible to minimize the wake-up time of the sensor node; (2) Redeveloping the software of the sensor node using Texas Instruments Real-Time Operating System (TI-RTOS) to exploit its tasks and semaphores capabilities to control the energy efficiency of the node operation; and (3) Using the Zigbee protocol to send the collected data from the sensor nodes and to send the actions to the actuators. This has low power consumption compared to WiFi which is widely used in literature [9–15]. We build a prototype of the proposed system to evaluate its performance. The experimental results show the ability of the proposed design to effectively control lighting and cooling devices in the home to the reduce the home's power consumption, and hence, reduce the electricity bills. Nevertheless, the sensor node has an 84.6 days lifetime which is approximately order of magnitude of the existing lifetime reported in the literature (9.4 days in [15]).

The rest of this chapter is organized as follows: Sect. 2 presents the layered architecture design for the IoT-based smart home system. Section 3 discusses the experimental results of the proposed system. Finally, Sect. 4 concludes the chapter.

2   Smart Home IoT Layered Architecture Design

The architecture design of the proposed IoT system is based on five layers adapted from [15] as shown in Fig. 1. The five layers are the physical perception layer, network layer, edge computing layer, cloud computing layer and finally security layer.

Physical Perception Layer: This layer consists of the sensor nodes (which collect the environment attributes and send the collected data to the next layer) and actuator nodes (which control the on/off operation of the Heating, Ventilation, and Air Conditioning (HVAC) and lighting system).

The proposed design and implementation of the sensor node differ from all previous related works [1–6, 9–15] by being entirely designed to consume the minimum possible amount of power. The core of the sensor node is the MicroController Unit (MCU). To minimize the consumed power of this node, we use Texas Instruments (TI) CC2650 SimpleLink multi-standard 2.4 GHz ultra-low power wireless MCU. This wireless MCU is suitable for energy harvesting and ultra-low power applications as it supports several flexible power modes including a shutdown mode [16]. In the shutdown mode, the CC2650 MCU only consumes 100 nA by switching off all hardware components, then it waits for an external interrupt event to wake up and switch to an active mode [17].

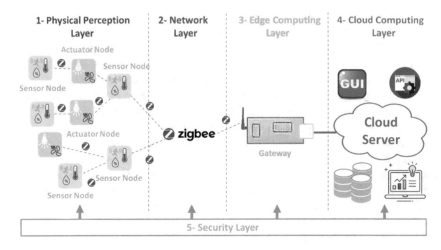

Fig. 1. Smart home layered IoT architecture.

The used sensors are temperature and humidity (HDC1010), ambient light (OPT3001), carbon dioxide (COZIR GC-0012), and Passive Infrared (PIR) motion detection (BOOSTXL-TLV8544PIR) sensors. Moreover, we use the TPL5110 external nano-timer to periodically wake up the microcontroller to collect the sensors' readings and pass them wirelessly to the gateway node. Note that, both this external timer and the PIR sensor waken the CC2650 MCU. However, the timer wakes the MCU based on a predefined period whereas the PIR sensor does so when motion is detected in the vicinity of the sensor node.

One important aspect of the designed node is its firmware. Therefore, we develop our sensor node's firmware in a way that minimizes the processor active time using TI Real Time Operating System (TI-RTOS) [18]. TI-RTOS is a compact and powerful embedded operating system that includes all the drivers of the devices in addition to software components required in development [19]. Furthermore, its simplicity and lightweight make it suitable for both fast system development and prototyping.

The actuator node is the other node implemented within the physical perception layer. The actuator node applies the desired changes in the smart home environment by controlling the operation of the HVAC devices and lights based on the decisions generated by the upper processing layers of the architecture. In the proposed system, the actuator node is connected to the air cooler, dehumidifier and LED lamp. The air cooler and dehumidifier are turned on or off. On the other hand, the intensity of the LED lamp is controlled by pulse width modulated (PWM) signal. Even though the actuator node is connected to devices that are powered by the mains electricity source, we also use the CC2650 MCU as the microcontroller of the actuator node to eliminate potential compatibility issues that could hinder the operation of the network connectivity layer.

Network Layer: This layer delivers the messages exchanged between the physical layer nodes and the network edge node that is implemented on the network gateway in our design. The network layer implementation includes the choice of a network protocol and a suitable transceiver that implements it. Unlike the related literature that use either cellular or WiFi protocols [1–15], we use the Zigbee protocol which is a high-level communication protocol that targets low-power applications. A distinguishing feature of the Zigbee protocol is its use of mesh network topology. Instead of connecting all the devices in the physical layer directly to the network coordinator implemented on the network gateway, Zigbee can connect the physical layer devices to one another and to the gateway by re-routing the messages received to a router to another router closest to the final destination of the message [20]. In our system, the actuator nodes have no power constraints as they are mains-powered, and hence, may be used as Zigbee routers.

Edge Computing Layer: This layer collects the measurements from the sensor nodes and sends the proper actions to the actuator nodes. Even though these tasks could be directly handled by the cloud computing layer, the existence of a local computing device at the edge of the smart home network eliminates any drops and latencies that may be caused by the Internet connection to the cloud servers. Furthermore, it can handle any time-critical actions and issue warnings.

The roles of this layer are implemented on the gateway implemented using a Raspberry Pi 3 Model B microcontroller. On top of the Raspbian OS installed on the Raspberry Pi, we use the Node-RED framework to build the monitoring and control software applications. To enable the gateway to act as the coordinator of the Zigbee network, XBee Zigbee S2C network module is interfaced to the Raspberry Pi 3 MCU and configured as a Zigbee network coordinator. This allows for the remote configuration of all Zigbee devices in the network through the gateway.

Cloud Computing Layer: This layer receives all the measurements available to the edge computing layer (i.e., the gateway) in real-time and stores them in a database implemented on a cloud server for historical archiving and advanced processing. Any authorized Internet-enabled device can access the database through the Internet. A Graphical User Interface (GUI) is implemented to allow the smart home users to have both the instantaneous measurements being received by the cloud server and a graphical illustration of the history of the home environment measurements. Advanced data analysis methods, such as machine learning and artificial intelligence, can be exploited to perform deep analysis of the data to find use patterns according to which the smart home power consumption and ambience are optimized. However, the implementation of such advanced analysis mechanisms is beyond the scope of this chapter. The results obtained by the data analysis methods are sent back to the gateway in the edge computing layer to optimize the control of the smart home environment.

Security Layer: This layer is concerned with how to secure the communications between the different devices in the different layers and to limit the access of the system to only the authorized users. Any communication between two devices in the architecture uses the security features provided by the used protocol. For example, the Zigbee protocol used in the network layer offers data encryption and central trust center [20] that we exploit in our system design. Also, all communications between the gateway in the edge computing layer and the cloud servers are protected using HTTPS protocol. The full implementation details of the security features are beyond the scope of this chapter.

3   Results and Discussion

In this section, we obtain the lifetime of the system and conduct several experiments to demonstrate the ability of a prototype of the system to control smart home environments.

System Lifetime: The only battery-powered node in the system is the sensor node. Therefore, its lifetime determines the lifetime of the whole system. The designed sensor node operates in two power modes: sleep mode and active mode. The node is designed to operate in the sleep mode most of the time and only wake up either periodically to perform routine measurements based on an interrupt signal issued by the external timer or when motion is detected in the vicinity of the node. The lifetime of the sensor node is obtained by first measuring the power consumption in both power modes, then calculating the time until the batteries of the node are drained using the measured power consumption values. Our average power measurements indicate that the node in the active mode operates in one out of three states, each consuming a different amount of power depending on which components of the node are switched on/off. In state 1, the microcontroller and the sensors are active. The communication module in this state tries to subscribe to a network through a power scan to find any available channel with the stored network identity. Meanwhile, in state 2, the microcontroller and the communication module are active. Finally, the microcontroller and the communication module perform a procedure to enter the sleep mode in state 3. The current consumption of the three active states are 49.16 mA, 30.96 mA and 11.16 mA. On the other hand, the node consumes on average only 1.46 mA when operating in the sleep mode. Our measurements also indicate that the sensor node needs only 500 ms to wake up, collect the measurements and send them to the gateway.

Considering that a typical smart home scenario implies that the sensor node wakes up every 20 minutes to measure the ambience and also wakes up when someone moves in front of it. This results in the node waking up 15 times per hour. The node needs 0.5 seconds to collect the measurements and send them to the gateway. Given that the node is powered by three 1200 mAh LiFePO4 batteries with 3.2 v output, the lifetime is 84.6 days which is order of magnitude if the 9.8 days lifetime reported in the literature [15]. It is worth noting that such a power consumption level entitles the node to operate using ambient energy harvesting techniques resulting in further extension in the lifetime.

Light Control: As a proof-of-concept, we implement two modes of light control: automatic and manual. In the automatic light control mode, the light is controlled automatically by the system based on the existence of residents in the room (using the motion detection sensor). The manual light control gives the smart home users the ability to manually change the light intensity using Internet-connected smart devices without relying on the motion detection signals received from the sensor nodes. Furthermore, a home security mode is allowed in which the lights are manually set to a dimmed intensity, but when motion is detected, an alarm siren will be fired to announce the existence of an intruder. The light intensity in the room and motion detection signal are both shown in Fig. 2 when using both the automatic (from 5:39 PM to 5:42 PM) and manual (from 5:43 PM on) control modes. Initially, the automatic light control is activated, and hence, lights are turned on if motion is detected in the room. At 5:43 PM, the profile is switched to manual light control. The light intensity is changed manually without depending on motion detection signal as shown in Fig. 2.

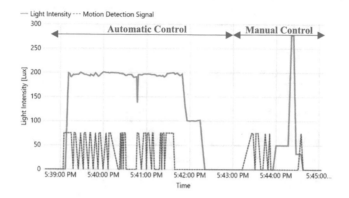

Fig. 2. Automatic and manual light intensity control in the smart home.

Humidity and Temperature Control: The proposed system's ability to control the relative humidity in the home is illustrated in Fig. 3. The value of relative humidity is kept in a pre-defined range (the red dotted horizontal lines in Fig. 3) set by the user using the system's GUI. The signal sent to the dehumidifier is also shown in Fig. 3 to demonstrate how the systems keeps the relative humidity within the desired pre-defined range. At time 5:29 PM, the relative humidity drops below the lower threshold. Accordingly, the dehumidifier is turned off. Then, the relative humidity exceeds the upper threshold at 5:33 PM. Consequently, the system turns the dehumidifier on again to lower the relative humidity. Similar results were obtained for the temperature control mechanism.

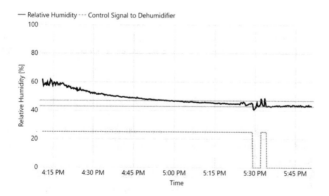

Fig. 3. Relative humidity control to operate within a pre-defined range (red lines).

Air Ventilation Control: In indoor environments, it is crucial to monitor and control the concentration of gas pollutants such as carbon monoxide (CO) and carbon dioxide (CO_2). In this experiment, we illustrate how the system controls a ventilation fan to push fresh air to balance the air quality inside the smart home. The ventilation fan is turned on when the concentration of pollutants exceeds a pre-defined upper threshold, and it is kept on until the concentration of pollutants is lower than a pre-defined low threshold. In the conducted experiment, the lower threshold is set to 1200 ppm and the upper threshold is set to 1300 ppm as illustrated in Fig. 4.

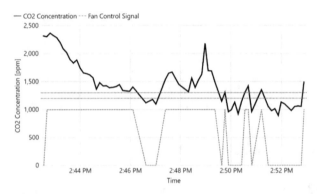

Fig. 4. Air ventilation control to keep pollutants within a pre-defined range (red lines).

4   Conclusions

An energy-efficient IoT-based smart home based on a layered architecture has been proposed. The IoT system design has aimed to minimize the energy consumed by its components to increase the lifetime of the system. Power analysis of the system has been carried out and showed remarkable lifetime extention of the system. Sample experiments that illustrate the ability of the proposed system to efficiently operate the smart home HVAC and lighting devices to reduce the electricity consumption has been presented.

Acknowledgement. This research was funded by the Egyptian National Telecom Regulatory Authority (NTRA) through a contract with the Electronics Research Institute (ERI).

References

1. Mao, X., Li, K., Zhang, Z., Liang, J.: Design and implementation of a new smart home control system based on Internet of things. In: International Smart Cities Conference (ISC2), Wuxi, China, pp. 1–5. IEEE (2017)
2. Ghayvat, H., Mukhopadhyay, S., Gui, X., Suryadevara, N.: WSN-and IOT-based smart homes and their extension to smart buildings. Sensors **15**(5), 10350–10379 (2015)

3. Hsu, Y.L., Chou, P.H., Chang, H.C., Lin, S.L., Yang, S.C., Su, H.Y., Chang, C.C., Cheng, Y.S., Kuo, Y.C.: Design and implementation of a smart home system using multisensor data fusion technology. Sensors **17**(7), 1631 (2017)
4. Gao, Y., Qin, Z., Zhang, R., Zhang, W., Duan, Y., Li, Z.: Research on data collection design based on Zigbee wireless technology smart home system. In: IOP Conference Series: Materials Science and Engineering, vol. 452, no. 4, pp. 042057. IOP Science (2018)
5. Balikhina, T., Maqousi, A.A., AlBanna, A., Shhadeh, F.: System architecture for smart home meter. In: International Conference on the Applications of Information Technology in Developing Renewable Energy Processes & Systems (IT-DREPS), Amman, Jordan, pp. 1–5. IEEE (2017)
6. Folea, S.C., Mois, G.: A low-power wireless sensor for online ambient monitoring. IEEE Sens. J. **15**(2), 742–749 (2014)
7. Brunelli, D., Minakov, I., Passerone, R., Rossi, M.: POVOMON: an ad-hoc wireless sensor network for indoor environmental monitoring. In: Workshop on Environmental, Energy, and Structural Monitoring Systems (EESMS), Naples, Italy, pp. 1–6. IEEE (2014)
8. Coelho, C., Coelho, D., Wolf, M.: An IoT smart home architecture for long-term care of people with special needs. In: 2nd World Forum on Internet of Things (WF-IoT), Milan, Italy, pp. 626–627. IEEE (2015)
9. Khan, A., Al-Zahrani, A., Al-Harbi, S., Al-Nashri, S., Khan, I.A.: Design of an IoT smart home system. In: 15th Learning and Technology Conference, Jeddah, KSA, pp. 1–5. IEEE (2018)
10. Jabbar, W.A., Alsibai, M.H., Amran, N.S.S., Mahayadin, S.K.: Design and implementation of IoT-based automation system for smart home. In: International Symposium on Networks, Computers and Communications (ISNCC), Rome, Italy, pp. 1–6. IEEE (2018)
11. Vishwakarma, S.K., Upadhyaya, P., Kumari, B., Mishra, A.K.: Smart energy efficient home automation system using IoT. In: 4th International Conference on Internet of Things: Smart Innovation and Usages (IoT-SIU), Ghaziabad, India, pp. 1–4. IEEE (2019)
12. Salman, L., Salman, S., Jahangirian, S., Abraham, M., German, F., Blair, C., Krenz, P.: Energy efficient IoT-based smart home. In: 3rd World Forum on Internet of Things (WF-IoT), Reston, VA, USA, pp. 526–529. IEEE (2016)
13. Mahmud, S., Ahmed, S., Shikder, K.: A smart home automation and metering system using Internet of Things (IoT). In: International Conference on Robotics, Electrical and Signal Processing Techniques (ICREST), Dhaka, Bangladesh, pp. 451–454. IEEE (2019)
14. Kumar, P., Pati, U.C.: IoT based monitoring and control of appliances for smart home. In: International Conference on Recent Trends in Electronics, Information & Communication Technology (RTEICT), Bangalore, India, pp. 1145–1150. IEEE (2016)
15. Alsuhli, G., Khattab, A.: A fog-based IoT platform for smart buildings. In: International Conference on Innovative Trends in Computer Engineering (ITCE), Aswan, Egypt, pp. 174–179. IEEE (2019)
16. Texas Instruments SimpleLink Solutions. http://www.ti.com/wireless-connectivity/simplelink-solutions/overview/overview.html. Accessed 07 Feb 2020
17. Texas Instruments CC2650 SimpleLink multi-standard 2.4 GHz ultra-low power wireless MCU. http://www.ti.com/product/CC2650. Accessed 07 Feb 2020
18. Texas Instruments TI-RTOS: Real-Time Operating System (RTOS) for Microcontrollers (MCU). http://www.ti.com/tool/TI-RTOS-MCU. Accessed 07 Feb 2020
19. TI-RTOS Texas Instruments Wiki. http://processors.wiki.ti.com/index.php/TI-RTOS. Accessed 07 Feb 2020
20. Zigbee Alliance. https://zigbee.org/. Accessed 07 Feb 2020

Smart City - Role of PMC in Circular Economy

Maged El Hawary[(✉)] and Aashley Bachani

Analytica Management Solutions, Dubai, UAE
maged.hawary@asgcgroup.com

Abstract. The world's population is increasing drastically. 68% of people are supposed to live in urban areas by 2050 (UN 2019); this could damage infrastructure and increase the need for intensive investments. Limited resources are big challenge. The Smart City approach is to modernize existing or new cities by combining Information Technology with people's intelligence. Smart City is established on the pillars of Transparency, Collaboration, Inclusion and Engagement among all stakeholders i.e. government entities, private entities, banks, developers, contractors etc. The Smart City plan addresses a city's economy, mobility, security, education and environment; the world is moving from a linear economy dilemma toward a circular economy paradigm aiming to efficiently utilize resources using sustainability concepts that apply to Economy, Environment and Society. This will be achieved by implementing Information Technology by applying IOT, cybersecurity and analytics. Smart City provides sustainable solutions, better quality of life and economic competitiveness. Project Management Consultants (PMC) play a key role in advising public and private entities for delivering the Smart City concept. The mission of this paper is to outline how the PMC's contribution in creating a Smart City ecosystem will benefit stakeholders, along with outlining the added benefits of adapting the circular economy concept within the construction industry.

Keywords: Smart City · Sustainability · Circular economy · Project management consultants

I. El Dimeery et al. (Eds.): JIC Smart Cities 2019, SUCI, pp. 178–188, 2021.
https://doi.org/10.1007/978-3-030-64217-4_21

1   Introduction

Urban Cities are witnessing an increase in population. Currently, urban areas account for 55% of the world's population (UN 2019) which is expected to increase to 68% by 2050 (UN 2019). There has been a rapid decline in rural populations due to improved standards of living in Urban cities. This alarming rate of migration has a toll on Urban Cities as they need to be Smart to manage the inflow of people.

Urban Cities are being converted to Smart cities or new cities are being developed as Smart cities. These cities reliant upon IoT, Blockchain, Artificial intelligence and other upcoming technologies. The pillars of Smart City are

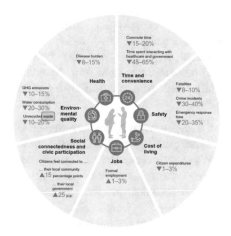

Fig. 1. Smart City

Smart Energy, Smart Transportation, Smart Data, Smart Infrastructure, Smart Mobility and Smart IoT devices.

A Smart City is a connected city which enhances the lives of citizens and visitors by deriving data through IoT sensors and other technologies. Refer to Fig. 1 (McKinsey and Company 2018).

A smart sustainable city is an innovative city that uses information and communication technologies (ICTs) to improve quality of life, efficiency of urban operations and services, and competitiveness, while ensuring that it meets the needs of present and future generations with respect to economic, social and environmental factors. (Government.ae 2019). A sustainable city is a community that is reliant upon renewable sources of energy and is overall designed to save energy. Smart City focuses upon social, economic and environmental impact and a sustainable city focuses specifically on environmental impact (Ahvenniemi et al. 2017).

In a Smart Environment people are provided with mobile applications to improve the quality of life by allowing them to do anything like finding a vacant parking spot and reporting an overflowing dumpster, pothole, crime or poorly lighted street.

A successful Smart city is created when there is cohesion between stakeholders i.e. private and public entities, and a common vision than benefits the government and the residents. In order to achieve this vision, stakeholders need the help of experienced private entities i.e. the PMC which plays a crucial role in the development of Smart City Ecosystem.

1.1   Smart City Ecosystem

A smart city ecosystem is a vibrant system comprising of people, organizations, businesses, policies, laws and processes integrated together. There are multiple value creators in a smart city ecosystem mainly focused upon the following (Chan 2019):

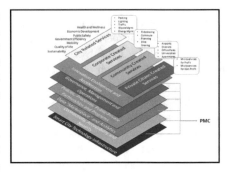

Fig. 2. Smart City ecosystem

- Services offered by the government i.e. Smart parking, Smart Lighting, Smart Waste Management etc. Along with these there are other value drivers which coexist in the Smart City Ecosystem.   Communities/residents, businesses and organizations.
- Business and organization will create Smart services to use and create data which is beneficial for the stakeholders. E.g. Uber for mobility, Google/Waze/SmartDrive for traffic and travel planning etc.
- Communities are miniature Smart Cities with specific smart needs according to their stakeholders. E.g. Commercial Spaces, residential communities, universities/schools, airports, financial districts etc.
- Residents/Citizens also play a crucial role as smart service providers. E.g. Residents can place air quality measurement sensors on their properties to provide information to fellow residents or government entities and responsible residents can take a picture of an accident and upload it on central police server, which benefits the city by sending emergency response units and another citizen through effective route planning.

Creation of a successful Smart City depends upon Public-Private Partnership. To overlook and support the stakeholders involved, Project Management Consultancies play a vital role in ensuring proper engagement and implementation of the innovative solutions. The PMC acts as an external consultant. Key dimensions provided by a PMC firm are knowledge-based economy, reduce-reuse-upcycle, life cycle cost optimization, smart utilization of buildings, smart waste management system and sustainability. The PMC implements innovative ideas in an ideal manner using best practices amalgamated from prior experiences. Currently, there are isolated smart service providers in the industry, the lack of cohesion could lead to failure in creating a smart city ecosystem, the PMC acts as an integrating body for all smart service providers maintaining the best interest of project stakeholders. Refer to Fig. 2 for realizing the role of the PMC at various level of a Smart City Ecosystem.

2 Key Dimensions of Smart City

Smart City aims at improving standard of living and digital management through data. The PMCs can suggest policies, procedures and optimum innovative solutions to the government and other stakeholders. PMCs can also help stakeholders realize the lifecycle impact or benefit of the implemented smart solutions.

Figure 3 highlights the multiple ways through which a PMC can add value to stakeholders.

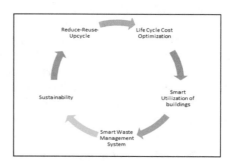

Fig. 3. Role of PMC in Smart City

2.1 Reduce-Reuse-Upcycle

Cities follow the concept of Reduce-Reuse-Recycle which reduces greenhouse gas emissions, mitigating a city's impact on the climate (Smart City 2019). Implementation of zero waste can conserve energy and reduce carbon emissions significantly. The benefit of this method is the positive environmental impact, but the financial benefits are often unrecognized. The PMC helps stakeholders realize the value of correctly implementing the waste management concept. There is potential in shifting from Reduce-Reuse-Recycle to Reduce-Reuse-Upcycle as explained in Fig. 4.

Recycling is a process in which a material is broken down to be transformed into something else whereas upcycling uses the integral pieces and parts of things to make new items that retain properties of original products and might end up being more valuable than the original product.

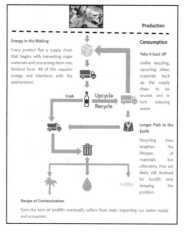

Fig. 4. Reduce-reuse-upcycle

Upcycling is important as it reduces the amount of waste and decreases the need for virgin material to be collected for production of new generations of product. In the case of metals, less mountains are mined; for paper less trees felled and overall this leads to less expended energy. E.g. Aluminum containers can be melted down to make new cans and consequently save over 90% of the energy required to produce new ones from scratch (Intercon 2019). Plastic bottles could be utilized for vertical gardening, food waste can be converted into compost, clothes can be reutilized and merged into emerging fashion trends etc.

The concept of Reduce-Reuse-Upcycle is correlated to circular economy i.e. a system of manufacturing and utilization centered around reusable and sustainable design. It eradicates waste from the current linear production system, where raw materials are obtained from the ground, and not long after are thrown in the trash. E.g. Timberland has tied up with a tire manufacturer to produce shoes from recycled tires; RAW for the Oceans manufactures clothes from recovering plastics found on the shoreline; Gamle Mursten reuses old bricks by cleaning them with vibration technology (Benzaken 2018).

2.2   Life Cycle Cost Optimization (LCC)

Smart City is an innovative and reactive ecosystem, due to fast technological changes and developing customer requirements, Smart City systems might have multiple modifications. Therefore, life cycle management is important to effectively manage development (Hefnawy et al. 2018).

In infrastructure or building the operation cost consumes majority of an LLC's budget. The life cycle costs of buildings are explained in Table 1 (Heralova 2014).

Table 1. Life cycle cost

Group costs	Cost item
Investment costs	Fees for design, cost of construction, cost of operation units, land cost, cost related to machines, equipment's, inventory, running cost for preliminaries and construction and other miscellaneous costs
Operation costs	Maintenance and renovation costs, power supply costs, water and wastewater costs, waste disposal costs, service fees, insurance, security costs, cleaning and maintenance costs and administrative fees
End of life cycle costs	Liquidation costs, cost of recovery of rubble etc.

One of the key challenges is whether a city can invest in the future while simultaneously reducing its energy consumption. From the government and other stakeholder's perspective constructing a smart city is not the sole purpose, there is a need to optimize the life cycle cost of facilities in a Smart City to help the stakeholders realize the benefit of their investments. This in turn benefits the residents of the city, reduces life cycle cost of city infrastructure and

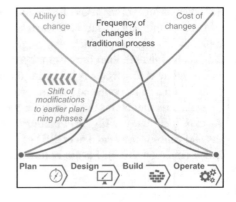

Fig. 5. Scope of change

buildings, and decreases the maintenance cost of buildings and services.

To optimize the LCC, stakeholders must have PMCs as part of the core team from the conceptualization stage of a Smart City project as they act as an integral part of project planning to optimize the LCC of a structure. Technologies that optimize the LCC must be introduced to the stakeholders during project conceptualization. If introduced later, it will have an increase project LCC cost as displayed in Fig. 5. As the construction of Smart City progresses the ability to change decreases and the cost of change exponentially increases (refer Fig. 5). Therefore, it is critical to have the right partners as part of the core team from the project conceptualization stage.

E.g. Smart Lighting system will automate the lighting and adjust the brightness of lights at various places according to multiple settings. This was implemented in Estonia, the municipality saved 65%–85% on energy consumption and they decreased the network maintenance budget by half (Laaboudi and D'Ouezzan 2016).

E.g. Optimized dispatching and synchronized traffic lights could cut emergency response time by 20–35%. Simultaneously it will provide peace of mind and freedom of movement to residents. This helps optimization and in turn reduces the budget for effective management of the city.

2.3   Smart Utilization of Buildings

There has been tremendous effort by governments to create infrastructure which is at par with the influx of population towards urban cities. City development authorities have built residential and commercial buildings and complexes, improved and constructed new infrastructure i.e. transportation, sewage, water and waste management etc. The Smart City concept has helped integrate smart technologies in buildings which has led to the collection of valuable data.

A rising constraint is lack of space to construct new buildings; city's house buildings constructed for specific events. These buildings become redundant after a few years of construction. E.g. Football Stadiums for FIFA World Cup 2022 Qatar, Commonwealth Games Village India etc. These structures are rigid and cannot be easily modified or used for other purposes. This leads to loss for the stakeholders as the projects would not be as profitable as initially planned.

Using its construction contracting background, the PMC will educate clients and other stakeholders in constructing buildings which can be later modified into more useful spaces. Buildings should also be able to adopt to the changing times. PMCs can advise the stakeholders and government entities on the creation of correct policies and procedures to be followed by the developers in the region.

Smart utilization of buildings has two components i.e. buildings should be multipurpose and there should be room for future modification in the building structure. E.g. Coca-Cola Arena Dubai, is an ideal example as this venue can host concerts, indoor sport events, conferences, seminars etc. This solves the problem of the world i.e. decreasing land space.

Dubai is leading the way in the world for innovation. The current mega project i.e. Expo 2020 comprises of multiple country pavilions which are temporary in nature or will be used for other purposes post completion of Expo 2020. 80% of the entire Expo 2020 area will be converted into District 2020 i.e. roughly 2.3 km^2 (District2020.ae

2019). Post Expo 2020, the sustainability pavilion at the Expo will be converted into the Children and Science Centre and the Dubai Exhibition Centre will be later used to host events and exhibitions.

District 2020 will integrate multiple neighborhoods that will include offices, collaborative workspaces, residential communities, social and cultural attractions, parks and multiple other business and leisure amenities (District2020.ae 2019).

Dubai has partnered with consultancies which have acted as value drives for the Expo 2020 management and Dubai government.

2.4 Smart Waste Management System

Currently 2.01 billion metric tons of Municipal Solid Waste are produced annually worldwide. The World bank estimates overall waste generation will increase to 3.4 billion metric tons by 2050. It is estimated that 40% of waste generated worldwide is not managed properly and instead dumped or openly burned (McKinsey and Company 2018).

The responsible entities need to rope in a waste management company for consultation to build a suitable solution for that city. The PMC's waste management team can study and implement effective waste management strategies which would be a sustainable solution, and beneficial to the residents and economically viable to the government.

Waste adds to CHG emissions in two ways, the first is through landfill and incineration. The second is through emissions produced by waste collection vehicles. Effective waste management strategies are to be adopted during city planning or redevelopment phase by the civic bodies of the city. This requires support from public and private stakeholders.

Solutions can be drawn from implemented strategies as follows (McKinsey and Company 2018):

- Digital tracking and payment for waste disposal, charges users exactly for the amount and type of trash they throw away. Helps in annual reduction of unrecycled solid waste by 30–130 kg/person and subsequently 10–15% potential reduction in CHG Emissions.
- Waste collection route optimization through tagging garbage trucks with RFID Solutions as implemented in Santander, Spain.
- South Korea's Songdo development is highly energy efficient and digitally connected. Household waste is sucked directly from homes and carried through a network of tunnels to processing centers for categorization and treatment, eradicating the need for garbage trucks.
- Barcelona, Spain- utilizes smart bins that use a vacuum and suck the waste into underground storage helps reduce the smell of trash waiting to be fetched and the noise and air pollution from collection vehicles.

The PMC is essential for deriving effective waste management strategies. PMC's can assist stakeholders for implementation of waste management solutions and help them understand the life cycle impact in the perspective of economics and environment.

2.5 Sustainability

The negative impact of cities on the environment is amplified by the increase in urbanization, industrialization, and consumption. Degradation of environment can have a drastic impact on residents' physical health and their quality of life—as well as on the long-term sustainability of the city itself. 70% of world's CHG emissions are generated by cities (McKinsey and Company 2018).

Irrespective of every city's unique factors, cities have a mutual goal to be more sustainable, smarter and more habitable for residents and visitors. Sustainability is a crucial aspect for creation of a habitable city. The drivers of sustainability are as follows:

- Decrease in consumption of energy and carbon emissions.
- Organic growth and development
- Savings of operational costs
- Life Cycle Cost optimization

A PMC consults stakeholder and educates them about sustainability of a project, the impact on environment and people and the long-term benefit to the stakeholders. While stakeholders' initial investments to achieve sustainable goals could be high, the long-term benefit is mutually beneficial for everyone.

Smart utilities have the advantage of assisting cities in construction and creating cleaner, healthier, and more pleasant environments while meeting sustainability objectives.

E.g. In the UAE, Masdar City is a master-planned live-work community with an emphasis on sustainability. An investment zone designed to attract a clean-tech innovation cluster, it is powered entirely by renewable energy and includes high-performance, low-carbon buildings (Elgazzar et al. 2017).

E.g. Grid Plus- "Smart Energy Agent" is powered by AI hardware, interconnected networks that use blockchain technology. The hardware is available at home and enables users to anticipate energy usage and can be utilized to automatically buy energy when it as at its lowest price. Smart energy agent accesses multiple energy markets and enables users to make decisions, this entire process is automated without involving the user based on pre-defined parameters. It helps residents sell excess energy generated by solar panels in their homes, residents can store energy and sell when prices at their peak to maximize the profit.

The UAE is among the top leaders in implementing sustainable infrastructure initiatives as they have access to the right talent and consultancy firms to help make the country meet its goals. To create a successful Smart Sustainable City, a team comprising the government and private stakeholders, residents and experienced consultants needs to be formed.

3 Smart City Strategies

As mentioned in ESWA-UN (2015) a city's government directs the vision for the smart city's and strategies that will be utilized in implementing the smart city. The city's existing infrastructure, along with pragmatic and realistic status i.e. economic, cultural, resources etc. will facilitate in formulating strategies. Every city has different capabilities and needs. Therefore, the concept of Smart City is different in various parts of the world. The role of the PMC in creating Smart City strategies is to adopt or create an already existing strategy or to create a bespoke solution for a city. But, some pillars of Smart Cities will remain at its core i.e. well-defined scope and infrastructure, measurable and overlap with multiple dimensions. E.g. Mobility and transportation are dimensions that make life better for residents by formulating ideal route from one place to another. E.g. The Barcelona Smart City strategy is focused upon improving the lifestyle of the people rather than capturing and monetizing the data whereas London's Smart City strategy is to improve digital services, open data, connectivity, digital inclusion, cyber-security and centers for innovation.

4 Role of PMC to Change the Fate of Failing Smart Cities

The main pitfalls which lead to failure of Smart City (Boorsma 2017) can be overcome by having PMC as a stakeholder in the project as explained:

- **Stuck in Silos:** Organizations i.e. private or public are organized in silos, confined environments that protect their own hierarchies, preserve their own systems and collect and retain their own data. This effort helps retain control over information but has an adverse effect on effective management of digitization. The role of the PMC becomes crucial as they will be responsible for collecting information from all stakeholders and for setting up guidelines for sharing policies, processes, assets and data to support the city in meeting its true potential.
- **No plan to Replicate or Scale:** Smart City projects start as pilot projects with no plan to scale the pilot implementation. The role of the PMC will be setting up guidelines and parameters that will enable the Smart City pilot project to be scaled and implemented throughout the city.
- **Top-Down Versus Bottom-Up Dichotomy:** Smart City is implemented through Top-Down dichotomy which is appropriate as it is backed by big corporates who possess capital investments. This approach could lead to a dead end due to lack of acceptance by resident's failure to produce user friendly design etc. The PMC's role will be to integrate bottom-up and top-down approaches. The core aim of Smart City is to make cities better and to improve the lifestyle for the citizens. The PMCs could lead as facilitators and as a communication channel from the residents to the big corporates. This is a win-win situation as big corporates will receive hands on advice and will decrease their R&D cost and residents will benefit as they get tailored solutions that meet their requirements.

5   Discussion and Conclusion

In this paper, we clarified and discussed topics related to the effective creation of a Smart City Ecosystem. Smart Service providers lead the technology component of the Smart City. However, there is a major gap to be breached i.e. effective implementation, management and life cycle benefit realization for the stakeholders and the residents of the city. This can be achieved if the stakeholders invest in the right project partners from the conceptualization stage itself i.e. consultancy organizations. It is emphasized, usually major projects have multiple organizations working together with a mutual objective i.e. to deliver a Smart City project but this often leads to lack of cohesion among organizations and consequently impacts the output of a Smart City.

Key points from this research paper are highlighted as follows:

1. Create a comprehensive Smart City Ecosystem based on the concept of circular economy.
2. PMCs should act as advisors or partners to the government or private entities or both for creation of a Smart City Ecosystem.

- PMCs will play a crucial role for developing synergy among stakeholders while facilitating the creation of a Smart City Ecosystem.
- PMCs are a part of the core team alongside the stakeholders from the conceptualization stage, delay in engagement could lead to decrease in the possibility of changes and increase in cost of changes as explained in Fig. 5.
- PMCs help determine life cycle cost (LCC) and optimizing it which is beneficial for both stakeholders and residents.
- PMCs will synergize all relevant project stakeholders, lead the scaling of the Smart City pilot project and be the core communication channel between residents and big corporations.
- A Smart City is sustainable, but all sustainable cities are not Smart. A city is Smart when the decisions, policies and procedures made are driven by data captured through innovative solutions powered through blockchain, AI, IoT or other IT based applications.
- Reduce-Reuse-Upcycle, as determined by PMCs, is a concept that can replace Reduce-Reuse-Recycle
- Consultancies can adopt Smart Waste Management technologies from across the globe and modify them to cater to project stakeholders.
- PMCs act as consultants as they help stakeholders understand the life cycle benefit of Smart Building designs and multipurpose buildings. This reduces land constraint, optimizes operation costs and reduces capital expenditure.
- Development of Smart City should be sustainable i.e. organic growth with decrease in carbon emissions.
- PMCs can help create plans for adopting renewable sources of energy and making Smart City an operational success.
- Every city in the world has distinctive needs, Smart City strategies are tailored according to each city's needs.

The PMC acts as a partner and advisor to public and private stakeholders while facilitating the development of a Smart City.

References

Ahvenniemi, H., Huovila, A., Pinto-Seppä, I., Airaksinen, M.: What are the differences between sustainable and smart cities?. Cities **60**(Part A) (2017)

Benzaken, H.: 5 Companies That Embrace the Concept of a Circular Economy - Goodnet. Goodnet (2018). https://www.goodnet.org/articles/5-companies-that-embrace-concept-circular- econo my. Accessed 7 Oct 2019

Boorsma, B.: A new digital deal. Boekscout BV (2017)

Chan, B.: The Smart City Ecosystem Framework - A Model for Planning Smart Cities. Create a culture of innovation with IIoT World! (2019). https://iiot-world.com/smart-cities/the-smart-city-ecosystem-framework-a-model-for-planning-smart-cities/. Accessed 9 Oct 2019

District2020.ae. Master Plan (2019). https://www.district2020.ae/masterplan.html. Accessed 9 Oct 2019

Elgazzar, R., El-Gazzar, R.: Smart Cities, Sustainable Cities, or Both? A Critical Review and Synthesis of Success and Failure Factors. SMARTGREENS 2017 Conference (2017)

Government.ae. Smart sustainable cities - The Official Portal of the UAE Government (2019). https://government.ae/en/about-the-uae/digital-uae/smart-sustainable-cities. Accessed 14 Oct 2019

Hefnawy, A., Bouras, A., Cherifi, C.: Relevance of lifecycle management to smart city development. Int. J. Product Dev. **22**(5), 351 (2018)

Heralova, R.: Life cycle cost optimization within decision making on alternative designs of public buildings. Procedia Eng. **85**, 454–463 (2014)

Laaboudi, K., D'Ouezzan, S.: Smart City Value Chain. White Paper, 3 (2016)

McKinsey and Company. Smart Cities: Digital Solutions for a More Livable Future. McKinsey & Company (2018). https://www.mckinsey.com/ ~ /media/mckinsey/industries/capital%20pr ojects%20and%20infrastructure/our%20insights/smart%20cities%20digital%20solutions% 20for%20a%20more%20livable%20future/mgi-smart-cities-full-report.ashx. Accessed 9 Oct 2019

Intercon. Recycling vs. Upcycling: What is the difference? (2019). https://intercongreen.com/ 2010/02/17/recycling-vs-upcycling-what-is-the-difference/. Accessed 7 Oct 2019

Smart City. How Zero Waste Management Can Bring Sustainability In Smart Cities (2019). https://www.smartcity.press/zero-waste-management/. Accessed 7 Oct 2019

UN DESA | United Nations Department of Economic and Social Affairs. 68% of the world population projected to live in urban areas by 2050, says UN | UN DESA | United Nations Department of Economic and Social Affairs (2019). https://www.un.org/development/desa/ en/news/population/2018-revision-of-world-urbanization-prospects.html. Accessed 10 Oct 2019

Self-X Concrete Applications in Smart Cities

Amin K. Akhnoukh[1($\boxtimes$)], Carol C. Massarra[1], and Pavan Meadati[2]

[1] Construction Management Department, East Carolina University, Greenville, NC, USA
akhnoukha17@ecu.edu

[2] Construction Management Department, Kennesaw State University, Atlanta, GA, USA

Abstract. Recent studies shows that the construction industry comprises 9% of the total gross domestic product (GDP) of the United States. This dynamic-2 billion dollar- industry is currently researching smart materials to be used in future construction projects, including smart cities construction. Self-X concrete represents a group of concrete products with superior characteristics to be used in different projects according to the required concrete performance. This paper presents different self-x concrete mix designs, their superior fresh and hardened properties, and potential applications in infrastructure and smart cities construction. Self-X concrete mixes are highly beneficial in developing structural members with superior strength and high durability, with lower environmental impact, lower life cycle cost, and improved construction site safety.

Keywords: Self-shaping self-compacting concrete · Photocatalytic concrete · Pervious concrete · Self-healing concrete

1 Introduction

Recent studies shows that the construction industry comprises 9% of the total gross domestic product (GDP) of the United States of America. This dynamic – 2 billion dollar – industry is currently researching smart materials to be used in future construction projects, including smart cities construction. Recent statistics shows that smart cities construction market will continue to grow during the next few years from roughly $310 billion in 2018 to $720 billion in 2023. The increase in demad is attributed to population growth and the need for efficient communications infrastructure.

Given its dominance in the construction market, new generations of concrete mixes are being investigated and developed to be used in different construction projects. New concrete mixes are required to have superior fresh and hardened characteristics as self-consolidating characteristics with ultra-high performance, high durability and resistance to environmental attacks, self-cleaning, and self-healing characteristics. The Self-X concrete mixes are highly beneficial in smart cities construction due to the ability of designing and constructing structural members with superior strength and durability, lower maintenance requirement, lower life-cycle cost, and increased projects safety. Self-X concrete properties are modified using different types of supplementary cementitious materials (SCMs) [1–4], the use of multi-wall carbon nano-tubes

I. El Dimeery et al. (Eds.): JIC Smart Cities 2019, SUCI, pp. 189–195, 2021.
https://doi.org/10.1007/978-3-030-64217-4_22

(MWCNTs) [2], and photocatalytic titanium dioxide particles [5, 6]. High strength concrete developed using SCMs could be used in pouring precast members fabricated using high diameter prestress strands [7]. The large diameter strands, used with end zone confinement, can provide members with increased flexure strength [8].

This paper presents the research investigations conducted to develop two self-x concrete mixes. First, self-consolidating high strength mixes to be used in pouring high strength deep members with high reinforcement. The developed mixes long-term performance will be investigated for alkali-silica resistivity (ASR) using accelerated mortar bar test (AMBT) [9], and SCMs positive effect on ASR mitigation is quantified [10]. Second, self-cleaning "photocatalytic" concrete mixes with minimal maintenance requirement for aesthetics.

2   Self-X Concrete Development

2.1   Self-consolidating Ultra-High Performance Concrete

The self-consolidating ultra-high performance concrete is a new generation of concrete developed in France in late 1990s [11–14]. The self-consolidating properties, also known as self-compacting, allows the concrete to fill the forms without external manual or mechanical compaction. The workability of the self-consolidating concrete (SCC) is evaluated by measuring the spread diameter of the concrete after flowing through the inverted Abrams cone [15].

In this research, self-consolidating concrete with superior performance characteristics are developed. The mix design is changed as follows: first, supplementary cementitious materials are used in partial replacement of cement. Micro-silica and class C fly ash are used in step wise replacement of Portland cement to increase the binder content and efficiency. Second, coarse aggregate is eliminated and replaced with well-graded fine sand. Third, high range water reducers, commercially known as super-plasticizers, are used to ensure high flowing ability using water-to-powder ratios less than 0.25. Finally, high energy paddle mixer, as shown in Fig. 1, is used in producing the concrete.

Fig. 1. High-energy paddle mixer for developing SCC mixes

Different mixing regimens were investigated in this research project. The successful mixing process was as follows:

- Granular materials including the binder content and dry sand were pre-blended (dry-mixing) for 2 min
- The water content and 50% of the HRWR were added and mixing continued for 13 min
- The remaining HRWR was added and mixing continued for 3 additional minutes. Total mixing time was 18 min

The afore-mentioned procedures were used in producing self-consolidating concrete mixes, with high early and final strength. An average binder content of more than 800 kgm and water-to-powder ratio smaller than 0.22 were used in mix production. Final mix constituents are shown in Table 1.

Table 1. Mix constituents (kgms) for SCC with ultra-high performance

Mix	Cement	Micro-silica	Fly ash	Fine sand	Water	HRWR
Mix A	630	90	180	1355	135	37
Mix B	625	80	80	1455	156	21
Mix C	670	145	145	1355	144	43

The strength results for the developed mixes range from 65 MPa to 85 MPa at 24 h, and final compressive strength range from 105 MPa to 118 MPa, as shown in Fig. 2.

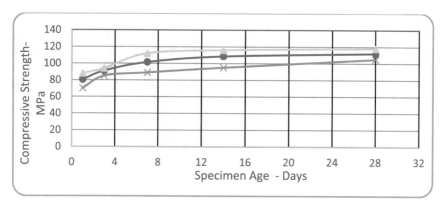

Fig. 2. Compressive strength testing results for SCC UHPC mixes

The long-term performance of SCC mixes are tested using the concrete capacity in resisting deterioration and cracking induced by long-term alkali-silica resistivity. ASR attacks concrete through the formation of expansive gel-like material as a result of the reaction of alkali-content within cement and reactive silica within specific types of aggregates in the presence of a high moisture content. In this research, different type and dosages of SCMs are introduced to regular cement and produced mortar bars for explansion test. According to ASTM Accelerated mortar bar test, an expansion smaller than 0.1% is an indication of non-reactive concrete mixes. Results of ASR expansion is shown in Fig. 3.

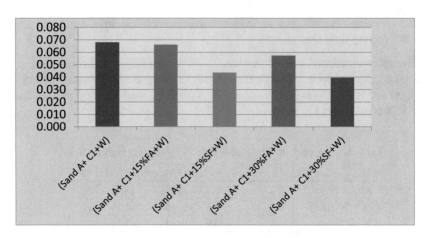

Fig. 3. Effect of silica fume and fly ash on expansion of concrete

2.2   Self-cleaning (Photo-Catalytic) Concrete

The experimental investigation is based on comparing a control specimen poured with ordinary Portland cement concrete and an additional specimen poured with the same concrete and capped with a 1.0 inch of TiO_2 infused mix for photocatalysis reaction investigation. The thickness of the photocatalytic layer is limited to 1.0 inch to minimize the additional cost of the infused nano-particles. The concrete mix was designed to attain a minimum compressive strength of 4,000 psi at 28 days to simulate normal concrete mixes used in highway construction projects. Mix proportions of the regular mix and the infused photocatalytic layer are shown in Table 2.

Table 2. Concrete mix proportions for tested specimens

Type	Standard mix	TiO_2 infused layer
Cement	26 lbs	1.80 lbs
Water	12 lbs	0.8 lbs
Limestone	35 lbs	2.4 lbs
Sand	71 lbs	4.9 lbs
TiO2	0	0.2 lbs
High range water reducer	8 oz	0.55 oz

Meltheylene blue dye was mixed with water and applied to the two concrete specimens surface to represent a dust/smog layer. The dye was left for three days to set on the concrete surface, and create permanent-stained concrete specimens, as shown in Fig. 4. Once settled, the light was applied on both specimens. The light intensity was calibrated at an average of 500 lx during the experimental phase to investigate the photo-catalytic efficiency of TiO_2.

Fig. 4. Specimens with permanent blue Meltheylene dye

The concrete specimens were left in room temperature for a time span of 18 h. Both specimens were visually inspected to determine the effect of TiO_2 in cleaning the blue dye stain under the continuous light. Figure 5a and 5b shows the result of the experimental investigation after 18 h have elapsed. Figure 5a shows the significant effect of the infused TiO_2 in cleaning the blue dye, as compared to regular specimen, shown in Fig. 5b, where no effect for the blue dye disintegration was recorded.

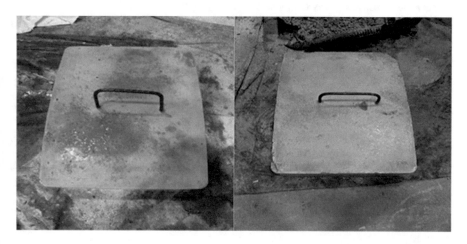

Fig. 5. (a) Concrete without TiO2 and (b) Concrete with TiO2

3 Conclusion

Different SCMs are used in mix designs of special types of concretes. SCMs including class C fly ash and silica fume are used in partial replacement of Portland cement to produce binder with high strength. Fine sand is used to optimize the mix gradation. In this research, silica fune and class C fly ash are used to produce ternary mixes with high final compressive strength. Similarly, micro and nano-sized TiO_2 is used in inducing photocatalytic reaction to clean concrete surface and results in concrete surfaces that doesn't require any additional maintenance (cleaning) for aesthetic purposes. Different types of self-X concrete mixes are currently used in different types of projects including federal government buildings, highway bridges, high strength utility pipes, and facades of different buildings.

4 Recommendation for Future Research

The development of new types of concrete requires the re-evaluation of design code charts and design curves to maintain a calibrated reliability indexes for new designs [16]. Additional research is required to decipher the overall characteristics of developed mixes to be incorporated in building information modeling softwares [17, 18].

References

1. Akhnoukh, A.K.: Overview of nanotechnology applications in construction industry in the United States. Micro Nano-syst. J. **5**(2), 147–153 (2013)
2. Akhnoukh, A.K.: The use of micro and nano-sized particles in increasing concrete durability. J. Part. Sci. Technol. **38**, 529–534 (2019)
3. Akhnoukh, A.K., Elia, H.: Developing high performance concrete for precast/prestressed concrete industry. J. Case Stud. Constr. Mater. **11** (2019)
4. Fahim, A., Moffatt, E.G., Thomas, M.D.A.: Corrosion resistance of concrete incorporating supplementary cementitious materials in Marine Environment. American Concrete Institute Special Publication SP-320-18 (2017)
5. Akhnoukh, A.K.: Implementation of nanotechnology in improving the environmental compliance of construction projects in the United States. J. Part. Sci. Technol. **36**(3), 357–361 (2018)
6. Elia, H., Ghosh, A., Akhnoukh, A.K., Nima, Z.A.: Using nano- and micro-titanium dioxide (TiO2) in concrete to reduce air pollution. J. Nanomed. Nanotechnol. **9** (2018)
7. Akhnoukh, A.: Prestressed concrete girder using large 0.7 inch strands. Int. J. Civil Environ. Struct. Constr. Architect. Eng. **7**(9), 613–617 (2013)
8. Akhnoukh, A.K.: The effect of confinement on transfer and development length of 0.7 inch prestressing strands. In: Proceedings of the 2010 Concrete Bridge Conference: Achieving Safe, Smart & Sustainable Bridges (2010)
9. ASTM C1260: Standard test Method for Potential Alkali Reactivity of Aggregates (Mortar Bar test), Annual Book of ASTM standards, American Society for Testing and Materials (2014)

10. Akhnoukh, A.K., Kamel, L.Z., Barsoum, M.M.: Alkali-silica reaction mitigation and prevention measures for Arkansas local aggregates. World Acade. Sci. Eng. Technol. Int. J. Civil Environ. Eng. **10**(2) (2016)

11. Akhnoukh, A.K.: Development of high performance precast/prestressed bridge girders. A Dissertation, University of Nebraska-Lincoln (2008)

12. Graybeal, B., Tanesi, J.: Durability of ultra-high-performance concrete. J. Mater. Civil Eng. **19**(10), 848–854 (2007)

13. Soliman, N.A., Tagnit-Hamou, A.: Using glass sand as an alternative for quartz sand in UHPC. J. Constr. Build. Mater. **145**, 243–252 (2017)

14. Koh, K.T., Park, S.H., Ryu, G.S., An, G.H., Kim, B.S.: Effect of the type of silica fume and filler on mechanical properties of ultra-high-performance concrete. In: Key Engineering Materials, vol. 774, pp. 349–354 (2018)

15. ASTM C143/143M: Standard test method for slump of hydraulic-cement concrete. Annual Book of ASTM Standards, American Society for Testing and Materials (2015)

16. Morcous, G., Akhnoukh, A.K.: Reliability analysis of NU girders designed using AASHTO LRFD. In: Proceedings of the American Society of Civil Engineers Structures Congress, California (2007)

17. Meadati, P., Irizarry, J., Akhnoukh, A.: Building information modeling implementation – current and desired status. In: Proceedings of the International Workshop on Computing in Civil engineering, Florida (2011)

18. Meadati, P., Liou, F., Irizarry, J., Akhnoukh, A.K.: Enhancing visual learning in construction education using BIM. Int. J. Polytech. Stud. (2012)

Infrastructure Systems and Management in Smart Cities

Hana Elgamal[✉] and Mahmoud El Khafif

German University in Cairo, New Cairo, Egypt
{Hana.elgamal,Mahmoud.elkhafif}@guc.edu.eg

Abstract. Nearly half of the world's population currently lives in cities, which has prompted serious concerns around scarcity of resources, deteriorating infrastructure and decline in living quality. To mitigate such concerns, an urgent global demand has risen to make cities "smart" through incorporating innovative information and communication technologies (ICTs) and capable smart solutions into city infrastructure developments. However, recent projects have shown that Egyptian smart city infrastructure developments are recurrently beset by critical challenges mainly attributed to Egyptian construction practitioners' lack of awareness of smart implementation protocols and "know-how". To contribute towards more successful Egyptian smart city infrastructure development projects, the aim of this paper is to enrich the existing body of knowledge on the applications, feasibility and benefits of ICTs integration into smart city infrastructure. It is apparent that this integration improves economic performance, safety, mobility and environmental sustainability for the benefit of all citizens and contributes towards more efficient infrastructure management.

Keywords: Information and communication technologies · Smart
infrastructure · Smart integration · Smart city · Resources sustainability

1 Introduction

1.1 Problem Definition

In recent decades, world population has increased significantly and so has the expectation of living standards in urban cities (Dobbs et al. 2011; Wang 2016). In fact, 10% of the world's population lives in the top 30 metropolis cities worldwide, and the biggest 600 cities accommodate one quarter of the world's population (Dirks et al. 2010). Currently, one half of the world's population lives in cities, and this figure is expected to increase to two thirds of the world's population by 2050 (Washburn et al. 2010; Nam and Prado 2011). In addition, cities are expected to accommodate one billion vehicles by 2020 (Marceau 2008). As a result of cities becoming overpopulated, energy consumptions are facing larger demands. At present, cities consume 75% of the world's resources and energy (Borja 2007). Moving forward, studies have forecasted even more severe environmental complications in the next few decades (Toppeta 2010). Focusing on Egypt, the country's population was expected to reach at least 96 million by 2020 (EEAA 2014), however, by 2019 its total population had already

I. El Dimeery et al. (Eds.): JIC Smart Cities 2019, SUCI, pp. 196–207, 2021.
https://doi.org/10.1007/978-3-030-64217-4_23

reached 99.2 million, clearly exceeding the initial forecast (Statista 2019). Of this total population, 42.6% live in urban cities (Fanack 2018). The north of Egypt is generally more prosperous and more vibrant with life compared to the south since the north hosts the major cities and businesses where the most economic activities takes place (Statista 2019). As such, significant migration takes place from the rural south to the urban north, especially directed towards Cairo, Giza and Alexandria (Statista 2019).

1.2 Paper Significance and Aim

Rapid urbanization increases the demand for high quality of living through the availability of critical resources, as energy, water supply and sanitation, and services as transportation, education and healthcare. This highlights the critical importance of using these resources and operating these services in a way that is more efficient or "smarter" than what is currently being carried out. This initiative can be furthered by developing fully operable "smart" cities that can sustain these resources and services over prolonged periods of time and meet the needs of citizens while absorbing future population growth (UNCSTD 2016).

Driven by advancements in Information and Communication Technologies (ICTs), as the Internet of Things (IoT), 5G, Big Data, Smart Sensors, Artificial Intelligence, cloud computing, network connectivity and other smart solutions, smart city projects are starting to become more of a reality, and aim to develop cleaner, better coordinated and economically flourishing cities (Siemens 2019). With Egypt being a developing country facing alarming population growth and energy demands, the importance and potential of smart city projects and smart infrastructure systems developments cannot be overemphasized.

On a global level, smart cities and smart infrastructure were recognized as two of the top priority subjects for the 2015–2016 period during the 18th annual session of the United Nations (UN) Commission on Science and Technology for Development (CSTD). These subjects were also perceived as key topics for successful human development and sustainable futures by the UN Addis Ababa Action Agenda, the UN novel Sustainable Development Goals campaign, the UN 2030 Agenda for Sustainable Development, and the Paris Agreement under the Framework Convention on Climate Change (UNCSTD 2016).

Against the presented problem statement, and driven by the discussed research significance, the aim of this research is to compile the findings of a thorough literature review to ultimately enrich the existing body of knowledge on smart cities and smart infrastructure systems definitions, benefits, technologies and implementations for the benefit of Egyptian and worldwide construction markets alike.

2 Infrastructure Systems

2.1 Definitions and Fundamentals

Infrastructure refers to the fundamental facilities serving a country, city, or other area including the services and systems necessary for its economy to function (O'Sullivan and Sheffrin 2003). A formal definition for infrastructure is: "the physical components of interrelated systems providing commodities and services essential to enable, sustain, or enhance societal living conditions" (Rice et al. 2010). Examples of such infrastructure systems are water distribution networks, communication networks, electricity grids and transport infrastructure. These systems create the foundation of modern societies as they provide the base for daily life and facilitate the flow of goods, information and services within urban and regional settings (Fulmer 2009).

2.2 Types of Infrastructure

Infrastructure systems can be classified into two types; hard or soft infrastructure. *Hard infrastructure* refers to the physical systems necessary for the functioning of a modern society (Fulmer 2009). These include *transportation*, *energy*, *water distribution*, *communication* and *solid waste* infrastructure systems. Examples of the different components that build up these systems will now be briefly touched upon. *Transportation* infrastructure system components include: 1) Road and highway networks and their structural components as bridges, tunnels, culverts and retaining walls; 2) Mass transit systems as subways, commuter rail systems, tramways, trolleys, bus transportation systems, city car sharing systems and canal ferries; 3) Seaport infrastructure and their components as dykes, quays, dry docks, spillways and lighthouses; and 4) Aviation infrastructure as aircrafts, airports, terminals, landsides, passenger buildings and air navigational systems. *Energy* infrastructure system components include: 1) Electrical power networks including generation plants, transformers, electrical grids, substations and distribution lines; 2) Natural gas networks including storage terminals and distribution pipelines; 3) Steam or hot water production facilities and distribution networks for district heating systems; and 4) Electric vehicle networks including charging stations and electric vehicles. *Water Distribution* infrastructure system components include: 1) Drinking water supply systems including storage reservoirs, filtration and treatment equipment and meters, pumps, valves and distribution pipes; 2) Sewage and wastewater collection, purification and disposal systems; 3) Drainage systems including storm sewers, manholes and ditches; 4) Major irrigation systems including reservoirs and irrigation canals; 5) Major flood control systems including levees, pumping stations and floodgates; and 6) Large-scale snow removal including fleets of salt spreaders, snow plows, snow blowers, and dedicated dump trucks. *Communication* infrastructure system components include: 1) Postal services and sorting facilities; 2) Telephone networks (land lines) including telephone exchange systems; 3) Mobile phone networks; 4) Television and radio networks including transmission and receiving stations and distribution networks; 5) The internet including the internet backbone, core routers, server farms, local internet service providers as well as the protocols and other basic software required for the system to function; and 6) Communication and global positioning system satellites and undersea cables. Finally, *Solid Waste* infrastructure system components include: 1) Municipal garbage and recyclables collection systems; 2) Solid waste landfills; and 3) Hazardous waste disposal facilities (Fulmer 2009).

Soft infrastructure refers to all the facilities and institutions that maintain the economic, health, social, and cultural standards of a society. These includes educational programs, parks and recreational facilities, law enforcement agencies, and emergency services. Another form of soft infrastructure are infrastructure services which depend on information technologies to support and manage other infrastructure systems. The adoption of infrastructure services is highly beneficial for a) governments to improve their management and maintenance of physical infrastructure systems through sparing increased costs, technical complexity and risks; and b) users to better operate their infrastructure services and optimizing their performances (Spacey 2018). A brief discussion on some common soft infrastructure services will be provided next. *Communication Services* include voice, email, messaging and collaboration tools. *Networking Services* can be provided through basic connectivity such as wired and mobile internet. *Data Processing*, through computing infrastructure as cloud computing platforms, can facilitate data processing and scaling. *Data Storage* can be performed through cloud storage and databases. *Knowledge Management* serves as a data analysis and document management tool. *Software Applications* enable managing the data processing and analysis procedures. *Internet of Things* is an integral part of soft infrastructure and relies on embedding computing devices within infrastructure components and consequently facilitate their interconnectivity via the internet. This enhances the ability to digitally manage the physical conditions of these components through automatic monitoring, maintenance and performance optimization. *Device Management* is important for controlling fleets of end-user devices such as laptops, tablets and phones. Finally, *Information Security* has been recently receiving notable attention and includes intrusion detection services and vulnerability management platforms (Spacey 2018).

3 Smart Cities

3.1 Definitions and Fundamentals

The smart city remains a concept, where a clear and reliable definition of the smart city is still absent among academia and practitioners. One of the formal definitions of the smart city found in literature is: "A city connecting the physical infrastructure, the information technology infrastructure, the social infrastructure, and the business infrastructure to leverage the collective intelligence of the city" (Techopedia 2019). Another formal and inclusive definition is: "A smart sustainable city is an innovative city that uses ICTs and other means to improve quality of life, efficiency of urban operations and services, and competitiveness, while ensuring that it meets the needs of present and future generations with respect to economic, social and environmental aspects" (Deloitte 2014). Evidently, there is a lack of a clear-cut definition for a "smart city" and therefore, a lack of a well-developed public perspective on the institution of a smart city. Since the aim of this research is to enrich the body of knowledge on smart cities development and implementation, a certain and clear definition for a smart city will be formed and adopted herein. Based on the definitions reported from various literature and regions, an all-encompassing and comprehensive definition for a smart city is: "A smart city is a designation given to a city that incorporates and employs ICTs, smart devices and other leading-edge technologies for facilitating the representation of the city as an interconnected and cohesive system of hard infrastructure, soft infrastructure and citizens/users, in order to 1) enhance the quality and performance of infrastructure systems such as energy, transportation and utilities; 2) facilitate better public urban services for citizens; 3) improve the quality of living for citizens; and 4) reduce resource consumption, environmental impact, waste production and overall costs."

3.2 Components and Characteristics

There are two main characteristics of smart cities: sustainability and smartness. The *sustainability* of a smart city reflects how well the different city infrastructure, governance and health systems are developed and maintained, and how well the city energy usages, waste and pollution emissions and global climate impact are controlled (Mohanty et al. 2016). The *smartness* of a city is conceptualized as the ability to boost economic, social and environmental standards of the city and its inhabitants (Mohanty et al. 2016). The five aspects of city smartness include smart people, smart economy, smart living, smart planning and smart environment (IEEE 2015). The first essential city smartness aspect is smart people who are the foundation of any smart city (Pribyl and Horak 2015). *Smart people* are citizens who have access to and are trained on the use of information and technology. These knowledgeable citizens embrace creativity and innovation to explore newer, smarter, faster and more efficient ways of carrying out their a) professional tasks such that they meet the needs of their employers; and b) social activities while preserving resources and protecting the environment (IEEE 2015). Second, a *smart economy* ensures high quality living and high paying jobs. By promoting business innovation, entrepreneurship and leadership, a smart economy fosters an environment where businesses are productive, efficient and globally competitive (IEEE 2015). Third, *smart living* is about improving the living standard of citizens which directly affects their emotional and financial well-being and their healthcare, education and safety quality. A smart living culture should nurture a healthy and safe lifestyle for citizens and should take pride in and even promote its community's diversity in terms of age, race, culture and faith (Pribyl and Horak 2015; Bureau 2016). Fourth, *smart planning* refers to the effective management of resources and services. Such planning entails optimizing city resources' usage and diligently planning the delivery of city services. Such services include a) maintaining roads, bridges and underground systems; b) providing safe drinking water and efficient waste management; and 3) making sure information services are available and accessible to everyone (IEEE 2015; Pribyl and Horak 2015). Finally, a *smart environment* is one that built on maintaining a balance between planning for growth and protecting resources. A smart city plans for future developments while protecting the natural environment, harmonizes living areas with workspaces, and balances between energy supply and energy use (Deloitte 2014).

4   Integration of ICTs into Smart City Infrastructure Systems

Integrating the hard infrastructure systems, that were previously discussed, with ICTs is what forms the *smart city infrastructure systems*, which are the core focus of this research. The hard infrastructure elements, as buildings, roads, railway tracks and power supply lines, represent the actual physical or structural entity of the smart city infrastructure system and are often termed the non-smart element of these systems. The ICT infrastructure is the core smart component of the smart city infrastructure that interconnects all system components together, thus acting as the nerve center of the smart city (Mohanty et al. 2016). Figure 1 (Inovatink 2018) demonstrates the different components of smart city infrastructure.

Fig. 1. Components of smart city infrastructure (Inovatink 2018)

4.1   Smart Traffic Management Systems

A smart traffic management system includes three main components which are traffic lights, cameras and/or road buried queue detectors and a central control system. The queue detectors identify and transmit the traffic flow status on all the city's main roads to the control system. Subsequently, the system controls the traffic lights to keep the city traffic-free. The system uses a model of real-world conditions in real-time simulation to decide whether to change the phasing of any of the lights. Advantages of smart traffic management include more punctual means of transportation, lower pollution rates at different locations, and fewer vehicles queuing on roads (Cambridge 2019).

4.2   Smart Parking Systems

A smart parking system process is shown in Fig. 2 (Thaker 2015). Smart parking systems use sensing devices such as cameras and sensors installed in pavements for

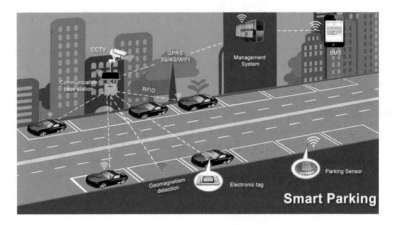

Fig. 2. Smart parking system (Thaker 2015)

counting vehicles and thus detecting parking lot occupancies and vacancies quickly and efficiently. Through these wireless sensors, drivers get a better idea of vacant parking spaces without spending much searching time. Ultimately, a larger number of more robust sensing devices can be installed to transmit this information in real time (Thaker 2015; Trafiksol 2018).

4.3 Smart Energy Grid

The smart energy grid uses computer technology to improve the communication, automation, and connectivity of the various components of the power network (Smartgrid 2014). Advanced metering infrastructure are critical to the successful functioning of the smart energy grid, where data is collected via smart meters installed to smart home controllers or end-consumer devices like thermostats, washers, dryers and refrigerators, which are considered a household's major energy consumers. Through analyzing this data, generation plants can better predict and respond to peak demand periods by improving the generation and distribution of energy, even to the extent of controlling the bulk transmission of power gathered from multiple plants. This entails reducing generation when less power is needed and to ramp up generation quickly when peak periods are approaching (Smartgrid 2014).

4.4 Smart Street Lights

Smart street lights are autonomously activated or deactivated depending upon whether there are people or vehicles passing in their spatial vicinity. Smart street lights will also communicate with neighboring street lights whenever passers are identified by sensors to automatically activate the surrounding lights. When no activity is detected the street lights will dim to low voltage lighting levels, and then brighten up to high voltage lighting levels when motion is detected. Overall, Smart Street Lighting systems implementation provides an efficient way to reduce energy consumption and CO_2 emissions (Intel 2019).

4.5 District Heating and Cooling

Smart district heating and cooling (DHC) introduces innovative thermal energy management solutions that aim to improve the performance of heating and cooling systems. Generally, heating and cooling systems are based on economies of scale, as heat generation in one large plant can often be more efficient than in multiple smaller plants than production (Stockholm 2017). The key smart features of in DHC are 1) smart heat load meters; 2) automated heat exchange monitor and control panels; and 3) complementary thermal storage and control systems (Stockholm 2017; Desmedt 2019). DHC unifies the energy generating sources in a city with buildings and facilities requiring heating and/or cooling (Desmedt 2019). In that manner, instead of each building having its own heating or cooling system, the energy is delivered to several buildings in a larger area from a central plant (Van den Ende et al. 2015).

4.6 Smart Bins

Smart bins can critically improve waste management services. Ultrasonic sensors are installed and placed on the inside of the bin lid (i.e. on the top side of the bin). This enables calculating the percentage of waste inside the bin through assessing the distance between the top level of the accumulated waste and the sensor. A decrease in the distance between waste and sensor indicates an increase in the percentage of waste inside the bin. This data is sent to the sensor's microcontroller in real time which processes this data and further transmits this as information to a smart application. This application visually displays the levels of waste inside bins to waste collection truck drivers and to management offices (McGrath 2017).

4.7 Smart Water Management

Smart water management systems use digital technology to save water, reduce costs, and increase water distribution reliability. Overlaying the physical pipeline network with a sensor-generated data network enables the analysis of flow and pressure data to determine real-time anomalies in the pipeline network, as leaks, for improved management of water flows (Polson 2013).

4.8 Smart Sewage System

Kansas City, Kansas, United States, used to spend hundreds of millions dollars on separating the combined sewer water and storm water from entering the Missouri River. These separation measures are an obligation as per the standards of the US Environmental Protection Agency (EPA) in order to conserve clean water. Through moving in the smart direction, today Kansas has the world's largest smart sewer network that is estimated to save $1 billion in the coming few years (Smart City Press 2018). Figure 3 (Rakshak 2017) illustrates how this smart separate sewer systems works.

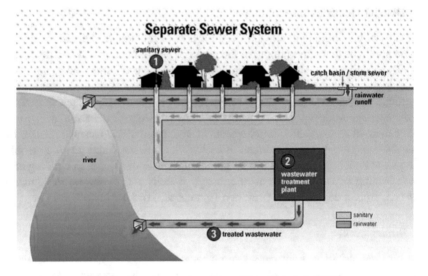

Fig. 3. Smart separate sewer system (Rakshak 2017)

4.9   Smart Building Management

Smart buildings can be described as buildings installed with smart systems (Wallace 2019), that make it possible for the buildings to learn and predict various needs as lighting, temperature, security and space availability, thus improving building energy efficiency. Several smart building technologies will be discussed within this section. Smart glass or electrochromic windows reduces glare and heat transfer while allowing visible light to pass through the windows. The darkness of the glass tint can be adjusted by passing an electric current through the glass. These windows can reduce the cost of air conditioning by 50% (Agarwal et al. 2010). Occupancy sensors detect when a room is occupied or unoccupied using passive infrared or ultrasonic sensors. They are connected to lighting and ventilation systems for energy conservation when rooms are empty (Schneider 2013). Indoor positioning systems provide users with indoor location data to simplify wayfinding on their mobile device. As these systems are deployed in smart buildings, visitors will be able to receive step-by-step instructions to their destination (Phumthai 2016). Other examples of smart building technologies include smart lighting and predictive heating, water, sanitation and fire detection.

4.10   Smart Health Care

Smart health care incorporates ICT solutions to improve and increase access to health care services. These improved services can remotely diagnose or prevent illness and can enable effective health care at lower costs. Examples of smart health care services are telemedicine, connected medical devices and illness spreading prevention methods. In Oslo, Norway, telemedicine systems have been developed making it easier for senior citizens to monitor and manage their own health care follow-ups from their homes. These systems include screens placed on home walls that older people can use to communicate with health care staff. These systems also incorporate wireless sensors that sound alarms if, for example, the oven is left on for too long or if a door is opened during the night. This system makes daily living easier for senior citizens and enables families to remotely check in on their loved ones and make sure everything is as it should be (Uddin et al. 2018).

5 Outlook and Future Potential

Rapid urbanization and population growth have led to cities currently consuming nearly 75% of worldwide resources and energy, and are, subsequently, considered main contributors to green-house gas emissions. Other complications include citizens' loss of basic functionalities, degradation in quality of life, traffic congestion, difficulties in waste management, declining infrastructure, scarcity of resources and environmental pollution are challenges faced by cities suffering rapid urbanizations and growing population.

With this alarming trajectory of rapid urban population growth and energy demands, there is an urgent need for developments of more sustainable or smarter cities that preserve and optimize the usage of natural resources. These smart cities with their smart infrastructure present the keys to more efficient urban services, enhanced living quality, flourished economic states, better resources sustainability to meet future generation demands, increased energy efficiency, and reduced waste and environmental emissions.

With Egypt lagging in smart city and infrastructure developments, this paper aims at enriching the body of knowledge on smart city and infrastructure systems definitions, benefits, smart approaches, technological integration, and feasibility, in an effort towards improving their incorporation into city-level infrastructure developments in Egypt.

To conclude, the paper illustrated that in order for Egyptian cities to smarten there is need for: 1) exploiting revolutionary ICTs and smart solutions to enhance or sustain its soft and hard infrastructure systems; and 2) engaging governments, citizens, visitors, and businesses in an intelligent, connected ecosystem.

As another important note, it was also revealed that most current research is focused on the applications and benefits of integrating ICTs with smart city infrastructure. Therefore, a recommendation for future research efforts is to focus on the technical, implementation feasibility and financial requirements of ICTs and illustrate how they are technically integrated with a city's infrastructure systems.

Another main issue is that a smart system needs smart people, not only to implement but also to maintain and utilize the system. Therefore, the educational aspects should be at the forefront of future considerations.

This would better benefit developing countries' engineers and users alike through offering them the proper implementation protocols, concepts and "know-how" which would ultimately facilitate the prospering of smart city developments in Egypt.

References

Agarwal, Y., Balaji, B., Gupta, R., Lyles, J., Wei, M., Weng, T.: Occupancy-driven energy management for smart building automation. In: Proceedings of the 2nd ACM Workshop on Embedded Sensing Systems for Energy-Efficiency in Building, pp. 1–6. ACM (2010). https://doi.org/10.1145/1878431.1878433

Borja, J.: Counterpoint: Intelligent cities and innovative cities. Universitat Oberta de Catalunya (UOC). Papers: E-Journal on the Knowledge Society (2007). http://www.uoc.edu/uocpapers/5/dt/eng/mitchell.pdf

Bureau, B.: Features of Smart Cities - Best Current Affairs: Best Current Affairs (2016). https://www.bestcurrentaffairs.com/features-of-smart-cities/

Cambridge: Smarter Cambridge Transport Smart Traffic Management (2019). https://www.smartertransport.uk/smart-traffic-management/

Deloitte: Communication infrastructure and its importance for broadband markets (2014)

Desmedt, J.: Heerlen, the Netherlands | Smartcities Information (2019). https://smartcities-infosystem.eu/scis-projects/demo-sites/heerlen-netherlands

Dirks, S., Gurdgiev, C., Keeling, M.: Smarter Cities for Smarter Growth: How Cities Can Optimize Their Systems for the Talent-Based Economy (2010). ftp://public.dhe.ibm.com/common/ssi/ecm/en/gbe03348usen/GBE03348USEN.PDF

Dobbs, R., Smit, S., Remes, J., Manyika, J., Roxburgh, C., Restrepo, A.: Urban World: Mapping the Economic Power of Cities (2011)

EEAA (2014). http://www.eeaa.gov.eg/portals/0/eeaaReports/SoE2007En/urban/09-urban9-FENG.pdf

Fanack: Urbanization in Egypt (2018). https://fanack.com/egypt/society-media-culture/society/urbanization

Fulmer, J.: "What in the world is infrastructure?": PEI Infrastructure Investor (July/August), pp. 30–32 (2009)

IEEE: Smart Cities (2015). http://smartcities.ieee.org/

Inovatink: Improved Smart Waste Management for Smart City (2018). https://medium.com/inovatink/improved-smart-waste-management-for-smart-city-7387a11f6204

Intel: Smart Street Lights for Brighter Savings and Opportunities (2019). https://www.intel.com/content/dam/www/public/us/en/documents/solution-briefs/smart-street-lights-for-brighter-savings-solutionbrief.pdf

Marceau, J.: Introduction: Innovation in the city and innovative cities. Innov. Manag. Policy Pract. 10(2–3), 136–145 (2008)

McGrath, J.: Smart home (2017). https://www.digitaltrends.com/home/barcelona-smart-city-technology

Mohanty, S., Choppali, U., Kougianos, E.: Everything You wanted to Know about Smart Cities (2016). https://www.researchgate.net/publication/306046857_Everything_You_Wanted_to_Know_About_Smart_Cities

Nam, T., Pardo, T.: Smart City as Urban Innovation: Focusing on Management, Policy, and Context (2011). https://www.researchgate.net/publication/221547712_Smart_city_as_urban_innovation_Focusing_on_management_policy_and_context

O'Sullivan, A., Sheffrin, S.M.: Economics: Principles in Action, p. 474. Pearson Prentice Hall, Upper Saddle River (2003)

Phumthai: IEI smart building solution coordinates the technology in your life into complete, brilliant experiences (2016). http://www.phumthai.com/smart-home

Polson, J.: Water Losses in India Cut in Half by Smart Meters: Itron (2013). https://www.bloomberg.com/news/articles/2013-03-15/water-losses-in-india-cut-in-half-by-smart-meters-itron

Pribyl, O., Horak, T.: Individual perception of smart city strategies. In: Proceedings of the 1st IEEE Smart Cities Conference, Guadalajara, 25–28 October (2015)

Rakshak, B.: Smart cities - what would you like to see (2017). https://forums.bharat-rakshak.com/viewtopic.php?t=7591

Rice, J.A., Mechitov, K., Sim, S.H.: Flexible smart sensor framework for autonomous structural health monitoring. Smart Struct. Syst. **6**(5–6), 423–438 (2010)

Schneider, D.: New Indoor Navigation Technologies Work Where GPS Can't (2013). http://spectrum.ieee.org/telecom/wireless/new-indoor-navigation- technologies-work-where-gps-cant

Siemens Intelligent Infrastructure (2019). https://new.siemens.com/global/en/company/topic-areas/intelligent-infrastructure

Smart City Press: How US Smart Cities Are Deploying Smart Sewer? (2018). https://www.smartcity.press/smart-sewer-in-united-states/

Smartgrid: The Smart Grid (2014). https://www.smartgrid.gov/files/sg_introduction.pdf

Spacey, J.: 12 Examples of an Infrastructure Service (2018). https://simplicable.com/new/infrastructure-services

Statista: Egypt - total population 2012–2022 | Statistic (2019). https://www.statista.com/statistics/377302/total-population-of-egypt

Stockholm: A Brief Introduction to District Heating and District Cooling (2017). https://stockholmdataparks.com/wp-content/uploads/a-brief-introduction-to-district-heating-and-district-cooling_jan-2017.pdf

Techopedia: What is a Smart City? (2019). https://www.techopedia.com/definition/31494/smart-city

Thaker, J.: Next Your parking can talk - smart parking is the solution (2015). https://www.softwebsolutions.com/resources/smart-parking-iot-solution.html

Toppeta, D.: The Smart City Vision: How Innovation and ICT Can Build Smart, "Livable", Sustainable Cities: The Innovation Knowledge Foundation (2010). http://www.thinkinnovation.org/file/research/23/en/Toppeta_Report_005_2010.pdf

Trafiksol: Save Time, space and fuel with Smart Parking Solution | Trafiksol (2018). http://www.trafiksol.com/blog/save-time-space-and-fuel-with-smart-parking-solution/

Uddin, M., Khaksar, W., Torresen, J.: Ambient Sensors for Elderly Care and Independent Living (2018). https://res.mdpi.com/sensors/sensors-18-02027/article_deploy/sensors-18-02027.pdf?filename=&attachment=1

United Nations Commission on Science and Technology for Development (UNCSTD): Issues Paper On Smart Cities and Infrastructure (2016). https://unctad.org/meetings/en/Sessional Documents/CSTD_2015_Issuespaper_Theme1_SmartCitiesandInfra_en.pdf

Van den Ende, M., Lukszo, Z., Herder, P.M.: Smart thermal grid. In: 2015 IEEE 12th International Conference on Networking, Sensing and Control (ICNSC), pp. 432–437 (2015)

Wallace, T.: Electrochromic Nano-Film Smart Coatings in Window Glass. Nanotechnology in City Environments (NICE) Database (2019). https://nice.asu.edu/nano/electrochromic-nano-film-smart-coatings-window-glass

Wang, G.: Smart cities in the city century (2016). https://www.huawei.com/en/about-huawei/publications/winwin-magazine/ai/smart-cities-in-the-city-century

Washburn, D., Sindhu, U., Balaouras, S., Dines, R.A., Hayes, N.M., Nelson, L.E.: Helping CIOs Understand "Smart City" Initiatives: Defining the Smart City, Its Drivers, and the Role of the CIO (2010). http://public.dhe.ibm.com/partnerworld/pub/smb/smarterplanet/forr_help_cios_und_smart_city_initiatives.pdf

Hydraulic Reliability Assessment of Water Distribution Networks Using Minimum Cut Set Method

Nehal Elshaboury[1]([⊠]), Tarek Attia[1], and Mohamed Marzouk[2]

[1] Housing and Building National Research Center, Giza, Egypt
nehal_ahmed_2014@hotmail.com
[2] Cairo University, Giza, Egypt

Abstract. Water Distribution Networks (WDNs) are one of the most valuable infrastructure assets worldwide. Reliability analysis plays an important role in the efficient planning and operation of a water network. It is classified into two main categories, namely mechanical reliability and hydraulic reliability. Mechanical reliability is defined as the ability of the network to function during unplanned events such as structural/mechanical failure. Whereas, hydraulic reliability is concerned with the ability of the network to cope with changes in demand and pressure head over time. This paper presents a methodology for evaluating hydraulic reliability of WDNs using the minimum cut set approach. The methodology involves the simulation of the network in normal and failure conditions. The simulation results from both conditions aid in obtaining the required as well as actual pressure head at demand nodes, respectively. The model is formulated with pressure conditions to evaluate the available demand at demand nodes. The available demand is then compared to the required demand for assessing the hydraulic reliability of the selected network. A case study of a network in Shaker Al-Bahery in Egypt is presented in order to demonstrate the process of hydraulic reliability assessment and illustrate the practical features of the proposed model.

Keywords: Water Distribution Network · Reliability assessment · Hydraulic model · Minimum cut set

1   Introduction

The typical Water Distribution Network (WDN) is articulated in terms of a graph with links/connections between pipes, hydraulic control elements, consumers, and sources (Ostfeld et al. 2002). It is considered one of the most significant elements in providing healthy drinking water. Besides, their associated investments constitute the largest weight (i.e. 80%) in maintenance budgets (Kleiner and Rajani 2000). The main task of a WDN is to provide consumers with an acceptable level of supply in terms of quality, pressure, and availability under both normal and abnormal conditions (Atkinson et al. 2014). In this manner, network reliability contributes significantly to the efficient planning, design, and operation of networks. There still doesn't exist a common definition or measure for the term "reliability" (Jalal 2008). Ciaponi et al. (2012) defined

I. El Dimeery et al. (Eds.): JIC Smart Cities 2019, SUCI, pp. 208–216, 2021.
https://doi.org/10.1007/978-3-030-64217-4_24

WDN reliability as "the ability to satisfy users taking into account the various working conditions to which it may be subjected during its operative life". Shuang et al. (2014) described reliability as "the probability that the WDN meet flow and pressure requirements under the possible mechanical failure scenarios (e.g. pipe breaks)". Therefore, the network reliability could be generally defined as the ability of the network to accomplish the assigned mission in an adequate way under stated conditions and for a prescribed time interval.

In every network, undesirable events or failures might occur that result in a decline in network performance (Ostfeld 2004). Therefore, it is difficult to find a network that is entirely reliable. Several studies have been emanated in the reliability modeling area of research to assess the hydraulic reliability of WDNs. Al-Zahrani and Syed (2005) evaluated the municipal water distribution system hydraulic reliability using the minimum cut-set method. Yannopoulos and Spiliotis (2012) presented a methodology to evaluate the mechanical and hydraulic reliability of a simple WDN using the minimum cut-set method. Gavrila et al. (2013) proposed a methodology to assess the hydraulic reliability of a large WDN using the minimum cut-set method. El-Abbasy et al. (2016) developed a mechanical reliability model of a small WDN using a series-parallel system. However, the authors highlighted the importance of incorporating a hydraulic model with the created model to determine the accurate water flow direction in the assessed network. Mohammed (2016) developed an integrated reliability assessment model that encompassed the mechanical and hydraulic condition of water main segments by pursuing the minimum cut-set approach.

The previously developed reliability assessment models for water networks had certain limitations: a) they applied the reliability techniques to small scale networks, b) they evaluated the network hydraulic performance without considering the effect of pressure on demands, and c) they rarely employed automation in the reliability models making the procedure tedious and calculations enormous. This research circumvents the limitations cited in the literature. The proposed model incorporates a hydraulic simulation model to reduce the computational time for reliability assessment and decrease chances for inadvertent errors. The model applies pressure-driven analysis which is suitable for networks under abnormal operating conditions since it takes into consideration the pressure dependency of nodal outflows. The proposed model will be tested on a medium scale WDN to check its viability on real applications.

In general, reliability analysis is classified into two main categories, namely mechanical reliability and hydraulic reliability. Mechanical reliability is defined as the ability of the network to function during unplanned events such as structural failure. Hydraulic reliability is concerned with the ability of the network to satisfy the consumers' requirements in terms of demand and pressure head at all times (Atkinson et al. 2014). Component failure may drop the pressure at demand nodes and consequently undermine the hydraulic integrity of a WDN (Gheisi and Naser 2015). Therefore, hydraulic reliability is also affected by mechanical reliability (Cullinane 1989). Some authors have argued about the water quality reliability assessment in terms of selected parameters (e.g., chlorine concentration). However, quality reliability assessment is often studied independently. The minimum cut-set technique is presumed to be the most efficient technique in evaluating network reliability (Tung 1985). The minimum cut-set could be defined, according to (Su et al. 1987), as a minimum set of network segments whose failure causes the failure of the network. But, the non-failure of one segment of the set doesn't lead to network failure (Billinton and Allan 1983). Subsequently, the overarching aim of this research is to develop a hydraulic reliability assessment model for water networks using the minimum cut-set method. The reliability assessment contributes in tackling the problem of optimizing maintenance and repair works for deteriorated water assets worldwide. This study is expected to help water municipalities for efficient budget allocation and scheduling of needed intervention strategies based on the condition of the network. Moreover, it aids decision-makers in enhancing the quality of life for people living and working in the city besides creating an economic development and improving sustainability for achieving a smarter infrastructure.

2   Research Methodology

The mechanism of the hydraulic reliability assessment model of a WDN is illustrated in Fig. 1. The model commences with estimating the total water demand and allocating the nodal demands. The next step comprises simulating the network in normal and failure conditions. The simulation results from both conditions aid in obtaining the required as well as actual pressure head at demand nodes. The model is formulated with pressure conditions to evaluate the available demand at demand nodes. The available demand is then compared to the required demand for assessing the hydraulic reliability of the selected network.

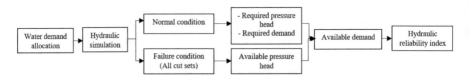

Fig. 1. Hydraulic reliability model flowchart

3   Model Development

The purpose of the hydraulic reliability assessment model is to evaluate the hydraulic performance of WDNs. This model is implemented through several stages, as described in the next sub-sections.

3.1   Water Demand Allocation

The average water demand is defined as the amount of water drawn within a specific period and it is expressed as a flow in m^3/day. The average water demand (Q_average) is computed by multiplying per capita demand by the forecasted population in the targeted year, as per Eq. 1.

$$Q_{average} = Per\ capita\ demand \times Pop_t \tag{1}$$

Where; $Q_{average}$ refers to the average water demand, and Pop_t refers to the population in the targeted year t.

The next step is to estimate the population at the target year. The Egyptian code dedicated for water and sewage networks (Egyptian code 2010) implies that the future population could be determined by knowing the current population as well as the annual rate of geometric increase using Eq. 2.

$$Pop_t = Pop_o(1+r)^n \tag{2}$$

Where; Pop_t is the estimated population at the target year, Pop_o is the current population at the time of design, r is the population growth rate, and n is the number of years between the design time and target time.

The firefighting demand (Q_{fire}) shall be taken into consideration when attempting to design a water distribution system. Although the fire demand is very less on an average basis, the rate at which the water is required to resist fire is very large in short periods. The rate of fire demand is treated as a function of population and type of location (e.g. industrial, commercial, etc.). The average total water demand is computed as the summation of the average water demand and firefighting demand. The network won't be able to meet the monthly, daily and hourly fluctuations if the average water demand is supplied at all the times. Therefore, an adequate supply of water must be provided so as to meet the peak demand.

3.2 Hydraulic Simulation

The hydraulic model comprises different network elements, which are pipes, junctions/nodes, and reservoir. The required attributes of the network are material, length, diameter, C-factor, hydraulic grade, and elevation. Each component of the WDN must be capable of meeting the required pressure and demand under both normal and failure conditions. The hydraulic model is said to be in a normal condition when all segments function well and satisfy the nodal demands at required pressure. On the other side, the hydraulic model is said to be in a failure condition when at least one of the segments is non-functional and halts the flow of water through it. It should be noted that a segment failure doesn't necessarily lead to network failure as water can propagate through other segments to reach the nodes. However, the unavailability of a segment might lead to a reduced water supply level of at nodes. In this research, Bentley WaterGEMS is employed to simulate the assessed water distribution system. The simulated hydraulic model in normal and failure conditions results in the required and available/actual pressure head at nodes, respectively. The model is then formulated with pressure conditions (minimum, maximum and required pressure) which aid in achieving the available demand at nodes as per Eq. 3 (Wagner et al. 1988).

$$Q_{j_{avl}} = \begin{cases} 0 & H_j < H_{j_{min}} \\ Q_{j_{req}} \sqrt{\frac{H_j - H_{j_{min}}}{H_{j_{req}} - H_{j_{min}}}} & H_{j_{min}} \leq H_j \leq H_{j_{req}} \\ Q_{j_{req}} & H_{j_{req}} < H_j \leq H_{j_{max}} \\ 0 & H_j > H_{j_{max}} \end{cases} \qquad (3)$$

Where; $Q_{j_{avl}}$ refers to the available demand at node j, $Q_{j_{req}}$ refers to the required demand at node j, H_j refers to the actual pressure head at node j, $H_{j_{min}}$ refers to the minimum pressure head at node j, $H_{j_{req}}$ refers to the required pressure head at node j, and $H_{j_{max}}$ refers to the maximum pressure head at node j.

The hydraulic network reliability can be expressed as the ratio of the available demand to the required demand, as shown in Eq. 4 (Zhuang et al. 2013).

$$R_{H(x)} = \frac{\sum_{j=1}^{N_{Node}} Q_{j_{avl}}}{\sum_{j=1}^{N_{Node}} Q_{j_{req}}} \qquad (4)$$

Where; $R_{H(x)}$ refers to the hydraulic network reliability in a failure condition x.
Finally, the actual hydraulic reliability of a WDN can be assessed by Eq. 5 as the average of hydraulic reliability in all failure conditions.

$$R_{NH} = \frac{\sum R_{H(x)}}{Number\ of\ minimum\ cut\ sets} \qquad (5)$$

Where; R_{NH} refers to the hydraulic reliability of a network.

4   Data Collection

According to the Egyptian code (2010), the total per capita water demand in new cities is estimated to be 300 L/capita/day. Concerning the firefighting demand, the fire demand for a city with population ranges between 25,000–50,000 capita should be 30 L/second with the duration of fire of 3 h. According to the code, the water distribution system must be designed to meet the maximum daily demand and firefighting demand or maximum hourly demand, whichever is greater. The maximum hourly and daily demands are described as a percentage of the average total demand. The maximum daily demand is estimated to be 1.6 of the average total demand. The maximum hourly demand for a city with a population census of less than 50,000 capita is estimated to be 2.25 of the average total demand. The code implies the pressure requirements for assessing the hydraulic performance of the selected network. It states that the minimum pressure in water distribution systems ranges between 20–30 m in residential districts. Moreover, the maximum pressure should be limited to 60 m of water. The velocity shall not be lower than 1 m/sec to prevent sedimentation and shall not be higher than 2 m/s to prevent erosion and high head losses. Finally, the head loss gradient shall not exceed 5 m/km. These specifications were used in assessing the hydraulic reliability model.

5 Model Implementation

The developed model is intended to assess the condition of WDN in Shaker Al-Bahery in El Gabal El Asfar, Egypt. The network layout is illustrated in Fig. 2. The hydraulic assessment model of the network is determined in major stages, as described below.

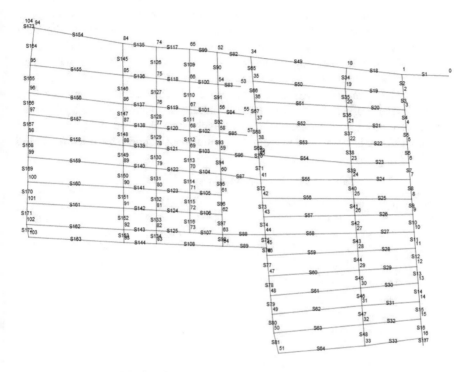

Fig. 2. Shaker Al-Bahery water network layout

It was reported that the population census in 2005 is about 15,464 capita. The annual rate of population growth for the years 2005–2025, 2025–2035 and 2035–2045 is estimated to be 2.3%, 2.15%, and 2%, respectively (Senior consulting office 2005). The population census at year 2045 is calculated to be 36,748. After estimating the population at the target year (i.e. 2045), the average water demand (Q_{avg}) is calculated: $(300 \times 36,748)/1000 = 11,024.4$ m³/day. In addition, the obtained fire demand (Q_{fire}) is equivalent to 2,592 m³/day. By calculating the maximum daily demand ($Q_{maximum\ daily} = 17,639.04$) and the maximum hourly demand ($Q_{maximum\ hourly} = 24,804.9$), the water demand used in design (Q_{design}) is computed as follows:

$$Q_{design} = maximum \left\{ \begin{array}{c} (1.6 \times 11,024.4) + 2,592 = 20,231.04 \\ 2.25 \times 11,024.4 = 24,804.9 \end{array} \right\} = 24,804.9 \text{ m}^3/\text{day}$$

This hydraulic reliability model utilizes the demand control center tool which assumes equal nodal demands at nodes/junctions based on the number and type of supplied consumers. The nodal demand is assumed to be 24,804.9/104 = 238.5 m³/day. The nodal demand allocation marks the completion of equipping hydraulic model. The hydraulic model of the WDN is now configured for simulation. In this model, the steady-state simulation which is used to determine the operating behavior under static conditions is adopted for simplification purposes. After simulating the constructed hydraulic model, the head loss gradient, water flow velocity, and pressure head at nodes are obtained in normal condition. Moreover, the hydraulic model for failure condition is simulated by considering the closure of minimum cut sets because these segment(s) causes the isolation/disconnection of a node and the failure of a network. The simulated hydraulic model in failure condition provides the available/actual pressure heads at nodes, as depicted in Table 1. The available/actual demand received by consumers at nodes which are calculated by comparing the available pressure heads with the required pressure heads at nodes are listed in Table 2.

Table 1. Available pressure head at demand nodes

Demand node	Available pressure head (m)						
	83 closed	16+33 closed	163 +172 closed	4+5+21 closed	6+7+23 closed	9+10 +26 closed	12+13 +29 closed
1	59.1	59.1	59.1	59.1	59.1	59.1	59.1
2	59.45	59.45	59.45	59.45	59.45	59.45	59.45
3	59.32	59.32	59.32	59.32	59.32	59.32	59.32
4	59.03	59.03	59.03	(N/A)	59.03	59.03	59.03
5	59.5	59.5	59.5	59.5	59.5	59.5	59.5
–	–	–	–	–	–	–	–
100	56.89	56.89	56.89	56.89	56.89	56.89	56.89
101	57.12	57.12	57.12	57.12	57.12	57.12	57.12
102	57.45	57.45	57.45	57.45	57.45	57.45	57.45
103	57.35	57.35	(N/A)	57.35	57.35	57.35	57.35
104	54.91	54.91	54.91	54.91	54.91	54.91	54.91

It is clear from Table 2 that the failure of segments "163+172" disconnects the demand node (103) from the source node, hence resulting in zero demand at this node. Apart from segments "163+172", the combination of segments "16 and 33" is found to be the most critical minimum cut-set than others because it causes zero demands at two demand nodes "16" and "17", indicating the impact of disconnection caused by the minimum cut set.

After finding the available demands at all nodes, the hydraulic reliability of the selected network is found to be equal to 0.99.

Table 2. Available demand at demand nodes

Demand node	Available demand (m³/day)						
	83 closed	16+33 closed	163 +172 closed	4+5+21 closed	6+7+23 closed	9+10 +26 closed	12+13 +29 closed
1	238.5	238.5	238.5	238.5	238.5	238.5	238.5
2	238.5	238.5	238.5	238.5	238.5	238.5	238.5
3	238.5	238.5	238.5	238.5	238.5	238.5	238.5
4	238.5	238.5	238.5	238.5	238.5	238.5	238.5
5	238.5	238.5	238.5	238.5	238.5	238.5	238.5
–	–	–	–	–	–	–	–
100	238.5	238.5	238.5	238.5	238.5	238.5	238.5
101	238.5	238.5	238.5	238.5	238.5	238.5	238.5
102	238.5	238.5	238.5	238.5	238.5	238.5	238.5
103	238.5	238.5	0.0	238.5	238.5	238.5	238.5
104	238.5	238.5	238.5	238.5	238.5	238.5	238.5

6 Conclusion

This research presented a methodology to assess the hydraulic condition of water networks in both normal and failure conditions. The simulation results in obtaining the required and actual pressure head at demand nodes, water flow direction, and velocity as well as head loss gradient. The model is then formulated with pressure conditions to evaluate the available demand at demand nodes which in comparison to the required demand aids in predicting the hydraulic reliability of the selected network. The proposed model is able to: a) construct hydraulic reliability of a water network considering the effect of pressure on demands, b) apply the hydraulic reliability assessment technique to medium scale network, and c) automate the reliability models to curtail the tedious procedure and enormous calculations. The capabilities of the developed model are demonstrated via Shaker Al-Bahery case study. The developed model is expected to help municipalities and decision-makers to plan for maintenance and rehabilitation actions of water networks.

References

Al-Zahrani, M.A., Syed, J.L.: Evaluation of municipal water distribution system reliability using minimum cut-set method. J. King Saud Univ. Eng. Sci. **18**(1), 67–81 (2005)

Atkinson, S., Farmani, R., Memon, F., Butler, D.: Reliability indicators for water distribution system design: comparison. J. Water Resour. Plan. Manag. **140**(2), 160–168 (2014)

Billinton, R., Allan, R.: Reliability Evaluation of Engineering Systems. Plenum Press, New York (1983)

Ciaponi, C., Franchioli, L., Papiri, S.: Simplified procedure for water distribution networks reliability assessment. J. Water Resour. Plan. Manag. **138**(4), 368–376 (2012)

Cullinane, J.: Determining availability and reliability for water distribution systems. Reliability Analysis of Water Distribution Systems, pp. 190–224, New York, USA (1989)

Egyptian code: Egyptian code for design and implementation of pipelines for drinking water and sewage networks. Housing and building national research center, Giza, Egypt (2010)

El-Abbasy, M.S., Chanati, H.E., Mosleh, F., Senouci, A., Zayed, T., Al-Derham, H.: Integrated performance assessment model for water distribution networks. Structure and Infrastructure Engineering, pp. 1–20 (2016)

Gavrila, C., Vartires, A., Gruia, I., Ardelean, F.: Reliability analysis of water distribution systems. Recent Advances in Energy, Environment, Economics and Technological Innovation, pp. 198–203 (2013)

Gheisi, A., Naser, G.: Multistate reliability of water distribution systems: comparison of surrogate measures. J. Water Resour. Plan. Manag. **141**(10), 1–9 (2015)

Jalal, M.: Performance measurement of water distribution systems (WDS). A critical and constructive appraisal of the state-of-the-art. M.Sc. thesis, University of Toronto, Toronto, Canada (2008)

Kleiner, Y., Rajani, B.: Considering time-dependent factors in the statistical prediction of water main breaks. In: Proceedings of American Water Works Association Infrastructure Conference, American Water Works Association, Denver, USA (2000)

Mohammed, A.: Integrated reliability assessment of water distribution networks. M.Sc. thesis, Concordia University, Montreal, Canada (2016)

Ostfeld, A.: Reliability analysis of water distribution systems. J. Hydroinform. **6**(4), 281–294 (2004)

Ostfeld, A., Kogan, D., Shamir, U.: Reliability simulation of water distribution systems - single and multiquality. Urban Water **4**(1), 53–61 (2002)

Senior consulting office: Initial report: Renewal of the water network in Shaker Al-Bahery region (2005). Accessed 10 Dec 2018

Shuang, Q., Zhang, M., Yuan, Y.: Performance and reliability analysis of water distribution systems under cascading failures and the identification of crucial pipes. PLoS ONE **9**(2), 1–11 (2014)

Su, Y., Mays, L., Duan, N., Lansey, K.: Reliability-based optimization model for water distribution systems. J. Hydraul. Eng. **113**(12), 1539–1556 (1987)

Tung, Y.K.: Evaluation of water distribution network reliability. In: Hydraulics and Hydrology in the Small Computer Age, Proceedings of the Specialty Conference, Florida, USA, pp. 359–364 (1985)

Wagner, J., Shamir, U., Marks, D.: Water distribution reliability: simulation methods. J. Water Resour. Plan. Manag. **114**(3), 276–294 (1988)

Yannopoulos, S., Spiliotis, M.: Water distribution system reliability based on minimum cut-set approach and the hydraulic availability. Water Resour. Manag. **27**(6), 1821–1836 (2012)

Zhuang, B., Lansey, K., Kang, D.: Resilience/availability analysis of municipal water distribution system incorporating adaptive pump operation. J. Hydraul. Eng. **139**(5), 527–537 (2013)

Trenchless Pipeline Rehabilitation in Smart Cities

Alan Atalah[✉]

Bowling Green State University, Bowling Green, OH, USA
aatalah@bgsu.edu

Abstract. Smart cities can be newly developed or existing old cities. The infrastructure, specially the gravity sewers, of new cities should be installed using the open cut technique before buildings and roads are constructed. However, the sewer lines of existing old smart cities cannot be replaced/constructed using the open cut technique without significant inconvenience and additional cost. For economic and social reasons, the smart cities attract population migration from rural areas. In developing countries like Egypt, this migration combined with high birth rate increase the current and future water and wastewater flow rapidly exceeding the capacity of these existing utilities causing environmental and health problems. Consequently, these smart cities need to install many new utilities, replace and rehab many existing pipelines to support the needed demand.

Gravity lines are the most challenging because of their depth, accuracy, and size. Most of these pipelines are located underneath very congested and narrow roads. Pipeline rehabilitation techniques can be economical alternative to open cut that reduces disturbance to business and residents. This paper presents an overview of the available rehabilitation techniques that can be employed in smart cities to meet the flow demands.

Keywords: Pipeline rehabilitation · CIPP · Open cut · Sliplining · Deform and reform

1 Introduction

In a city like Cairo, many water and wastewater pipes have been in service for 50 to 100 years well exceeding their anticipated service life. Most of these underground systems are severely deteriorated and in need for costly maintenance and repair. Common problems of these old sewers involve corrosion and deterioration of their materials, failure or leakage of pipe joints, and reduction of flow capacity due to build up debris inside the pipe. Ground movements from adjacent construction activity, soil consolidation, differential settlement, or other ground instability damage these lines over time. These damages compounded by poor design, poor construction, corrosion, and aging. This damages increase in infiltration of ground water and storm water inflow into sewer systems and ex-filtration of sewage into the ground water aquifer. The infiltration and inflow unnecessarily increase the discharge flow to the wastewater treatment plant; the ex-filtration pollutes the aquifer, which is the source of drinking water to many people. These problems manifest themselves in the sewer system in the form of overflows, structural deterioration, maintenance problems, hydraulic bottlenecks, and odors. Aging networks require more expensive maintenance and repair.

For repair and replacement, the conventional techniques have involved open cut excavation to expose the pipe, followed by either replacement of pipe sections and/or service connections. This process requires lane or street closure for a lengthy period, which magnify the negative impact on nearby residents, business, and road users. Since open cut construction is not a viable and economical solution to replace many of these infrastructures, trenchless solutions are needed. In many situations, trenchless pipeline rehabilitation provide a plastic liner that adds another 50 years of service to the pipe.

2   Pipeline Rehabilitation Techniques

There are several pipelining technologies available in the marketplace, such as cured-in-place-pipe (CIPP), reform and refold, and slip lining. The lining techniques can increase the flow by about 20% through the smooth surface, but they slightly reduce the flow through the smaller diameter netting a small increase in capacity. Most of the lining methods do not need access excavation to the pipeline. This is a significant advantage in narrow streets with poorly design and construction buildings, which is common in random growth suburbs of Cairo, Egypt. These buildings were never structurally designed and were illegally build with many other floors added on over the years. The open cut excavation to install or replace deep sewers require sheeting installation and removal, dewatering, and backfilling, which is very risky to these nearby buildings endangering the lives of hundreds of people.

Before pipe lining, a closed-circuit television (CCTV) filming the line is helpful in determining the existed pipe conditions and the lateral connections locations. CCTV is video cameras mounted on small tractor devices or sleds and driven or pulled through a pipeline commonly used to inspect repairs and to assess the condition of the pipe. Condition assessment is identifying and locating features that may affect the integrity and performance of a utility including defects, obstructions, leaks, infiltration, inflow, etc. (ISTT 2019).

If the pipeline is in reasonably good conditions; then the best action is to continue to maintain, monitor, and collect information over time until the conditions change; at which point, rehabilitation such as point repairs, short length repairs, or full length lining from manhole to manhole can be implemented. Maintenance actions to clean and videotape the line can be taken at intervals suitable to the line conditions. If point repair is required, cementitious grout, chemical grout, or point repair CIPP are available options. Full length lining from manhole to manhole is the more suitable solution if the pipe has multiple problems at several locations along the run from manhole to manhole.

When considering the full-length repair, the designer needs to consider the conditions of the soil surrounding pipe and the degree of line deterioration (partial or full) to determine the thickness of the liner according to the ASTM 1216, installation in wet or dry conditions, and the needed flow bypass.

2.1   Sliplining

Sliplining include continuous, segmental, and spiral wound. Sliplining is the installation of a new pipe (commonly used materials are HDPE, glass fiber reinforced polyester (GRP), and PVC) within the existing pipe. After the sliplining pipe is pulled (continuous fused pipe) or pushed (segmental pipe) into place, grouting fills the annular space between the host and sliplining pipes as shown in Fig. 1. Continuous sliplining is a method of lining with a pipe made continuous for the length of the section to be renovated prior to insertion, and which has not been shaped to give a cross sectional diameter smaller than its final diameter after installation. Segmental lining as a method of lining with pipe sections made of at least two pieces with both longitudinal and circumferential joints (ISTT 2019).

Fig. 1. Grouting the annular space between old pipe and lining pipe (Utah DOT)

The sliplining footprint includes insertion/pulling shafts (depends on system), pipes storage space and pipe handling equipment. The annular space, or area between the existing pipe and the new pipe, is typically grouted to prevent leaks and to provide structural integrity. If the annulus between the sections is not grouted, the liner is not considered a structural liner (EPA 1999).

As shown in Figs. 2 and 3, the typical steps for sliplining can be summarized as follow:

- Inspect the existing pipe.
- Bypass the flow from the section to be replaced.
- Clean and clear the host of obstructions
- Identifying the locations of lateral connections. The Egyptian standards of practice prohibits connecting the service pipes to the main line. This is a very favorable condition for the lining techniques.
- Provide access to host pipe, usually, by saw cutting the existing pipe at the spring line and removing the top half.
- Join the pipe (butt fusion, gasketed bell and spigot).

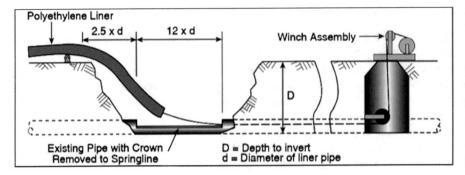

Fig. 2. Slipling continuous HDPS (Plastic Pipe Institute 2019)

Fig. 3. Sliplining segmental pipes (Infra S.A. 2019)

- Insert liner pipe and position it inside the existing pipe.
- Grout the annular space.
- Restore the lateral connections.
- Connect the terminal points to the rest of the network.

2.2   Spiral Wound

Spiral Wound PVC liners are extruded at the factory and coiled onto a large drum, then shipped to the job site. Once onsite, a continuous strip of PVC material is fed into the winding machine, and the liner is constructed onsite within the host pipe as shown in Fig. 4. The PVC profiles interlock with each subsequent strip of material during the winding process. The profiles include gasket materials that, once wound into place, form a tight fit mechanical lock. Manufacturers offer tight fit lining systems for diameters 8" – 60" that do not require annular space grouting and fixed diameter grouted solutions for 36" to 200" and larger including circular and non-circular applications, utilizing a wide range of structural and non-structural grouts (Jaques and Shallenberger 2018). Spiral wound tend to be more cost competitive than other lining techniques in larger pipes where the liner has to be thicker increasing the amount and cost of the required felt and resin.

The major advantages of sliplining include simplicity, economics, and speed. The major disadvantages of sliplining include the need for an insertion pit and reduced pipe

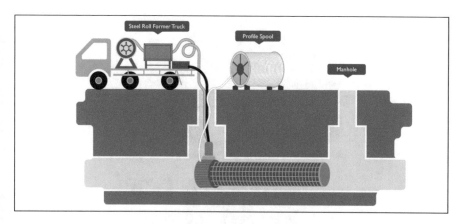

Fig. 4. Spiral wound PVC liner method (PUB, Singapore's National Water Agency 2018)

diameter. It is not well suited for small diameter pipes due to the significant diameter reduction.

2.3   Cured-in-Place Pipe

CIPP is a method of lining with a flexible tube impregnated with a thermosetting resin that produces a pipe after resin cure (ISST 2019). The thermosetting resin system can be unsaturated polyester, which is economic solution with high chemical resistance to sewage, good physical properties, and excellent working characteristics. The resin also can be vinyl ester, which has higher chemical and thermal resistance and suitable for industrial and pressure pipeline applications. Epoxy resin is required in lining water pipelines. All these thermosetting resins provide corrosion protection, structural enhancement, leakage abatement, and other benefits. The impregnated-with-resin flexible tube is a fabric that can be made from felt, fiberglass or carbon fiber.

CIPP can be a purely trenchless method that requires little or no digging, and takes significantly less time to complete than other construction methods. The crews insert a flexible liner inside the old pipe. The liner is inflated (by compressed air or water) to press firmly against the inside wall of the old pipe as shown in Fig. 5. The liner is exposed to heat or ultraviolet light (UV) to "cure" the resin. During curing, the liner gradually hardens forming a rigid, smooth surface that seals cracks and restores the old pipe to near-new condition with at least 50 years of design life (City of Portland, Oregon 2019). The impregnated CIPP tube can be installed using the winched insertion or the water inversion method. After the liner inflation and curing, dimples occur in the line where the laterals exist; these dimples can be found by TV inspection or robotic equipment. A Tee can be placed at the junction before or after rehabilitation. Laterals are typically reinstated with robotic cutting devices, or manually for large-diameter pipes, by cutting the liner (EPA 1999).

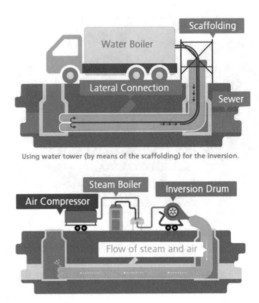

Fig. 5. The CIPP lining process.

2.4   Modified Cross Section Lining

The modified cross section lining methods include deformed and reformed methods, Swagelining™, and rolldown. These methods either modify the pipe's cross-sectional profile or reduce its cross-sectional area so that the liner can be inserted inside the existing pipe as shown in Fig. 6. The liner is subsequently expanded to snug tightly against the wall of old pipe (EPA 1999).

Fig. 6.   The modified cross section lining

2.5   Deform/Reform Liner Pipe Technique

The HDPE liner pipe is extruded in a round shape during the manufacturing process then deformed using a combination of heat and pressure and wound onto spools. Prior to installation, the spooled pipe is heated until it becomes pliable. During installation, the deformed pipe is pulled off the spool and inserted into the existing pipe through a manhole using a winch. Once in place, hot water or steam under pressure is fed into the inside of the deformed pipe to soften and reform the pipe and push the liner against the wall of the existing pipe as shown in Fig. 7. Lateral connections are installed in a similar manner to the CIPP. The deform/reform liner pipe is more economical than CIPP, but it has the major disadvantage of contracting after cooling down (due to the high thermal coefficient of HDPE) shifting the location of the lateral connection (Kung'u 1999). Waiting (a few hours to a day) until the HDPE rebounds significantly reduces this problem.

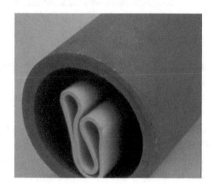

Fig. 7. The deform/reform liner pipe

Fold and form liner is similar to the deform/reform liner except that the pipe is made of polyvinyl chloride (PVC) instead of HDPE. PVC does not shrink/expand as much as HDPE reducing/eliminating the dislocation problem, but the liner may collapse over time shortening the design life of the liner (Kung'u 1999).

The ideal conditions for the application of CIPP and deform and reform liners include:

- Pipeline has adequate flow capacity during the expected life of the liner.
- Pipe diameter and length of run are within the capabilities of the CIPP or FFP
- Point source repairs are not practical
- The host pipe is not severely damaged, broken, or deformed.
- Flow bypassing is feasible or unnecessary due to low flows.

The limitations of these techniques include the need to bypass the flow, the reduction of the pipe diameter, which limit the flow increase, shrinkage or unfolding of the liner after expansion, and the limited structural support that the liner provides.

Another form of deform and reform liner is the rolldown technique; in which, the HDPE is compressed using mechanical means (series of rollers) to reduce the pipe liner's diameter as shown in Fig. 8. The temporary reduction of the pipe diameter allows the pipe insertion inside host pipe. After pulling the liner through the existing pipe, heat and pressure are applied to expedite the liner's restoration to its original pipe diameter and form tight fit against inside wall of host pipe (EPA 1999).

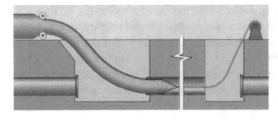

Fig. 8. Rollers rolling down the liner pipe

The Swagelining™ is a typical drawdown process; the new liner is heated and subsequently passed through a reducing die that temporarily reduces the liner's diameter by 7 to 15% to allow pulling the liner through the existing pipe. As the new liner cools, it expands to its original diameter. In both the rolldown and drawdown techniques, it is critical to allow the HDPE pipe to relax and return to its original length before terminating the line or making lateral connection (EPA 1999). Some disadvantages of these two techniques are the need to bypass or divert the flow, the reduction of the pipe diameter, which limit the flow increase. Table 1 presents a concise comparison of various sewer rehabilitation techniques available in the market place.

3   Trenchless Technology in Egypt

When the above-cited-rehabilitation techniques are properly selected, designed, and implemented in the US and EU, they were more economical than open cut in most urban areas. Due to the higher labor wages in the US and EU, the construction methods relies more on construction machinery, skilled labors, and innovative technologies. In Egypt, the labor wages are small fraction of their counterparts in the US and EU making construction more labor intensive and less machinery based. Also, the trenchless equipment and material are more expensive in Egypt due to the additional cost of shipping, currency exchange, overhead, and so forth. In addition, the movements of people, equipment, tools, and material within the US and EU are much easier than they are in the Middle East. Consequently, responding to system problems is much more costly and time consuming in Egypt than within the US and EU. This reduces the advantage of trenchless construction over the open cut one in Egypt than that in developed countries. The author hypotheses that trenchless construction is more economical than open cut in Egypt as well. The magnitude of the advantage may shrink, but there is still a significant advantage when the liner is properly selected, designed, and installed over the long run. This advantage is more significant in the random growth areas where excavation and heavy equipment risks the lives of hundreds (if not thousands) of people due to risk to nearby buildings. Research to test this hypothesis is needed to compare the costs of open cut versus trenchless methods in Egypt.

Table 1. A comparison of various sewer rehabilitation techniques (EPS, 1999)

Method		Diameter range		Max. run length		Liner Material
		(mm)	(in)	(m)	ft	
In-line expansion	Pipe bursting	100–600	4–24	230	750	PE, PP, PVC, GRP
Sliplining	Segmental	100–4000	4–158	300	1,000	PE, PP, PVC, GRP (-EP & -UP)
	Continuous	100–1600	4–63	300	1,000	PE, PP, PE/EPDM, PVC
	Spiral wound	150–2500	6–100	300	1,000	PE, PVC, PP, PVDF
Cured-in-place product linings	Inverted-in-place	100–2700	4–108	900	3,000	Thermoset Resin/Fabric Composite
	Winched-in-place	100–1400	4–54	150	500	Thermoset Resin/Fabric Composite
	Spray-on-linings	76–4500	3–180	150	500	Epoxy Resins/Cement Mortar
Modified cross-section methods	Fold and form	100–400	4–15	210	700	PVC
	Deformed/Reformed	100–400	4–15	800	2,500	(thermoplastics) HDPE
	Drawdown	62–600	3–24	300	1,000	(thermoplastics) HDPE, MDPE
	Rolldown	62–600	3–24	300	1,000	HDPE, MDPE
	Thin-walled lining	500–1,100	20–46	960	3,000	HDPE
Internal point repair	Robotic repair	200–760	8–30	N/A	N/A	Epoxy Resins/Cement Mortar
	Grouting/Sealing & spray-on	N/A	N/A	N/A	N/A	Chemical Grouting
	Link seal	100–600	4–24	N/A	N/A	Special Sleeves
	Point CIPP	100–600	4–24	15	50	Fiberglass/Polyester, etc.

Note: Spiral wound sliplining, robotic repair, and point CIPP can only be used only with gravity pipeline.
All other methods can be used with both gravity and pressure pipeline.
EPDM = Ethylene Polypelene Diene Monomer
GRP = Glass fiber Reinforced Polyester
HDPE = High Density Polyethylene
MDPE = Medium Density Polyethylene
PE = Polyethylene
PP = Polypropylene
PVC = Poly Vinyl Chloride
PVDF = Poly Vinylidene Chloride

Egypt is a developing country with very high population density in the urban areas where the majority of governmental and business activities occur. This technology offers a better employment opportunities and quality of life for the typical Egyptian.

3.1 History

Many large tunneling projects took place in Egypt since the Pharaonic times. Several miles of large diameter tunnels were installed in Cairo to transfer the sewer flow to the main WWTP in El Gabal El Asfar in the 1980s. The feeding lines to these main sewer tunnels were constructed using the utility tunneling, pipe jacking and open cut techniques. Trenchless sewer installations started with the US Agency for International Development contract #20 that included 18 pipe-jacking crossings. The contract was extended to connect the WWTP in El Haram area to the rest of the network; the scope of work included several crossings underneath critical streets, railroads, and water streams that was constructed using pipe jacking and microtunneling later on in the early 1990s. Also during the early 1999s, the Cairo subway system was constructed using the tunneling and microtunneling techniques. These tunneling and pipe jacking installations were executed by foreign contractors, which aspired two Egyptian contractors to acquire microtunneling machines. They became successful, and they created a duopoly associated with good quality and high prices. Consequently, the healthy profit margins attracted other contractors who competed on price; sometimes, on the account of quality that caused significant third party damages and unsatisfied owner and engineer. However, the Egyptian trenchless technology community invested a significant effort to rectify the situation and restore the owner's confidence; such as the inclusion of trenchless technology specifications in the Egyptian code of practice.

The oil industry successfully utilized horizontal directional drilling (HDD) to install natural gas and oil pipelines. The Egyptian construction industry is slowly using HDD to install main pressure lines (water and force main) underneath difficult crossing. However, these trenchless applications are still limited in scope and used only in very limited situations where open cut is almost impossible. The low utilization of the technology make it costly in unit prices. These high prices are compounded further by limiting the installations to the minimum length possible. As a result, the utility market gets into a spiraling vicious cycle of high prices, low utilization, and apprehension from employing the correct solution. The proposed research attempts to compare the cost of trenchless applications versus the open cut ones in Egypt over the long run. The research will consider the advanced machinery that require high capital investments and the labor intensive open cut construction.

3.2 Challenges Facing the Trenchless Industry in Egypt

The competition over the well-trained operators and crewmembers from the Arabian Gulf area is very strong; these workers can five to ten times their wages in Egypt. The Egyptian employers cannot bind the workers to stay with the company after receiving extensive and expensive training on using the trenchless system. This situation leads to using the second tier workers after the departure of the trained workers; thus increasing the price per meter to recover the cost of training an excessive number of workers.

The utility construction market is dominated by a small number of mostly public companies and private large firms. These firms subcontract the fieldwork to a larger number of small contractor entrepreneurs who are short on capital and formal education. These small contractors have difficulty renting trenchless systems to perform the first few jobs for financial reasons. In addition, the suppliers require a form of guarantee to recover the cost of equipment in case it is damaged or lost. The insurance and surety system does not work in favor of small contractors in this situation, and consequently a letter of credit from their bank becomes almost the only solution. However, this solution freezes up the cost of the system in the bank adding up an undue financial burden on the contractor and further increasing cost.

The Egyptian utility owners and engineering designers offer almost all pipeline projects to be constructed using the open cut methods. They specify trenchless techniques for short lengths at critical crossings, which results in significant increase in cost per meter because the cost of the equipment and its mobilization and setup are spread over short length.

There is significant difficulty and cost in shipping construction equipment (trenchless systems) across Middle Eastern countries for political, security, and economical reasons. This limits the utilization of the system to only a few installations within the country for which it is purchased.

4   Conclusions and Recommendations

The pipeline rehabilitation techniques present good economical solution to extend the life of existing sewer lines because they provide some additional flow capacity with minimum physical disturbance. The installation without excavation is significant advantage in the random growth areas that are characterized by narrow street, high-rise buildings, and poor construction. These techniques have the limitations of the need for an existing pipe with adequate flow capacity. Research is strongly needed find economical rehabilitation and replacement systems relative to the costs of open cut in developing countries like Egypt.

References

City of Portland, Oregon: Cured-in-Place Pipe-Lining (CIPP) (2019). https://www.portlandoregon.gov/bes/article/467513

Plastics Pipe Institute: Slip Lining (2019). https://plasticpipe.org/municipal_pipe/advisory/sliplining/index.html

Atalah, A.: The safe distance between large-diameter rock pipe bursting and nearby buildings and buried structures. ASCE J. Transp. Eng. **132**, 350–356 (2006)

Atalah, A.: The need for trenchless pipe replacement techniques in developing countries. North American Society for Trenchless Technology 2007 No-Dig Conference & Exhibition, San Diego, California (2007)

EPA: Collection Systems O&M Fact Sheet - Trenchless Sewer Rehabilitation. United States Environmental Protection Agency Office of Water, Washington, D.C. (1999)

ForConstructionPros.com: A Technology Bursting with Possibilities; In the right conditions, pipe bursting can be a cost-effective pipe replacement solution (2011). https://www. forconstructionpros.com/equipment/underground/horizontal-bursting-boring-piercing/article/ 10228825/pipe-bursting-opens-utility-installation-options

Infra, S.A.: Technical description of a rehabilitation method segmental pipe sliplining made of GRP, PE, PVC AND PP according to PN-EN 752-5; 13689; ISO 11296-1; 13566-5 (2019). http://infra-sa.pl/de/dienstleistungen/renovation/segmental-pipe-sliplining.html

Iowa Statewide Urban Design and Specifications: SUDAS Design Manual. Iowa State University, Institute for Transportatio, Ames, IA (2019)

ISTT: Glossary (2019). http://www.istt.com/index/glossary/letter.P

Jaques, J., Shallenberger, R.: Spiral wound liners and the pipe rehabilitation industry. Tech Tips from NASSCO, p. 1, August 2018

KB Contractors: KB Contractors (n.d.). https://www.kbcontractors.co.nz/pipe-bursting

Kung'u, F.T.: Sanitary sewer system rehabilitation techniques vary (1999). https://www. wateronline.com/doc/sanitary-sewer-system-rehabilitation-techniqu-0002#def

Lindeburg, M.R.: Civil Engineering Reference Manual for the PE Exam, 15th edn. PPI, A Kaplan Company, Belmont (2015)

PUB: Singapore's National Water Agency, 2018. PUB, Singapore's National Water Agency. https://www.facebook.com/PUBsg/posts/1859458157478190

Utah DOT: Culvert Rehabilitation Practices, s.l.: s.n (n.d)

Evolution of a Smart City from the Challenge of Flood Disaster: Case Study of New Owerri Capital City, South East of Nigeria

N. U. Okehielem[✉], C. O. Owuama, and C. J. Enemuo

Federal University of Technology, Owerri, Imo, Nigeria
nellok8@yahoo.com

Abstract. Owerri city, south east of Nigeria is a city bedeviled with perennial flood disasters. These disasters usually come with attendant loss of lives and properties of inestimable values.

A critical study of Owerri city was conducted with respect to its flood plain, diverse range of drainages (drainage system), canals, rivers together with rain gauge values for the past 2 decades and other extenuating factors. All the varying factors were extrapolated to highlight possible cause or causes of constant flood disasters.

These highlighted causes were made spring boards of solutions to mitigate future occurrence and establish smart city status for new Owerri with respect to flood control.

Keywords: Smart city · Flood disaster · Evolution · Drainage system · Climate change

1 Introduction

The perennial flood disasters happening in Owerri city has progressively been of immense proportion in recent times. This has been adduced to be caused by factors like: poor drainage system; blockage of existing drainage system; urban population growth; climate change to increase of impervious surfaces arising from recent developments among others. Owerri urban city, southeast of Nigeria consists of two sections divided by the Nworie River, with Owerri I on the eastern section and Owerri II (New Owerri) on the western section. Whereas the impact of urban sprawl originated from Owerri I, the drift is gradually being felt in Owerri II. One of the principal negative impacts being focused in this study is flood and ways of mitigation through use of smart city techniques. Alawadhi et al. (2012) is of the opinion that the concept of "smart city" is such that it can be deployed towards mitigation and resolution of urban problems, while simultaneously making urban development more sustainable.

Smart city has been defined from diverse perspectives depending on the background of the definer. Alawadhi (2012) examined wide range of definitions of Smart city and highlighted the following findings: That smart city is defined as "a city well performing in a forward-looking way in economy, people, governance, mobility, environment, and living, built on the smart combination of endowments and activities

I. El Dimeery et al. (Eds.): JIC Smart Cities 2019, SUCI, pp. 229–237, 2021.
https://doi.org/10.1007/978-3-030-64217-4_26

of self-decisive, independent and aware citizens" (Giffinger et al. 2007). Whereas Garaliu et al. defines 'smart city' as "when investments in human and social capital and tradition (transport) and modern (ICT) communication infrastructure fuel sustainable economic growth and a high quality of life, with a wise management of natural resources, through participatory governance. Greater emphasis on the application of technology was held paramount by (Washburn et al. 2010) while defining smart city as "The use of smart computing technologies to make the critical infrastructure components and services of a city-which include city administration, education, healthcare, public safety, real estate, transportation, and utilities-more intelligent, interconnected, and efficient". Anavitarte and Tratz-Ryan (2010) further expatiated on the relevance of technologies (ICTs) as they defines it as "an urban area functioning and articulated by modern information and communication technologies in its various verticals, providing ongoing efficient services to its population". However, Vinordkumar (2016) stressed on the relevance of Global Information System (GIS) in upgrading existing cities into smart cities. He further explained that smart components include smart economy, smart governance, smart mobility, smart environment and smart living.

In all the definitions, the consistent strand of components for evolution and development of smart cities could be found in technological domain, whether in terms of planning, road transportation, drainage, lighting, waterways, communication, economy etc. The aggregate of negative impacts of flooding is as shown in Fig. 1 at the heart of Owerri city were water level rose to about 0.45 m.

Fig. 1. Flood incidence at the heart of Owerri city on July 4, 2018

2   Study

The study area covers Owerri II (New Owerri) which is located within latitudes 05″25′ and 05″32′ and longitude 06″57′ and 07″07′. This area spans from south east of Obinze towards Nekede, then along Nworie River slightly north-eastwards and then north-westwards towards Akwakuma, while looping Orogwe, Nzegwu and back to Obinze. The New Owerri has annual total mean rainfall of 2190 mm. (Ministry of Works & Transport, Imo State 1984). The rainfall intensity variance and standard deviation shown in Table 1 below indicates the fluctuation of intensity and high intensity recorded for the last few years indicated. The average height difference of +5.159 m between water table and assumed ground level at Owerri II (New Owerri) from Nworie river source North West to the point of confluence with Otamiri river (AC-Chukwuocha 2017). Figure 1 shows New Owerri on the west of Nworie River, while the old Owerri is located on the eastern part of the Nworie River.

Table 1. The rainfall intensity in New Owerri for 2004–2013 (Nwachukwu 2018)

Year	2004	2005	2006	2007	2008	2009	2010	2011	2012	2013
Intensity in mm/Hr	175.358	208.608	211.533	176	198.992	227.742	176.208	201.083	236.15	164.642

Figure 1 above shows the pattern of rainfall for a period of 10 years, from 2004 to 3013 at the new Owerri. In the table, the pattern indicates that the highest recorded amount of rainfall was in 2012 which was 236.15. Beyond that, there was preponderous of high records of amount of rainfall in later years as the data shows. This can be explained to be as a result of climate change trend. Hence, there is need to adopt new techniques and technologies to check this change. A critical look at the base map as indicated in Fig. 2 shows that the crisscrossing of the Nworie and Otamiri Rivers between new Owerri and old Owerri is a drainage advantage; however, a proper channeling of storm water towards these rivers is required for proper drainage efficiency. An open drain of about 6 m wide and 2.5 m deep which was constructed to drain water into the river has been blocked with silt and sand. This makes it difficult to drain water as required.

Most of this study was done by field observations, documentary evidences, and literature reviews. Case studies of techniques adopted in standard smart cities were carried out and this gave a clearer picture of the means for mitigating flood disasters in smart cities using recent technologies. The identified global best practices were compared with the existing practice being adopted in New Owerri II flood disaster control to discover shortcomings that needs to be improved upon.

Within New Owerri, the drainage system is largely of open channels which are inefficient for a smart city. In some areas, there are no drainage systems or where there are, they are completely blocked with sand or garbage. In some housing estates, the road networks are not more than 6.00 m in width between one property line (fence) and another, with no sufficient space for sidewalks or drainage channels. Residential Houses are mainly made up of single family bungalows, single family duplexes and multiple families' block of flats (medium rise) with no high rise apartment blocks. Due to non-existent/inefficient drainage systems the road spaces suffer serious degradation during rain in conveying storm water and in some cases vulnerable to development of pot-holes.

Fig. 2. Map of Owerri II (New Owerri) and Owerri I (Old Owerri)

2.1 Drainage System

Drainage system is categorized into artificial or natural. There are two main types of drainage systems. They are surface drainage system and subsurface drainage systems. The surface drainage system is usually made up of shallow ditches called open drains as indicated in Fig. 3 below. It could be of either natural or artificial drains. If natural, it means that the excess water runoff flows through farms to swamp or lakes and rivers (Brouwer et al. 1985).

Fig. 3. Open   Drain;   Source: SANIMAS (2005)

Fig. 4. Closed Drain; Source: SANIMAS (2005)

2.2   Surface Drainage

Excess water gathering on the surface of land as a result of rain could be removed by either natural or artificial means. The present scenario in new Owerri area is such that there is surface drainage and even in some areas the open drains have blocked and turning the scenario into surface drainage.

2.3   Subsurface Drainage (Deep Open Drains, Pipe Drains)

Cleaning out open drains whether deep or shallow is difficult to organize but utterly necessary. Blockages can cause spill-over and cause flooding. A solution could be to cover it with concrete slabs as shown in Fig. 4 above. These concrete slabs may be opened at intervals for cleaning. Alternatively, the conventional system of drainage system as indicated in Fig. 5 below may be adopted

Fig. 5. Conventional Drainage System; Source: SANIMAS (2005)

3   Techniques and Technologies for Smart City Drainage Evolution

There are wide ranges of technologies and tools that may be deployed as part of technique for flood disaster control in smart city. They include bioswale, covered drain, concrete pavers, road median, curb grate, sidewalk planters etc. Also pervious concrete pavers over structural soil could be used to direct stormwater runoff to drains through curb grate as shown in Fig. 7 while at same time permit water sippage into soil strata below. The road median with hedges at intervals or sidewalk planters as indicated in Fig. 6 below may also be useful tool for flood control.

Fig. 6. Road Median/Sidewalk Planters Hawkins, K. H. (2009)

Fig. 7. Curb Grate Hawkins, K. H. (2009)

3.1   Bioswale

Bioswales are linear channels designed to concentrate and convey stormwater runoff while removing debris and pollution. Bioswales can also be beneficial in recharging groundwater. Bioswales are typically vegetated, mulched, or xeriscaped. They consist of a swaled drainage course with gently sloped sides (less than 6%). When it is in the green parking lot, it contain single- and multi-stemmed deciduous and evergreen trees, shrubs, and herbaceous plants (Scharenbroch et al. 2016).

3.2   Flood Plain

The definition of floodplain largely depends on the goal in the mind of the definer (Department of Regional and Environmental Executive Secretary for Economic and Social Affairs Organization of Americana States 1991). The goal of the definer could range from topographical, geo-morphogenical, to hydrological and even more. However, Leopold et al. (1964) took a comprehensive perspective in defining it as, "a strip of relatively smooth land bordering a stream and overflow at a time of high water". The floodplain along Nworie River in new Owerri II fluctuates seasonally. Therefore, it will be erratic to use the old or fixed flood plain in working for a solution towards solving the existing flood problem. Hence, there is need to log for flood plain in Owerr II over a period of time to determine the current trend with respect to climate change and be informed on the right strategy to adopt for a solution.

3.3   IOT Flow Sensor

Internet of things (IOT) flow sensor is a device that could be used in monitoring free flow of storm water through the drainage system. It sends signal to the central unit on sensing blockage within the drainage system which in turn triggers alert well ahead of blockage. It also identifies the location of the blockage along the drainage system. This is usually part of automated drainage system. Indeed, most IoT applications are modeled as data transformation workflows that consist of different parts (Nardelli et al. 2017) (Fig. 8).

FLOW SENSOR

Fig. 8. IOT Flow sensor

Table 2. Principal roads in Owerri I and 11 with their flood control measures

	WHETHERAL ROAD	DOUGLAS ROAD	OKIGWE ROAD	PORTHARCOURT ROAD	BANK ROAD	TETLOW ROAD	WORLDBANK ROAD	CONCORD BOULEVARD/ YARADUAH DRIVE	ZUMA DRIVE	SAM MBAKWE AVENUE
CLEARED DRAINAGE SYSTEM			✓							
SURFACE DRAINAGE	✓	✓			✓		✓			
DEEP OPEN DRAINAGE							✓			
IOT FLOW SENSOR										
BLOCKED DRAINAGE SYSTEM	✓	✓		✓	✓		✓	✓	✓	✓
COVERED DRAINAGE										
SUB-OPEN DRAINAGE					✓	✓		✓	✓	✓
ENORMOUS ADJACENT IMPERVIOUS SURFACES	✓	✓	✓	✓	✓	✓				
CURB GRATE										

3.4 Result and Discussions

It has been discovered from the foregoing that the flood incidences in New Owerri II have been lately progressively degenerative. This is as a result of aggregate of numerous challenges ranging from blocked drains; increase of impervious surfaces triggered by uncontrolled urban development; non-supporting of the original design level of the drains run-off/discharge to the present challenges; non usage of existing tertiary collector open drain to non automation of the drainage system in response to smart city status.

Therefore, introduction of curb grates, where there is none and more curb grates, where there are few will go a long way to forestall drain blockages; reduction of overall impervious surfaces will also help in reduction of storm water through seepage into subsoil; climate change will have affected the original flood plain, therefore, determination of recent flood plain is essential to ensure compliance to new techniques and strategies so as to check recent challenges. Result shown in Table 2 indicates that Owerri II falls short of flood control measures in many ways and requires contingency approach to be put back on track.

In the Table 2 referred to in Owerri I and II, there exist no completely covered drains; complete absence of IOT Flow sensor and curb grates in all the roads assessed; 50% of the drains were blocked; only one road has well established linkage to open deep drain; there were enormous adjacent impervious surfaces to highly crowded developed properties which are mainly densely populated.

4   Recommendation and Conclusion

It is recommended that a thorough assessment of the causes of acute flood disaster in New Owerri II is carried out and documented. This will disclose all existing drainage deficiencies in the city. Also to be reviewed are all the techniques in place for drainage vis-à-vis the global best practice requirements for smart cities, so as to infuse them into the building by-laws, where necessary. If the government intends to realize the dream of Owerri II as an upcoming smart city with respect to flood control, the identified measures for control of flood in smart cities must be put in place, while the inhabitants should be sensitized in same direction. The plan for evolution of the existing trend towards smart city status should be clearly drawn. Thereafter, implementation of the drawn plan would be allowed to gradually evolve overtime.

References

AC-Chukwuocha, N., Ngah, S.A., Chukwuocha, A.C.: Vulnerability studies watershed area of Owerri South East Nigeria using digital elevation model. J. Geosci. Environ. Protect. 5 (2017)

Alawadhi, A., Aldama-Nalda, A., Chourabi, H., Gil-Garcia, J.R., Leung, S., Mellouli, S., Nam, T., Pardo, T. A., Scholl. H. J. Walker, S.: Building understanding of smart city initiatives. In: International Federation for Information Processing (2012)

Anavitarte, L., Tratz-Ryan, B.: Market insight: 'smart cities' in emerging markets. Gartner (2010). http://www.gartner.com/id=1468734

Brouwer, C., Goffeau, A., Heibloem, M.: Irrigation water management: training manual no 1- introduction to irrigation. Food and Agriculture Organization of the United Nations, Chapter 2 and 6 (1985)

Department of Regional and Environmental Executive Secretary for Economic and Social Affairs Organization of Americana States: Primer on Natural Hazard Management in Integrated Regional Development Planning, Washington, D.C (1991)

Giffinger, R., et al.: Smart Cities: Ranking of European Medium-Sized Cities, Centre of Regional Science (SRF). Vienna University of Technology, Vienna, Austria (2007)

Hawkins, K.H.: Deadrick Green: Creating Tennessee's First Green Street, Hawkins Partners, Inc., Nashville (2009)

Nardelli, M., Nastic, S., Villari, M., Ranjan, R.: Osmotic flow: osmotic computing + IoT workflow. Blue Skies Department (2017)

Nwachukwu, M.A., Alozie, C.P., Alozie, G.A.: Environmental and rainfall intensity analysis to solve the problem of flooding in Owerri urban. J. Environ. Hazard **1**, 107 (2018)

SANIMAS: SAIMAS-Informed Choice Catalogue (2005)

Scharenbroch, B.C., Morgenroth, J., Maule, B.: Tree species suitability to bioswales and impact on the urban water budget. J. Environ. Qual. **45**, 199–206 (2016)

Vinordkumar, T.M.: Geographical Information System for Smart Cities, p. 3. Copal Publishing Group (2016)

Washburn, D., Sindhu, U., Balaoras, S., Dines, R.A., Hayes, N.M., Nelson, L.E.: Helping CIOs Understanding "Smart City" Initiatives: Defining the Smart City, its Drivers and the Role of the CIO. Forrester Research Inc., Cambridge (2010)

Correlation of Non-destructive with Mechanical Tests for Self-Compacting Concrete (SCC)

Jeovany Amgad, Nancy Hammad$^{(\boxtimes)}$, and Amr M. El-Nemr

Civil Engineering Program, German University in Cairo, Cairo, Egypt
nancy.saad@guc.edu.eg

Abstract. The innovation of self-compacting concrete (SCC) has greatly reduced the high level of energy consumed in vibration to compact the fresh concrete. SCC has the ability to flow under its own weight and achieve full compaction even in congested reinforcement area or in case of complex shapes of concrete structures saving time, labour and energy. This paper investigated SCC mixes produced using kaolin and silica fume as additives to assess the mechanical properties in terms of destructive and non-destructive testing. Four SCC mixtures with different percentage of kaolin and silica fume were considered in this study. Compressive, splitting tensile and flexural strengths as destructive testing were carried out and determined. Furthermore, non-destructive testing such as Schmidt hammer and ultra-sonic pulse (UPV) were assessed and correlated with those of the destructive testing. From results, it was deduced that SCC with 10% kaolin and 15% silica fume achieved the highest compressive and flexural strength after 28 days. Moreover, the results provided the possibility of using non-destructive tests to asses the compressive strength of SCC.

Keywords: Self-Compacting Concrete (SCC) · Mechanical properties · Non-destructive tests · Schmidt hammer and ultrasonic pulse velocity

1 Introduction

Compaction of concrete during placement is a key factor for achieving the design strength and durability. Inadequate compaction could negatively affect the performance of concrete in fresh and hardened state [1]. Innovation of self-compacting concrete (SCC) was in a great need to cope with the evolution of reinforced concrete designs and complex structural shapes [2]. SCC has introduced more benefits rather than conventional concrete such as increased productivity, enhanced construction quality and improved environment on site. The changing properties of concrete required enhanced and reliable testing methods, both in laboratory and in situ. One of these enhanced testing methods are non-destructive tests such as Schmidt hammer and ultrasonic pulse velocity (UPV) [3].

The main advantages of non-destructive tests are avoiding the damage of concrete, simple and quick. Furthermore, based on the fact that some physical properties of concrete could be related to the compressive strength. Schmdit hammer and UPV tests

© The Author(s), under exclusive license to Springer Nature Switzerland AG 2021
I. El Dimeery et al. (Eds.): JIC Smart Cities 2019, SUCI, pp. 238–246, 2021.
https://doi.org/10.1007/978-3-030-64217-4_27

could be used to correlate their reading with the compressive strength of concrete and be an effective tool for concrete inspection [4].

2 Experimental Program

2.1 Materials

Ordinary Portland cement confirming to the ECP 203/2018 [5] with specific gravity of 3.15 and blaine fineness of 378 kg/m^2 was used in this study. Kaolin and silica-fume were added to the concrete mix as partial replacement of cement. The chemical and physical properties of silica-fume and Kaolin are illustrated in Table 1 and Table 2 respectively.

Table 1. Properties of Silica-fume

Property specification	Limits	Silica fume	Reference method
Moisture content	$\leq 3.0\%$	0.6%	AS 3585.2
Loss on ignition	$\leq 6.0\%$	3.4%	AS 3583.3
Sulfuric anhydride	$\leq 3.0\%$	0.3%	AS3583.8
Total silica content SiO$_2$	$\geq 85.0\%$	94.7%	AS 2350.2

Sand obtained from mountain quarries of maxiumum size of 4.75 mm was used and tested according to ECP 203/2018 [5]. The fineness modulus of the sand was 2.6, whereas the water absorption and specific gravity were 1.21% and 2.5 respectively. Coarse aggregate of maximum nominal size of 20 mm was used in this study with water absorption 1.2%, specific gravity 2.5, resistance to abrasion 33.3% and impact value 15.3%. High range water reducing naphthalene sulphonate based superplasticizer was used for production of free flowing concrete. The used superplasticizer is commercially known by Sikament-NN with density of 1.185 kg/L at 20 °C. It complies with ASTM C494 type F [6].

2.2 Mix Proportions

Four SCC mixes were prepared containing different content of kaolin and silica-fume as a partial replacement of cement. The main purpose of those mixes to study the of kaolin and silica fume addition on the SCC. The first and second SCC mix composed of 10% and 20% of kaolin (SCC-10 k/0sf & SCC-20 k/0sf), while the third one contains 10% kaolin and 15% silica-fume (SCC-10 k/15sf) and the fourth one has higher percentage of kaolin (20%) and 15% silica-fume (SCC-20 k/15sf) as a partial replacement (Table 3).

Table 2. Properties of Kaolin

Constituents	Percent by weight
SiO_2	48.0%
Al_2O_3	36.0%
Fe_2O_3	1.0%
K2O	1.5%
Na_2O	0.3%
MgO	0.4%
DO2	0.3%
CaO	0.1%
F_3O_3	0.2%
L.O.I	12.2%
Temperature	1185 °C (8 h)
Contraction	5.5%
Water absorption	14.0%
Modulus of rupture	380 kg/cm^2

Table 3. Mix proportions

Mix	Cement Kg/m^3	Water Kg/m^3	Fine Agg. Kg/m^3	Coarse Agg. Kg/m^3	Koalin Kg/m^3	Silica-Fume Kg/m^3	Superplasticizer Kg/m^3	W/B
SCC-10 k/0sf	495	231	909	612	55	0	10.9	0.42
SCC-20 k/0sf	440	231	909	612	110	0	10.9	0.42
SCC-10 k/15sf	412.5	231	909	612	55	82.5	16.4	0.42
SCC-20 k/15sf	357.5	231	909	612	110	82.5	17.3	0.42

2.3   Test Methods for Assessing the Fresh Properties of SCC

To evaluate the fresh state of the SCC, several tests shall be carried out such as:

1. J-ring test: this test aims at investigating both the filling ability and the passing ability of SCC. It can also be used to investigate the resistance of SCC to segregation. The test was performed in accordance to EN 12350: part 12-2010 [19]. The J-ring test measures three parameters: flow spread, flow time T50J (optional) and blocking step. The J-ring flow spread indicates the restricted deformability of SCC due to blocking effect of reinforcement bars by measuring the diameter of the flow in the two perpendicular directions. While the flow time (T_{50J}) indicates the rate of deformation within a defined flow distance.

2. V- Funnel test: this test is used to determine the filling ability (followability) and the viscosity of SCC. Stability of the mix is determined by measuring the time required to empty the V funnel completely 5 min after filling. This test was carried out in accordance to EN 12350: part 9-2010 [20].

2.4   Tests on the Hardened SCC

The hardened concrete was evaluated through testing the compressive, splitting tensile and flexural strengths. Then non-destructive tests such as schmdit hammer and ultrasonic velocity were used to assess the quality of SCC and correlate the results revealed with those obtaained by the destructive tests.

3   Results and Discussion

3.1   Fresh Properties of SCC

The result of the self-compatibility tests conducted on the four SCC mixes are given in Table 4. The results of the SCC are found to be within the acceptance limit in accordance to the ECP 203/2018. Furthermore, It was revaled that the addition of Kaolin and silica-fume did not affect the timing of the J-ring. However, the diameter of SCC-20 k/0sf showed the largest diameter among tested mixes due to increase in the percentage of kaolin to 20%. Koalin is known by its positive effect on the workability of the SCC [21]. The results of the v-funnel test has indicated that SCC-20 k/0sf and SCC-10 k/15sf have the highest filling ability.

Table 4.   Fresh properties of different mixes

	J-ring			V-funnel
	T_{50J} (s)	D_1/mm	D_2/mm	Time (s)
SCC-10 k/0sf	2	600	620	10
SCC-20 k/0sf	2	730	800	8
SCC-10 k/15sf	2	600	630	8
SCC-20 k/15sf	2	600	600	11
Acceptance	2–5	600–800	600–800	6–12

3.2   Hardened Properties of SCC

3.2.1   Compressive Strength

The results of the compressive strength test of SCC mixes after 7 and 28 days are illustrated in Table 5. The results showed that the addition of kaolin did not enhance the compressive strength after 7 and 28 days. On contrary, the addition of silica-fume had a positive effect on the compressive strength of the SCC after 28 days that reached 34.50 MPa with an increase of 21.0%. It is clear that the addition of silica-fume promoted the consumption of $Ca(OH)_2$ released during the hydration process leading to formation of calcium silicate hydrate (C-S-H). The formation of C-S-H contributed to the strength of interfacial transition zone between the matrix and the aggregate particles [3].

Table 5. Hardened properties of different mixes

Mix	Compressive strength (MPa)		Rebound number	Tensile strength (MPa) after 28 days	Flexural Strength (MPa) after 28 days	Ultrasonic pulse velocity (m/s)
	Compressive test (7 days)	Compressive test (28 days)				
SCC-10 k/0sf	22.76	28.51	39	2.26	6.90	4530
SCC-20 k/0sf	16.18	23.73	35	2.31	5.94	3750
SCC-10 k/15sf	25.66	34.50	42	2.97	8.72	4651
SCC-20 k/15sf	21.91	34.39	41	2.3	9.05	4450

Schmidt hammer test was applied on concrete cubes to correlate between the averaged compressive strength by crushing the specimens and the corresponding averaged rebound hammer number of the SCC. Figure 1(a) shows a good correlation between the compressive strength after 28 days and the rebound number. Power regression was used for establishing the most suitable relationship between the compressive strength of the SCC and the corresponding rebound number as illustrated in Eq. (1) with coefficient of determination $R^2 = 0.9651$. The obtained relationship is compared with the previously proposed relationships by other reaserchers as shown in Fig. 1(b). The comparsion shows that the obtained model followed the same pattern of the previously developed ones. The slight variation between these different models are the different concrete mixes and admixtures, concrete surface and different regression models.

$$f_{ck} = 0.0105 \, N^{2.1679} \tag{1}$$

Where f_{ck} is the compressive strength in MPa and N is the rebound number.

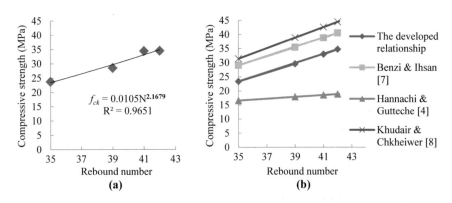

Fig. 1. (a) Correlation between compressive strength and rebound hammer number of the SCC (b) Comparison between the developed compressive strength and rebound number model and the other reaserchers models

3.2.2 Splitting Tensile Strength

Table 5 presents the splitting tensile strength of SCC mixes after 28 days. It is clear that increasing the percentage of kaolin addition slightly enhanced the splitting tensile strength. In contrast, the addition of silica-fume by 15% has significantly increased the splitting tensile strength of SCC-10 k/15sf mix by 31.4% to reach 2.97 MPa. In order to obtain a suitable relationship between the averaged compressive strength and the corresponding averaged tensile strength of the SCC, polynominal regression (Eq. (2)) was selected after several trials to increase the degree of accuracy and reaching a coefficient of determination (R_2) of 0.9141 (see Fig. 2(a)).

$$f_{ct} = 0.0085f_{ck}^2 - 0.4561f_{ck} + 8.2495 \qquad (2)$$

Where f_{ct} is the splitting tensile strength in MPa and f_{ck} is the compressive strength in MPa.

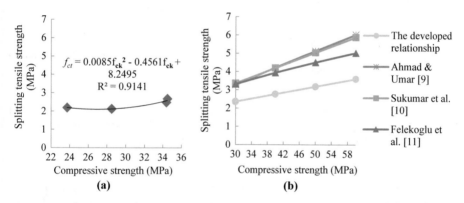

Fig. 2. (a) Correlation between the compressive strength and the splitting tensile strength of SCC (b) Comparison between the developed the splitting tensile strength and the compressive strength model and the other reaserchers models

The relationship developed between the compressive strength and splitting tensile strength of the SCC after 28 days showed a good agreement with the relationships developed by other researchers with some variation as shown in Fig. 2(b). This variation is attributed to different SCC mix components such as kaolin, silica-fume, different strength grade and the different regression models adopted by the other researchers [9].

3.2.3 Flexural Strength

The results revealed that the addition of silica fume as a partial replacement of cement of 15% has enhanced the flexural strength after 28 days by 26.3% and 52.3% for SCC-10 k/15sf and SCC-20 k/15sf mixes respectively (see Table 5). Based on the results of the compressive and the flexural strengths tests, the graph shown in Fig. 3(a) was plotted to develop a suitable non-linear relationship as shown in Eq. (3).

$$f_{ct} = 2.3704e_{ck}^{0.0382f} \tag{3}$$

Where f_{cf} is the flexural strength in MPa and f_{ck} is the compressive strength in MPa.

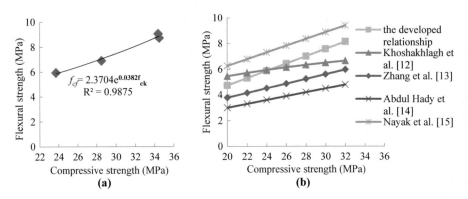

Fig. 3. (a) Correlation between the flexural strength and the compressive strength of SCC (b) Comparison between the developed the flexural strength and the compressive strength model and the other reaserchers models

The proposed relationship is agreed with the relationships given by other researchers (see Fig. 3(b)). The percentage of error of the proposed equation with respect to the tested results is found to be 1.25% ($R^2 = 0.9875$).

3.2.4   Ultrasonic Velocity Test (UPV)

UPV test usually provides an indication of the concrete homogeneity through measuring the wave propagation in the concrete specimen. UPV is extermly affected by the presense of pores, flaws and voids. For that defects could be easily recogonized by using the UPV test [16]. SCC mixes showed high UPV after 28 days which reflects the high quality of the mixes as illustrated in Table 5. The percentage of voids is minimized with addition of kaolin and silica-fume that improved the interfacial trasition zone of the concrete [3]. SCC-10 k/15sf recorded the highest UPV which was 4651 m/s.

In order to correlate the averaged compressive strength to the averaged UPV readings, exponential regression (Eq. (4)) was used with determination coefficient $R^2 = 0.7427$. Figure 4(a) shows the relationship between the compressive strength after 28 days and the UPV. This developed relationship is compared with the previously propoesed relationship by other reasearchers as illustrated in Fig. 4(b). The obtained relationship is found to be within the range of the other's researchers relationships.

$$f_{ck} = 5.7455e^{0.0004v} \tag{4}$$

Where f_{ck} is the compressive strength in MPa and v is the UPV in m/s.

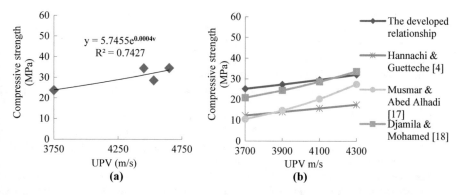

Fig. 4. (a) Correlation between compressive strength and UPV of SCC (b) Comparison between the compressive strength and UPV model and the other reaserchers models

4   Conclusion

On the basis of the experimental work presented in this study, the following conclusions could be drawn out:

1. The addition of kaolin by 20% has positively affected the workability of the SCC.
2. The addition of silica-fume motivates the formation of calcium silicates hydrates which promotes the compressive strength of the SCC.
3. The addition of kaolin by 10% and silica-fume by 15% as a cement replacement produced the highest compressive and splitting tensile strengths of SCC after 28 days.
4. Forward and useful mathematical relationships were developed to correlate between the compressive and the splitting tensile strengths as well as the compressive strength and the flexural strength of the SCC with high coefficient of determination
5. Non-destructive tests such as schmdit hammer and UPV could be used to statisfactorily asses the compressive strength of SCC through using the derived equations.

References

1. Ahmad, S.: Effect of glass and polyvinyl alcohol fibres on fresh and hardened properties of self-compacting concrete. M.Tech, Department of Civil Engineering, Aligarh (2014)
2. Okamura, H.: Self compacting high performance concrete – ferguson lecture. Concr. Int. **19** (7), 50–54 (1996)
3. Ulucan, Z.C., Turk, K., Karatas, M.: Effect of mineral admixture on correlation between ultrasonic velocity and compressive strength for self-compacting concrete. Russ. J. Nondestr. Test. **44**(5), 367–374 (2008)
4. Hannachi, S., Guetteche, M.N.: Application of the combined method for evaluating the compressive strength of concrete on site. Open J. Civ. Eng. **2**, 16–21 (2012)

5. ECP 203-2008: Egyptian Code of Practice for Design and Construction of Reinforced Concrete Structures. Ministry of Housing, Utilities & Urban Development, Egypt (2008)
6. American Society for Testing and Materials 2011, Standard Specification for Chemical Admixtures for Concrete, ASTMC494, USA
7. Benzi, D.K.H., Ihsan, M.A.: Estimating strength of scc using non-destructive combined method. In: The Third Conference on Sustainable Construction Materials and Technologies, Japan (2013)
8. Khudair, J.A., Chkheiwer, A.H.: Non-destructive tests on self compacting concrete made with local materials in basrah. Wasit J. Eng. Sci. 4(1), 163–177 (2016)
9. Ahmad, S., Umar, A.: Characterization of self-compacting concrete. In: International Symposium on Plasticity and Impact Mechanics, Implast (2017)
10. Sukumar, B., Nagamani, K., Raghavan, R.S.: Evaluation of strength at early ages of self-compacting concrete with high volume fly ash. Construct. Build. Mater. 22(7), 1394–1401 (2008)
11. Felekoglu, B., Turkel, S., Baradan, B.: Effect of water/cement ratio on the fresh and hardened properties of self-compacting concrete. Build. Environ. 42(4), 1795–1802 (2007)
12. Khoshakhlagh, A., Nazari, A., Khalai, G.: Effects of Fe2O3 nanoparticles on water permeability and strength assessments of high strength self-compacting concrete. J. Mech. Sci. Technol. 28(1), 73–82 (2012)
13. Zhang, Y., Che, J., Xia, G., Wang, D., Guo, W.: Relationship between compreesive strength and flexural strength of PVA-DSECC. Hans J. Civ. Eng. 7(6) (2018)
14. Abdul Hady, M.A., A, A.A., El-Ghazaly, H.A.: The effect of coarse aggregate types on properties of self compacting concrete. J. Mech. Sci. Technol. 14(3), 53–57 (2017)
15. Nayak, R.R., Alengaram, U.J., Jumaat, M.Z., Yusoff, S.B., Alnahhal, M.F.: High volume cement replacement by environmental friendly industrial by-product palm oil clinker in cement – lime masonry mortar. J. Clean. Prod. 190, 272–284 (2018)
16. Heba, A.: Effect of fly ash and silica-fume on compressive strength of self-compacting concrete under different curing conditions (2011)
17. Musmar, M.A., Abed Alhaldi, N.: Relationship between ultrasonic pulse velocity and standard concrete cube crushing strength. J. Eng. Sci. 36(1), 51–59 (2008)
18. Djamila, B., Mohamed, G.: The use pf non-destructive tests to estimate self-compacting concrete compressive strength. In: MATEC Web of Conference, vol. 149, Algeria (2018
19. EN 12350-12:2010: Testing fresh concrete - Part 12: Self-compacting concrete - J-ring test
20. EN 12350-9:2010: Testing fresh concrete - Part 9: Self-compacting concrete - V-funnel test
21. Lenkaa, S., Panda, K.C.: Effect of metakaolin on the properties of conventional and self-compacting concrete. Adv. Concr. Constr. 5(1), 31–48 (2017)

Investigating Barriers to Implement and Develop Sustainable Construction

Mostafa Namian[1]([✉]), Ahmed Al-Bayati[2], Ali Karji[3],
and Mohammadsoroush Tafazzoli[4]

[1] East Carolina University, Greenville, NC, USA
Namianml9@ecu.edu
[2] Lawrence Technological University, Southfield, MI, USA
[3] Penn State University, State College, PA, USA
[4] Washington State University, Pullman, WA, USA

Abstract. The construction industry plays a crucial role in a sustainable development. The United States Green Building Council (USGBC) has traced almost 40% of Carbon Dioxide emissions in the United States to the construction industry. As the world's population grows, the demand for housing, infrastructures, and other facilities grows as well. If trends continue as they are now, CO_2 emissions initiated by the construction industry will continue to grow exponentially significantly endangering our planet. Therefore, implementing and developing sustainable practices in construction is critical to curbing the rising CO_2 emissions, global warming, and environmental crisis to build sustainable and smart cities. Sustainable design and construction aim for efficiency and low operating costs at the life cycle of projects. Sustainable buildings and sustainability in construction have received much attention in recent years. Although statistics show a steady increase in the popularity of sustainable projects, the construction industry is lacking a large movement in sustainability. There are many obstacles and barriers in the way of future development for green living in smart cities. In order to widely implement sustainable practices and construct sustainable buildings, these barriers need to be identified, investigated, and tackled. The current study aims to investigate the barriers to the implementation and development of sustainable construction. Thirty construction professionals in the US were interviewed to find the most critical barriers of the sustainable construction. Fifteen obstacles were identified among which financial constraints, design constraints, inadequate technology are among the top factors. The results of this article can be used by construction professionals and policymakers to accelerate broad implementation of sustainability in construction projects. By implementing sustainability, the construction industry will take a huge step to protect our environment for the next generation and also to the development of smart green cities.

Keywords: Construction sustainability · Smart cities · Sustainable development · Sustainable design · High-performance buildings · LEED · Sustainable barriers

I. El Dimeery et al. (Eds.): JIC Smart Cities 2019, SUCI, pp. 247–252, 2021.
https://doi.org/10.1007/978-3-030-64217-4_28

1 Introduction

Today, sustainable development has become one of the global concerns around the world. Based on World Commission on Environment and Development (WCED), sustainable development is "a development that meets the needs of the resent without compromising the ability of future generation to meet their own needs" [1]. Sustainable development has been particularly tangled to the construction industry [2]. Sustainable construction has a great impact on our society by playing an important role in the environmental protection and the development of smart cities [3]. The United States Green Building Council (USGBC) has reported that 39% of Carbon Dioxide emissions in the United States are produced by buildings [4]. Furthermore, buildings are responsible for 70% of electricity consumption, and 40% of the world's raw materials [5]. Without immediate interventions, CO_2 emissions from buildings will continue to grow exponentially. Sustainable construction practices are critical to curbing the rising CO_2 emissions and building smart cities. Therefore, implementing and developing sustainable practices in construction are critical to curbing the rising CO_2 emissions, global warming, environmental crisis and to the development of smart cities [6]. Statistics dating back to 2009 show a steady increase in popularity of projects achieving LEED accreditation. By 2009, more than 200 cities, counties, and towns in the US had implemented some form of LEED initiative [7].

In fact, sustainable construction efforts have received much attention in recent years. However, green building initiatives are still new [8]. The construction stakeholders aiming to adopt green buildings are now facing new challenges associated with sustainable construction. These emerging challenges must be resolved; otherwise, they can hinder or eliminate the accomplishment of the primary project goals [9]. In order to foster sustainable construction, it is critical to first identify the barriers and risks that may inhibit the acceptance and execution of green building designs and construction projects. Previously researchers have mentioned challenges associated with the construction of sustainable buildings. Rafindadi et al. [10] showed that political conditions pose a high risk to sustainable construction projects as a general condition. They also determined factors within the project life cycle that play key roles in sustainable construction projects are project complexity, designers' skills, financial resources, and human performance. Bamgbade et al. [11] found that regulatory factors are one of the main influencers on sustainable construction and design in Malaysia. Despite previous efforts, a comprehensive study to find the critical obstacles of expanding sustainable construction has not been studied yet. This study presents the first empirical effort to identify key obstacles that must be tackled to accelerate sustainable construction to build smart cities.

2 Research Methods

In order to identify key obstacles of sustainable construction, the research was conducted in two phases. In Phase I, an extensive literature review was accomplished to identify potential barriers to achieve the desired sustainability in construction which is essential to expand smart cities [11, 12]. After integrating the data, the researchers identified fifteen elements (see Table 1). These 15 elements were used in Phase II in order to be validated using construction professionals' perceptions. The researchers contacted 30 construction experts currently employed in the US to ask their insights to identify contemporary challenges that the construction industry is facing to achieve sustainable construction.

Table 1. Identified barriers to sustainable construction

#	Sustainable construction obstacle
1	Design constraints
2	Financial constraints
3	Improper contract method
4	Inadequate proactive plans
5	Inefficient legal framework
6	Inefficient technology
7	Insufficient commitment of upper-level management
8	Insufficient environmental competencies
9	Lack of awareness amongst stakeholders
10	Lack of employee welfare package
11	Lack of sustainable waste management
12	Lack of worker's training on sustainable operations
13	Management
14	Political impacts
15	Preferences of suppliers/institutional buyers

The experts were contacted via email. One reminder was sent to encourage all experts to participate in the study. The participants' responses were obtained within six weeks. Overall 25 out of 30 invited experts participated in the study (response rate = 83.33%). All the participants identified themselves knowledgeable about sustainability in construction. In addition, three out of 25 participants asserted they have a valid LEED certificate.

The participants filled out a questionnaire including a list of 15 obstacles. The participants were asked to evaluate each obstacle and validate the applicability of each barrier to the construction industry based on their experience and expertise. Moreover, the participants were asked to suggest any other obstacles that were important in their opinion but were not included in the questionnaire. Although the construction experts in a few cases, provided specific comments on some of the barriers, no new obstacle was recommended by the experts to be considered. In other words, the interviewed experts endorsed the comprehensiveness of the list of barriers.

3 Results and Analysis

As mentioned before, 25 construction experts participated in the study. They expressed their opinion using a binary scale. The participants rated each identified barrier either "1: applicable barrier" or "0: not applicable barrier." The number of responses selected to be "1: applicable barrier" was added for each barrier and the percentage of applicability was calculated using Eq. 1.

The data were integrated and analyzed using Microsoft Excel. The results show that there is unanimous agreement among experts on design and financial constraints along with the inefficiency of technology. In other words, design, monetary, and technological barriers are critical to enhancing sustainability in construction.

In contrast, "improper contract method", "inefficient legal framework" and "lack of employee welfare package" are not highly applicable as sustainable development barriers in construction (AI <50%). Figure 1 shows the AI of all studied elements.

$$AI_i = N_i/T \times 100 \tag{1}$$

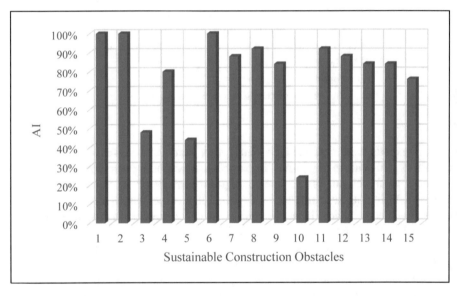

Fig. 1. Applicability index of obstacles to contemporary construction

Where AI_i is the applicability index expressed as a percentage for obstacle$_i$;
N is the number of participants who asserted that obstacle$_i$ is applicable; and
T is the total number of participants which is 25.

4 Discussion

The participated experts endorsed the comprehensiveness of the list of identified barriers. An AI equal to or above 80% was assigned to eleven barriers. Only three obstacles received an AI below 50%. However, three different participants raised concerns regarding "Inadequate proactive plans", "Lack of employee welfare package", and "Management" that they are not specific enough in order to be evaluated by them. One expert mentioned that Lack of employee welfare package cannot be addressed by the construction practitioners "due to requirements by law". In other words, the expert asserted that "Lack of employee welfare" is out of the scope of this study. Finally, another construction expert evaluated "Improper contract method" as a "not applicable barrier" and further explained that "contracts can be written with the same framework as a typical build[ing]".

Further research must be conducted to study the relative importance of the identified factors in order to able construction professionals and practitioners to focus on the most influential factors to accelerate the expansion of sustainable practices to more construction projects.

5 Conclusions

Smart cities are built of sustainable construction projects. Although sustainable construction has received much attention in recent years, there are not widespread. There are several challenges and barriers that hinder the expansion of sustainable construction. The first step to aim for smart and sustainable cities is to identify barriers to sustainable construction. The current study is an empirical effort to comprehensively identify barriers to implement and develop sustainable construction.

In order to achieve the objective of the study, an extensive literature review along with conducting interviews with 25 construction experts currently employed in the US revealed the most critical obstacles that must be tackled to foster sustainability in construction practices. Overall, 15 barriers were identified out of which design, monetary, and technological constraints are critical to enhancing sustainability in construction. The results of the current study can help construction professionals and practitioners and also managers and government to find the most effective strategies to address current barriers to help and foster sustainable construction. Further studies must be conducted to identify the relative importance of the identified factors given the characteristics of each particular construction projects.

References

1. WCED: World Commission on Environment and Development: Our Common Future. Oxford University Press, Oxford (1987)
2. Karji, A., Woldesenbet, A., and Khanzadi, M.: Social sustainability indicators in mass housing construction. In: 53rd ASC Annual International Conference Proceedings on Proceedings, pp. 762–769. ASC, Seattle (2017)

3. Karji, A., Woldesenbet, A., Khanzadi, M., Tafazzoli, M.: Assessment of social sustainability indicators in mass housing construction: a case study of mehr housing project. Sustain. Cities Soc. **50**, 101697 (2019)

4. Abergel, T., Dean, B., Dulac, J., Hamilton, I.: Towards a zero-emission, efficient and resilient buildings and construction sector. Global Alliance for Building and Construction. 2018 Global Status Report. United Nations Environment Programme, WGBC, Katowice, Poland (2018)

5. Holowka, T.: USGBC: LEED–immediate savings and measurable results. Environ. Design Constr. **10**(7), 13–18 (2007)

6. Bortscheller, M.: Equitable but ineffective: how the principle of common but differentiated responsibilities hobbles the global fight against climate change climate law reporter. Sustainable Development Law & Policy **10**(2), 49–69 (2009)

7. To LEED or not to LEED. http://search.proquest.com/docview/195915980. Accessed 15 Oct 2019

8. Tafazzoli, M., Nochian A., Karji, A.: Investigating barriers to sustainable urbanization. In: International Conference on Sustainable Infrastructure 2019 on Proceedings, pp. 607–617. ASCE, Los Angeles (2019)

9. Andrea, O., Owusu-Manu, D., Edwards, D., Holt, G.: Exploration of management practices for LEED projects: lessons from successful green building contractors. Struct. Surv. **30**(2), 145–162 (2012)

10. Rafindadi, A., Mikic, M., Kovačić, I., Cekić, Z.: Global perception of sustainable construction project risks. Proc. – Soc. Behav. Sci. **119**, 456–465 (2014)

11. Bamgbade, J., Kamaruddeen, A., Nawi, M., Qudus, A., Salimon, M., Ajibike, W.: Analysis of some factors driving ecological sustainability in construction firms. J. Clean. Prod. **208**, 1537–1545 (2019)

12. Rieckmann, M.: Education for sustainable development goals: learning objectives. UNESCO, Paris (2017)

Potential of Straw Block as an Eco - Construction Material

Manette Njike[1(✉)], Walter O. Oyawa[2], and Silvester O. Abuodha[3]

[1] Pan African University Institute for Basic Sciences,
Technology and Innovation, Nairobi, Kenya
Manette.njike@yahoo.fr
[2] Jomo Kenyatta University of Agriculture and Technology, Nairobi, Kenya
[3] University of Nairobi, Nairobi, Kenya

Abstract. Interest of using straw bale as construction material has increased worldwide. This result from the need of developing building envelopes which are climate responsive and can significantly reduce building's energy consumption. Research on straw bale has shown that straw bale has good thermal conductivity while plastered straw bale assemblies has good mechanical properties. Up to date, straw bale construction consists of stacking straw bale in a running bond and use different techniques to push down straw bale wall before plastering them. No clue has been given if this method is structurally beneficial than to stabilized single straw bale before assembling them into a structural panel. This paper presents a method of construction that consist of manufacturing straw blocks before using them in masonry. Blocks of dimension $29 \times 14 \times 14$ mm were manufactured using chopped straw with a natural binder. The average compressive strength and density of blocks are respectively 1.25 MPa and 522 kg/m^3; which are respectively 73 and 5 times greater than that of straw bale. Also the average thermal conductivity of straw block and straw are similar (0.06 W/mK). Thus the use of straw blocks will improve the structural performance of straw houses.

Keywords: Straw block · Compressive strength · Density · Thermal conductivity · Water absorption

1 Introduction

Straw bale is made of residue material from agricultural crop (rice, wheat). The bales are used for construction of load bearing houses. The construction of this type of house generally consist of stacking straw bale on top of each other in a running bond, using different technique to compress the assemblies. After roofing, lime-cement or earth based plaster is applied. Straw bale is an environmental friendly material; it provides significant benefit in term of cost and human health [1]. Moreover, straw bale walls offer excellent thermal insulation; their thermal property is around 0.06 W/mK [2]. For a 450 to 500 mm thick walls, the thermal transmittance (U-values) is about 0.13 to 0.19 W/m^2 K depending on bale density and thermal conductivity [3]. The thermal properties and thermal transmittance show how effective a given construction material

© The Author(s), under exclusive license to Springer Nature Switzerland AG 2021
I. El Dimeery et al. (Eds.): JIC Smart Cities 2019, SUCI, pp. 253–261, 2021.
https://doi.org/10.1007/978-3-030-64217-4_29

can prevent heat transfer in the building. The lower these values, the more energy performant that material will be. In addition, the reported maximum load for plastered straw bale walls is 66 KN per running meter [4].

Made of organic material, straw bale durabilities properties (fatigue, decay under certain condition) can influence the durability of structure. Under its own weight and roof weight, load bearing straw bale wall can lose its original height in few weeks thus, resulting in susceptible damage in the structure. In response to this, techniques suchs as the use of threaded rod through the bales at intervals of 1.8 m, use of stucco mesh sheet on the faces of wall [5] or use of tensioning cable [6] to tie the top plate to the foundation while pushing down straw bale assemblies are employed. However, the mechanic of straw bale wall is complex in a way that; when the cables or wire are removed, straw bale wall gain its initial height; but if straw reach the fatigue stage, the tensioning wire will no longer perform its function, thus will be useless in the structure.

It is also known that straw bale has a low compressive strength and low density. These are the parameters that make straw not to be accepted by many people, thus the difficulty to obtain permit for straw bale construction [7]. Most studies in straw bale house are focused on the strength of plastered straw bale assemblies. There are few data on the strength of straw bale unit. Lina Nunes [8] mentioned that compressive strength of straw bale is around 0.017 MPa; 2000–5000 times lower than the compressive strength of concrete. Generally, platered straw bale forms a composite structure that works like a sandwich panel. Plastered straw bale assemblies gain its strength and stiffness from the render coats [9, 10]. Thus the strength of plastered straw bale panel is between 0.8 and 1.03 MPa [11].

From the above, pre-compression is an important step in the process of straw bale house construction. In our knowledge, the structural performance of masonry is based on the characteristics or the properties of its constituents material (brick and mortar, straw bale or block). Further research is required to investigate the potential of stabilized straw bale (straw block) in solving the problem of low strength, low density and displacement in the straw bale assemblies. This study consists of manufacturing straw blocks of improved density and compressive strength compared to that of straw bale. Straw bale is made of straw stems tie down with two or three strings while straw block is made of chopped straw stems mixed with natural binder to form a solid material.

2   Materials and Methods

2.1   Materials

Material used to manufacture straw blocks were wheat straw, Rhodes grass, saw dust, gum Arabic and tap water.

Wheat straw bale and Rhodes grass (Chloris gayana kunth) from Nakuru, the western part of Kenya were used in this study. The bales were kept in the laboratory under a temperature of 24 °C. The moisture content were recorded using straw bale moisture meter. The physical properties of straw bale are presented in Table 1. Saw dust was collected from a local carpenter and particles passed through sieve size 1.2 mm.

Gum Arabic is a natural gum from the hardened sap of acacia trees. Acacia Senegal's Gum from Isiolo, northern part of Kenya was used. Gum Arabic was crushed and passed through a sieve size of 1.2 mm. Chemical test was conducted on gum Arabic using BRUKER S1TITAN 600. Results of this test are presented in Table 2.

Table 1. Properties of wheat straw and Rhodes grass bale

Properties	Wheat straw	Rhodes grass
Average moisture content (%)	10	13
Average dry density (kg/m^3)	85.07	103.76
Average comp. strength (MPa)	0.011	0.017

Table 2. Chemical properties of gum Arabic

Parameters	Value in %	Parameters	Value in %
MgO	27.312	Fe	0.385
Al2O$_3$	2.179	K$_2$O	26.165
CaO	43.396	Mn	0.226

2.2 Method

Binder Preparation

The binder preparation consisted of producing homogenous glue from the mixture of hot water and gum Arabic. Four ratio (1:0.5, 1:0.75, 1:1 and 1:1.25) per volume were investigated. Ratio 1:0.5 flew easily through straw stems but could not hold them well together and after 3 days, mold appeared on the blocks. Ratio 1:0.75 also flew freely through straw stems, could hold straw stem together but during curing straw particle could come out. Ratio 1:1 flew well through straw stems and the block looks better that made of ratio 1:0.75. Ratio 1:1.25 did not flow well through straw stems, the mixture was thick and made it difficult to mix with straw.

Straw Preparation

Straw stems were chopped in small length (1 to 8 cm). A bucket of 10 L was used for measurement. Three mixtures were made with water to gum ratio of 1:1. Mixture A was made of 2 buckets (0.658 kg). Mixture B was made of 2.5 buckets (0.823 kg). Mixture C was made of 3 buckets (0.987 kg). The volume of mixture B was more accurate to the mould's volume thus was used to manufacture blocks.

Straw Block Manufacturing

A cubic meter for this mixture requires 215 litter of water, 244 kg of gum Arabic and 89 kg of chopped straw.

Three types of blocks were manufactured. Blocks type 1 were made using wheat chopped straw; chopped Rhodes grass were used for blocks type 2. For blocks type 3, wheat straw were partially replaced with saw dust in order to include fine particle in the blocks; this type of blocks were made of 80% wheat straw and 20% saw dust. Blocks were manufactured using hand compression machine. Twelve blocks were made for each types of blocks and they were kept in the open air for curing. After 1 month and half, the density and weight (W1) of each block were recorded. In order to verify that straw blocks were dry, the blocks were then put in drying oven (105°c) for 24 h then the dry weight (W2) was also recorded. Primary engineering properties of materials (density, Compressive strength, water absorption and thermal conductivity) were recorded. For each type of block and each test, 3 blocks were tested and the average values are reported in Table 3.

Testing Methods
compression test was conducted as per BS EN 12390 part 3: 2002. AMSLER'S compression testing apparatus was used. Also a Load cell of type CLP-50B, capacity 50tf and sensitivity 2 mV/V; 2 dial gauge displacement transducers type DDP-30AS2; of capacity 30 mm to measure transversal and longitudinal displacement in the blocks. Data were recorded from a data logger in an interval of 5 s. 2 methods were used to record displacement in the blocks. The first recording was automatically with the displacement transducers. When the transducer reached its maximum (30 mm), further displacements were recorded manually at 3 different loads: 20 KN, 50 KN and 100 KN (Fig. 1). Loading was carried out manually in small increment.

Water absorption test was conducted to determine the moisture absorption percentage of straw blocks samples after 24 h. The test was conducted as per BS EN 772 part 11:2010.

Thermal conductivity test was conducted in order to investigate the block's insulation capacity. Due to the current environmental and energy concerns (the extreme outdoor temperature and its related indoor temperature), a good indoor climate is important for the success of any building; it makes its occupants comfortable, and also decides on its energy consumption, thus influences building's sustainability [12]. The method of hot plate was used under a tension of 50 volts. The intensity (A), cold temperature (°C) and time (s) were recorded at every interval of 20 °C (hot temperature).

3   Results

Table 3 represents the characteristiques of straw blocks while Fig. 1 and 2 are respectively load-displacement and stress-strain curve.

Table 3. Characteristiques of different type of straw block

Block sample	Average bale/block dimensions (mm)	Average weight (Kg)	Average density (Kg/m^3)	Average. Water absp. (%)	Average Thermal cond. (W. m^{-1}K^{-1})	Average Compressive strength* (MPa)
Wheat straw block	290 × 140 × 140	3.08	541.87	5.1	0.06	1.25
Rhodes grass block		2.988	525.69	11.2		
Wheat straw – saw dust block		3.445	606.09	4.3		

* compressive strength at the point where the blocks or bale gain stiffness at 50KN for straw blocks and 10KN for straw bale as shown on stress strain curve.

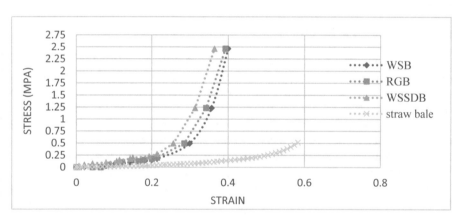

Fig. 1. Stress-strain curve

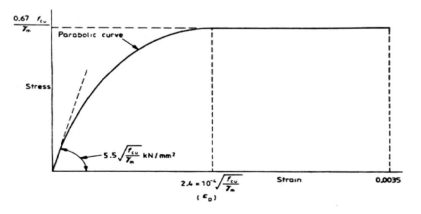

Fig. 2. Strsess-strain curve of concrete [13]

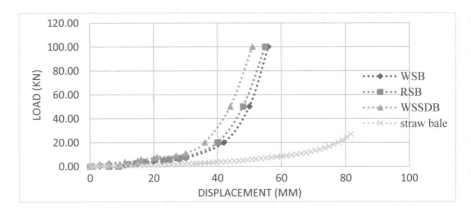

Fig. 3. Load-displacement curve

4   Discussion

4.1   Compressive Strength

Compressive strength was conducted in order to measure the displacement and resistance of straw block under compression. Result obtained shows that in the first stage of loading wheat straw saw dust block (WSSDB) has high compressive strength and small displacement compare to those of wheat straw block (WSB) and Rhodes grass block (RGB). WSB and RGB exhibited almost similar strength and displacement. In the second stage of loading, the three types of blocks have the same strength for the same load. Moreover, it is recorded that as the load increase straw block gains stiffness and becomes more compact or harder with small displacement in the block. Zhang [14] reproded that when the curve start going upward vertical axis (Fig. 3), straw bale became compact and harder to be crushed. At this point there is no voids in the bale.

As shown on Fig. 2, the compressive strength of straw blocks is 1.25 MPa when the blocks gain stiffness. This value is greater than 1 MPa found from similar study by Colsom et al. [15]. They stated that the compressive strength of straw cube made of chopped straw mix with bio binder is around 1 MPa. furthermore, the compressive strength of straw block is 73 times greater than the strength (0.017 MPa) of straw bales currently used in construction. It is atated that the strength of plastered straw bale panel is between 0.8 and 1.03 MPa [11]. This means that plaster plays a significant role for structural capability of straw bale houses. Besides plastered straw bale assemblies behaves as a composite panel which gain its strength and stiffness from the render coats [9, 10]. Since the required strength for an ordinary single-storey building wall is about 0.2 MPa, straw blocks can be used in this type of construction. Furthermore, depending on the designer or the design strength of straw block house, the width of straw block can be increased or the method of solid wall (masonry wall with 2 layers tickness) can be used in load-bearing straw block wall. The strength of masonry depends also on the thickness of wall, thus increasing the thickness of the wall will also increase its load carrying capacity.

4.2   Load Displacement Characteristics

The stress-strain curve represents the relationship between the resultant stress calculated from the applied load and deformations in different types of straw blocks. Stress-strain curve represented on Fig. 1 shows the behavior of straw blocks under compression loaded up to 100 KN. Under compression load, straw block behavior is different from that of concrete or soil block under the same condition. Generally at maximum stress concrete or soil block crushed under compression (Fig. 2) while straw blocks gain strength as the load increases (Fig. 1). Zhang [14] reported that when the curve start going upward vertical axis, straw bale became compact and harder to be crushed. As stated by Sutton et al. [9], the low compressive strength of straw block is governed by the displacement rather than the block failure. In reality straw stem is like pipes and a straw bale is made of many stems. Displacement in straw bale is due to voids between straw stems and then those in each straw stems. In straw block, most of the voids between straw stems are filled with the binder. Straw block or bale became compact and stiffer when voids are removed during the compression. For the serviceability consideration, further study is required to investigate the deflection in a plastered straw block assemblies and compare it with allowable deflection in a structural wall.

4.3   Failure Patterns

The failure patterns of Wheat Straw-Saw Dust Block is presented on Fig. 4. No vertical cracks appeared throughout the test, however at the middle of all types of blocks, horizontal crack were observed. This may represent the layer of material during block manufacturing. In fact, chopped straw stems are like fiber, to ensure that it was well sprayed, the mix design was put in 4 layers in mould before compression. Straw bale is a flexible material, it does not crush under compression, compression push air outward and the bale become compact with reduced height. But when the load is removed on the straw bale, it gains its initial height in a short time. This behavior is not observed in the straw block. The binder kept straw stems together and formed a solid material. The displacement in transversal direction was higher in wheat straw saw dust block (7 mm) than in wheat straw block and Rhodes grass block (3 mm).

(a)   (b)

Fig. 4. Wheat straw saw dust block blocks (a) and respective failure pattern (b)

4.4 Water Absorption and Durability

The aims of water absorption test was to investigate the resistance of straw blocks to water. Results obtained from this test shown that there is little difference between WSB (5.1%) and WSSDB (4.3%) water absorption. RGB absorbed more water (11.2%) than other types of blocks. It was observed that Straw stems in the straw block did not absorb water until all the gum Arabic around them dissolved. The water absoption of wheat straw – saw dust block and wheat straw block is smaller the minimum water absorption (7%) for class B Engineering bricks as per BS EN 771-1: 2003. In the other hands, the water absorption of the three types of straw blocks are lower than 12%, the minimum value for class A engineering bricks as per IS 3495: 1992.

As mentioned by Faine and Zhang [16], water proofing, fire and pest resistance of straw block are observed with a variety of techniques like construction on conventional trip footing, raising first bottom course [16–18]. Adequate plaster is also used to protect straw block walls against weather and enhance it durability. vapor permeable based plaster like cement-lime or earth based plaster are used in the straw bale construction industry. Under variable weather conditions, the plaster mixture can include additives to enhance the resulting performance of plastered straw bale house [3].

4.5 Thermal Conductivity

It was recorded from this study that there is no significant variation on the behavior of the three types of straw block to heat flow. The thermal conductivity of the different types of blocks were around 0.06 W/m K. This value is similar to that of straw bale reported by Costes et al. and Nunes [2, 19]. It is known that material with small thermal conductivity provides good indoor thermal comfort. The thermal resistance of straw block is an indicator that straw block house will be energy efficient.

5 Conclusions

The average compressive strength and density of blocks are respectively 1.25 MPa and 522 kg/m^3; which are respectively 73 and 5 times greater than that of straw bale. The presence of fine particle (saw dust) in straw block reduces greatly displacement in the blocks and increases their strength. From this study, the use of straw blocks in construction will not required pre-compression of walls before plastering them. It will greatly improve the structural and thermal performance of straw houses. Thus used for construction of single storey houses. Moreover, using straw block as construction material will reduce environmental burden of buildings. However, further study is required to investigate high density straw block.

References

1. Cascone, S., Rapisarda, R., Cascone, D.: Physical properties of straw bales as a construction material: a review. Sustainability 11, 3388 (2019)

2. Costes, J.P., Evrard, A., Biot, B., Keutgen, G., Daras, A., Dubois, S., Lebeau, F., Courard, L.: Thermal conductivity of straw bales: full size measurements considering the direction of the heat flow. Buildings **7**, 11 (2017)
3. Walker, P., Thomson, A., Maskell, D.: Straw bale construction. Elsevier (2016)
4. Walker, P.: Compression load testing straw bale walls (2004)
5. King, B: Straw bale construction (1998)
6. Lecompte, T., Le Duigou, A.: Mechanics of straw bales for building applications. J. Build. Eng. **9**, 84–90 (2017)
7. Vardy, S.P., MacDougall, C.: Compressive testing and analysis of plastered straw bales. J. Green Build. **1**, 63–79 (2006)
8. Nunes, L.: Nonwood bio-base materials (2017)
9. Sutton, A., Black, D., Walker, P.: Straw bale: an introduction to low-impact building materials (2011)
10. Vardy, S.P.: Structural behaviour of plastered straw bale assemblies under concentric and eccentric loading (2009)
11. Vardy, S., MacDougall, C.: Concentric and eccentric compression experiments of plastered straw bale assemblies. J. Struct. Eng. **139**, 448–461 (2012)
12. Thapa, S., Panda, G.K.: Energy conservation in buildings – a review. Int. J. Energy Eng. **5**, 95–112 (2015)
13. BS8110-3 1985 Structural use of concrete (1985)
14. Zhang, J.Q.: Load carrying characteristics of a single straw bale, pp. 1–16 (n.d)
15. Colsom, V., Dalmais, M., Le Cunff, T., Jadeau, O.: Rigid insula + on panel from hemp (2019)
16. Faine, M., Zhang, J.: A pilot study examining and comparing the load bearing capacity and behaviour of an earth rendered straw bale wall to cement rendered straw bale wall. In: International Straw Bale Building Conference, Wagga Wagga, December 2002, pp. 1–19 (2002)
17. Bou-Ali, G.: Straw bales and straw bale wall systems (1993)
18. Hodge, B.G.: Building your straw bale home. CSIRO Publishing (2006)
19. Douzane, O., Promis, G., Roucoult, J.M., Le Tran, A.D., Langlet, T.: Hygrothermal performance of a straw bale building: in situ and laboratory investigations. J. Build. Eng. **8**, 91–98 (2016)

The Role of Energy Modelling in the Development of Sustainable Construction Regulations for Al-Madinah City Central District

Alaa Kandil[1(✉)], Mostafa Sabbagh[2], and Samia Ebrahiem[1,2,3]

[1] Menoufia University, Shebin Elkom, Egypt
alaa.kandil@sait.ca
[2] King Abdulaziz University, Jeddah, Saudi Arabia
[3] South Alberta Institute of Technology, Calgary, AB, Canada

Abstract. This article discusses part of the procedure used to develop the energy requirements in the new green standards for the central district of Al-Madinah city. A comprehensive energy auditing was conducted for sample existing buildings. Two buildings were selected representing both high-end and medium quality typical construction in the district. Field-collected energy use data were used to calibrate energy models to compare the performance of sample buildings with reference buildings complying with ASHRAE 90.1, 2010. It was found that current regulations, the type of ownership in the district, and the increase in utility cost lead to energy performances that matched buildings complying with ST90.1, even though no energy efficiency requirements are currently mandated. These results paved the road for proposing the application of mandatory energy performance regulations to the district that matches international standards and are applicable in the meantime.

Keywords: Sustainable regulations · Measurement and verification · Sustainable development · Energy-efficient cities

1 Introduction

Various jurisdictions across the world regulated energy efficiency goals using different means. Good literature of global best practices in energy efficiency policies could be found by CTPD [1] or by GBPN [2]. In general, polices could be classified under three classifications:

1. Energy efficiency regulations based on prescriptive requirements. The energy requirements in the Saudi Building Code is an example of this method as it identifies minimum required insulation for walls and roofs and glazing type. Several other municipalities across the world adopted prescriptive requirements defined in national or international codes, e.g. parts 3 to 9 of ASHRAE 90.1 [3] and parts 3 to 7 of NECB [4].
2. Energy efficiency regulations based on the total performance of the building compared to a reference building using energy modelling simulation. Examples of

I. El Dimeery et al. (Eds.): JIC Smart Cities 2019, SUCI, pp. 262–269, 2021.
https://doi.org/10.1007/978-3-030-64217-4_30

that include complying with ASHRAE 90.1 part 10 [3], NECB part 8 [4], and California Title 24 part 6 [5]. It is also the method used by some voluntary certifications, e.g. LEED [6] that uses ASHRAE 90.1 Appendix G for compliance.

3. Energy efficiency regulations based on the total performance of the building compared to absolute performance metrics. Examples of these policies include the Britch Colombia Step code [7] and Toronto Green Standards [8]. It is also the method used by some voluntary standards, e.g. Passivhaus [9].

While the first approach is the easiest to verify by authorities having jurisdiction (AHJ) because of its "checklist" notion, it puts lots of restrictions on innovation and diversity. On the other hand second and third methods require the use of energy modelling which is much harder for AHJ to verify but gives more freedom for innovative ideas and technologies. Because of that reason, several jurisdictions allow compliance with both method one and two. However, with the globe's growing interest in reaching net-zero construction by 2030, it was found that using reference models as the bases of comparison is not the best way to go. Reference models typically have the same geometry of proposed designs. That may let a proposed project with a very poor mass design (e.g. a starfish-like building) comply with energy regulations because the reference model will have the same form. Finally, the third method is the method used by AHJ with a clear road map toward net-zero targets. It has the least amount of design restrictions, but it has a very strict energy use intensity cap. Designers and builders need to work together in an integrated approach to meet that cap. Not only that, some AHJs, e.g. City of Vancouver, have three caps, one for total energy use intensity (TEUI), one for thermal energy demand intensity (TEDI) and another one for greenhouse gases intensity(GHGI) [10].

This study was a part of a broader study that aimed to develop new green building standards for the central district of Al-Madinah City. One of the main concerns in regard to what new requirements to apply in the standards was the applicability of the requirements. The green standards are intended to push the boundaries of sustainability in the city, but if it is not applicable, nothing will happen. So the goal of this part of the study was to investigate the current status of existing building stock in the central district of the city by conducting a comprehensive energy auditing for sample buildings. The result should reveal if available consultation and construction capacities are capable of complying with the new regulations.

2 Study Methodology

This study utilized ASHRAE level 3 energy auditing [11]. This method provides the highest level of accuracy and it can account for the special operating characteristics of the holy area of Al-Medina. A simpler method, e.g., using utility bills to compare energy performance of the hotels in central district with international benchmarks will not put into consideration the special operating schedules in this area due to special prayer times during peak seasons like Hajj and Ramadan.

The study used Appendix G of ASHRAE 90.1 (ST 90.1) [3] as a benchmark for comparing the performance of the buildings under study. ST 90.1 is the most widely used standard across the world for the determination of energy efficiency of buildings. It is used by hundreds of municipalities in developed countries for buildings' permit approval and by most major green certification agencies for granting various green certificates, e.g. LEED [6].

The energy modelling for the projects was conducted using ENERGYPLUS [12] Ver.8.6. ENERGYPLUS is approved by ASHRAE 140 [13] to be used for compliance for ST 90.1.

This part of the study included the following activities:

- Collecting as-built building information. That included collecting record drawings, conducting walk-through, and interviews.
- Collecting energy use information using utility bills and sub metering.
- Building a set of as-built energy models using the best available information.
- Calibrating the as-built models against actual energy use.
- Building a set of reference models based on ST 90.1.
- Running both as-built models and reference models using standard operating conditions.
- Comparing the performance of the two sets of models and performing analysis.

2.1 Selection of the Sample Buildings

The selection process of appropriate samples aimed to select buildings with available information and with size and quality that represent the majority of current buildings stock in the central district.

Based on these criteria, two buildings were selected:

- Building-A: a three stars hotel with a total gross area of 12,902 m^2 distributed over two public floors, fourteen guest rooms' floors, plus three underground floors.
- Building-B: a five stars hotel with a total gross area of 36,104 m^2 distributed over three public floors, ten guest rooms' floors, plus three underground floors.

2.2 Identification of Energy Models Thermal Zones

The thermal zones for the energy models were defined to reflect:

- The differences between system types
- The effect of orientation
- The operating schedule.

Repeated typical floors were modelled as one floor with a zone multiplier and with adiabatic masses to substitute omitted floors.

2.3 Building Envelope Specification

Based on the field investigations, both buildings have very similar building envelope characteristics for both transparent and opaque assemblies that could be summarized in Table 1 and Table 2:

Table 1. Buildings envelope thermal properties for sample buildings

Walls	Un-insulated mass wall: CMU base wall with 20 mm rain screen granite cladding
Windows	Double light blue 6 mm glazing with 13 mm air gap
	Aluminum metal framing with no thermal break
Roofs	50 mm EPS insulation installed above the concrete deck

Table 2. Window to wall ratio for sample buildings

	Total	North	East	South	West
Building-A Window to wall [%]	14.01	35.24	12.08	0.00	8.25
Building-B Window to wall [%]	25.62	23.30	27.73	24.84	26.24

Building-A facades have a set of exterior traditional shading system *"Mashrabia"* that was simplified in the models as exterior louvres system.

Thermal performance characteristics of reference models were defined based on values in tables 5.5-1 (climate zone 1 (A, B)) of ST 90.1. ST 90.1 specifies the maximum window to wall ratio (WWR) of 40% for climate zone 1, however, if the as-built building WWR is less than 40%, the code requires using the as-built ratio, which was the case for both buildings. ST 90.1 also specifies that reference models should have no exterior shading components. All elements that may act as exterior shades were removed from the models, including exterior columns. The only elements that may result in exterior shading were the projected floors over the pedestrian sidewalk.

2.4 Mechanical Systems Definition

The main mechanical system for Building-B is a fan coil system served by 3 + 1 air-cooled chillers providing cooling, and a set of makeup units with heat recovery on the roof providing ventilation. The main mechanical system for Building-A is a Variable Refrigerant System (VRV) providing cooling, and a set of fresh air and exhaust fans with heat recovery on the roof providing ventilation.

Reference buildings' systems were defined based on table G3.1.1A, B of St 90.1 based on building size and use.

2.5 Lighting Systems Definition

A comprehensive room by room field survey for installed lighting was conducted for both buildings and as-built models' lighting used data collected by the survey. The study team conducted sample measurements for actual lighting levels for several spaces and the results showed that lighting levels in most of the locations were below average Saudi standards (see sample table below). Reference lighting power densities (LPD) for various uses were defined based on table 9.6.1 of ST 90.1 (Table 3).

Table 3. Measured lighting level (LUX) at sample locations in Building-B

Location	Average Saudi Code	Measured	
		Max	Min
Entrance halls	100	142	45
Restaurant at floor level	200	174	38
Restaurant at table level	200	216	51
Corridors floor	100	88.8	8.3
Stairs	150	100	65
Bathrooms at room	200	285	194
Offices	500	485	285
Car parks P	75	63.2	23.2

2.6   Identification of Weather Data

The project team didn't find any published weather files for the Madinah location. Energy modelling requires hourly weather information for a typical year to model standard conditions and for the measurement year for calibration. In order to generate a proper weather file, a climate information database was acquired from Meteoblue Point + service [14], an international weather data provider. A weather file was developed based on the newly available data and missing information, e.g. solar radiation data, was completed using the closest location with valid weather data. This method is believed to provide accurate weather information for the purpose of this study.

2.7   Calibration of Energy Models Using Measured Data

Field measurements were conducted using both fixed data loggers connected to current transmitters and hand-held measurement stations to check the performance of installed systems. The field measurements included measuring and verifying the following systems:

- Energy use of central cooling systems (VRV system and central chillers)
- Energy use of central ventilation systems
- Energy use of mechanical pumping systems
- Energy use of domestic water pumping systems
- Energy use of service zones (kitchens and laundry areas)
- Energy use of elevators
- Total facility hourly and monthly energy use
- Total facility water consumption
- Typical operation schedules.

Data collected by field measurements were used to adjust inputs for energy models. Some corrections were straightforward, e.g. peak equipment power densities for laundry zones in Building-B were corrected from the default of 54 w/m^2 to a significantly higher value of 190 w/m^2 based on actual collected data.

Some other corrections were based on observing hourly energy use pattern, rather than identifying a single input. For example, the hourly cooling energy use pattern was plotted against simulated cooling energy use and outside temperature. It was found that measured energy used is higher than expected simulated use in general and it tends to be more constant across the day, even though there is a significant outdoor temperature change between day and night. This energy use pattern could be partially due to problems with the control logic of the central cooling system, however, a significant portion of this discrepancy was fixed by fixing the following factors:

- Underestimating the infiltration rates which reduced the overall energy use in the simulated models.
- The use of higher cooling set point in the simulated models compared to actual set points.
- Underestimating the effect of thermal mass, which will make the energy model more responsive to changes in outdoor temperature.

3 Results and Discussion

Simulated Energy Use Intensity (EUI) for the two buildings were 241 and 162 kWh/m^2, which are close to the median benchmark for some developed countries, e.g. the United States median for hotels is 231 kWh/m^2 [15]. It should also be noted that this median includes all hotels and motels of all ages and sizes and covers a wide range of climate conditions. As expected, EUI for the five stars hotel was higher due to the type and quantity of available service spaces.

Based on design documents, field verifications, and surveys, both buildings are incorporating several energy efficiency measures including:

- Low window to wall ration with some exterior shading
- High thermal mass
- High efficient central cooling systems, (VRV or screw chillers)
- Heat recovery for ventilation and exhaust systems
- Low lighting power density (LPD).

These measures were driven by three factors:

- The type of ownership of the properties: buildings that are designed and constructed for the same owner - who also will operate the property - are typically designed with considering long term running cost.
- A big portion of hostelry businesses belongs to multinational corporates that enforce international sustainable standards in their hotels regardless of local regulations. That was the case of Building-B where the designers were required to comply with IHG Green Engage certification system [16].
- Current construction regulations for the central district zone that dictates a special style that promotes traditional facades treatment with arcades and exterior shading.

The above factors lead to total annual energy uses for both sample projects that are lower than reference models built based on ST 90.1 (Table 4).

Table 4. Comparison between total annual energy use for both sample buildings and reference building based on Appendix G of ASHRAE 90.1 2010.

Building-B	Total energy [kWh]	EUI [kWh/m^2]
ASHRAE 90.1 2010	7760825	260.03
	7,202,943	243.03
Percentage savings	**7%**	
Building-A	Total energy [kWh]	EUI [kWh/m^2]
ASHRAE 90.1 2010	1906054	190.2
	1686774	168
Percentage savings	**13%**	

4   Conclusions and Recommendations

Based on the results of the simulation activities of both the calibrated as-built models and reference models following ST 90.1 guidelines, the following conclusions could be obtained from the study:

The energy performance of both case studies were better than the reference performances based on ASHRAE ST 90.1 2010. Both cases also surpassed the reference performance by a comfortable margin which may even qualify them to for LEED-certified.

The current performance is driven by: 1) the type of ownership and operation of hotel buildings that makes owners care about running cost as well as initial cost, 2) the sustainable commitment of several hotel international operators, and 3) the current regulations controlling the construction in Medina central district.

The previous sustainability drivers cannot be taken 100% for granted as a wider survey for the façades in the central district found that: although current regulations require façade design and window to wall ration to reflect orientation of the façade to minimize excessive solar gains, and although the overall average WWR was 35%, the WWR varied between 1% to 76% and that variation was unrelated to the orientation. The type of ownership is not consistent as well since a significant number of hotels in the kingdom is not operated by international chains, especially low-end hotels.

It is fair to say that: it is very applicable to mandate complying with ASHRAE ST 90.1 2010 or any equivalent energy performance standard without adding significant challenges to investors, designers and builders.

The use of Reference-Based energy regulation should be only the first step as the sustainable community is moving toward more rigorous Absolute Metrics-Based regulations. However, it is an important step toward educating and training the construction and consultation community in the kingdom about energy efficiency.

References

1. City of Toronto Planning Division (CTPD): Zero emissions buildings framework, Toronto TRF (2016)
2. GBPN, LBNL: Building Energy Efficiency: Best Practice Policies and Policy Packages. Berkeley LAB Publications, Berkeley (2012)
3. ASHRAE-ANSI: Energy standards for buildings except low-rise residential buildings. American Society of Heating, Refrigeration and Air Conditioning Engineering, Atlanta (2010)
4. NRCC: National Energy Code of Canada for Buildings. National Research Council of Canada, Ottawa (2017)
5. California Energy Commission: Title 24. https://www.energy.ca.gov/programs-and-topics/programs/building-energy-efficiency-standards. Accessed 29 Sep 2019
6. U.S Green Building Council: LEED. https://new.usgbc.org/leed. Accessed 29 Sep 2019
7. Government of British Colombia: Energy step code. https://energystepcode.ca. Accessed 29 Sep 2019
8. City of Toronto: Toronto green standards. https://www.toronto.ca/city-government/planning-development/official-plan-guidelines/toronto-green-standard. Accessed 29 Sep 2019
9. Passive House Institute: Passive House Certificate. https://passivehouse.com. Accessed 29 Sep 2019
10. City of Vancouver: Green buildings policy for rezoning. Planning, Urban Design and Sustainability Department, Vancouver (2016)
11. ASHRAE: Procedure for commercial building energy audits. American Society of Heating, Refrigeration and Air Conditioning Engineering, Atlanta (2011)
12. U.S. Department of Energy: EnergyPlus. https://energyplus.net. Accessed 29 Sep 2019
13. ASHRAE-ANSI: Standard method of test for the evaluation of building energy analysis computer programs. American Society of Heating, Refrigeration and Air Conditioning Engineering, Atlanta (2017)
14. https://www.meteoblue.com/en/pointplus. Accessed 29 Sep 2019
15. ENERGY STAR: Portfolio manager technical reference. https://www.energystar.gov/buildings/tools-and-resources/portfolio-manager-technical-reference-energy-star-score. Accessed 29 Sep 2019
16. https://www.ihgplc.com/responsible-business/environmental-sustainability/ihg-green-engage-system. Accessed 29 Sep 2019

The Impacts of Climate Zone, Wall Insulation, and Window Types on Building Energy Performance

Abid Nadeem[1]([✉]), Yerzhan Abzhanov[1], Serik Tokbolat[1,2],
Mohamad Mustafa[2], and Bjørn R. Sørensen[2]

[1] Department of Civil and Environmental Engineering, Nazarbayev University,
Nur-Sultan, Kazakhstan
`abid.nadeem@nu.edu.kz`
[2] Institute of Buildings, Energy and Material Technology, UiT The Arctic
University of Norway, Narvik, Norway

Abstract. Building energy consumption tends to increase over the next few decades due to the increasing level of urbanization and population. These days much attention has been paid to the enhancement of energy performance of residential and non-residential structures. One should consider various factors for proper building thermal design and assessment. In this study, a simulation-based investigation is applied to analyze the influence of building envelope, climate region, and window's physical features on energy performance. Building's energy consumption and amount of CO_2 emissions are studied. EnergyPlus tool interfaced with DesignBuilder software was used to perform energy simulations. Annual energy analyses are carried out on the reference house model over the five climate regions from the Koppen-Geiger climate classification map. According to results obtained, climate condition, wall envelope, window type, and window to wall ratio can significantly influence a building's energy performance. Application of insulating materials and the use of specific window type results in considerable energy savings and reduction of CO_2 emission amounts.

Keywords: Building energy simulation · Insulating materials · Climate regions · Energy performance · Energy-saving potential

1 Introduction

In recent times, 30–40% of the total energy produced worldwide is consumed by building sector where heating, ventilation and air conditioning (HVAC) system utilize a significant portion of the energy [1]. It is also responsible for 8% of energy production related to CO_2 emissions. This value may rise since it is anticipated that by 2050, there will be a 50% increase in global energy consumption [2]. Consequently, there will be more harmful effects on the environment in the future due to increased building energy consumption and growth in fossil fuel demand associated with a higher amount of CO_2 emissions.

I. El Dimeery et al. (Eds.): JIC Smart Cities 2019, SUCI, pp. 270–277, 2021.
https://doi.org/10.1007/978-3-030-64217-4_31

During the design stage of buildings, the designers consider many factors that influence the energy performance of buildings. Active techniques of improving the building's energy consumption are about the advancement of HVAC and lighting systems, while passive methods imply the development of building envelope [3].

Heating and cooling energy reductions, and improvement in the thermal comfort of residents may be achieved by introducing modifications into the building envelope [2]. This approach results in significant outcomes since building envelope design influences 20–60% of building energy consumption [4]. For instance, techniques such as the use of insulation materials and phase change materials in building envelope design can be used [2]. Simona et al. [5] investigated that the inclusion of thermal insulation materials into building envelope leads to the reduction of heating and HVAC energy consumption. Fang et al. [6] performed experimental studies and determined that insulated building envelopes consume less energy in contrast with envelopes without insulation.

Besides, substantial attention has been paid to the effect of climate zones to the energy performance of buildings. Nadeem [7] investigated the relationship between location and energy consumption of a two-story residential building. The results showed that the use of insulated envelopes tends to decrease the heating requirement and increase the cooling needs of buildings. Aldawi et al. [8] have investigated the performance of two current, and four new wall envelopes among six climate zones of Australia which showed that polyurethane insulation materials performed 40% better than polystyrene insulation in energy savings.

Windows are an essential component of the building envelope. Amaral et al. [9] evaluated the windows' effect on thermal comfort, heating, and cooling energy consumption of the reference room. They concluded that optimal window characteristics could be determined and applied to enhance its energy performance. These characteristics relate to the building's geographical location, orientation, and physical properties. In a study by Gasparella et al. [10], the effect of window features on cooling and heating energy demand, under the climatic conditions of Paris, Milan, Nice, and Rome, were investigated for 2-storey insulated residential building. They also provided some solutions for the impaired performance of windows in summer in the form of utilization of shading systems. Based on the above literature review, it is evident that several factors may influence a building's energy consumption. However, it is not convenient to reflect all of them in one research study. Therefore, this paper evaluates the impact of the location of the structure and window features that are necessary for the assessment of energy performance of buildings. This study will provide evidence to the importance of these factors in design and assessment of building's energy consumption) and amount of CO_2 emissions.

With growing population of major cities and pressing demand for smart solutions, energy efficiency of residential buildings can be both a driver of smart cities as well as the positive outcome. Buildings which are energy efficient as a result of building envelope optimization and adjustment to local climate conditions will undoubtedly allow cities to become smarter.

2 Methodology

2.1 Locations (Climate Zones)

In this study, five cities from different climate zones were selected. These cities are Nur-Sultan, Beijing, Chicago, Singapore, and Valencia. Nur-Sultan is in central Asia and, in general, has a semi-continental climate with hot summers, freezing and dry winters. Beijing and Chicago have a humid continental climate, with hot and humid summers, cold and dry winters, while the first is affected by monsoons from the East. Singapore has a humid tropical climate throughout the year with slight variations in temperature and abundant rainfall. The environment of Valencia is considered as arid with little rain, warm winters, and hot, dry summers. The climatic characteristics of these cities are summarized in Table 1.

2.2 Building Characteristics and Energy Simulation

3D model and interior partitioning of the building is shown in Fig. 1. This reference building is a wood-framed, two-story single-family residential house that is oriented at $0°$ to the north. Kitchen, master bedroom, and living room are on the first floor whereas three more bedrooms are on the second floor and the total built-up area is 247 m^2. The analysis assumed that the building envelope parameters were same for different locations. The models were built mainly focusing on external walls which were selected from Design Builder software's library, the data on roofs, floors and doors, and other components of the building envelope were assumed to be same regardless of the construction norms and standards of respective countries.

Table 1. Climate characteristics of selected locations

	Nur-Sultan	Beijing	Chicago	Singapore	Valencia
Koppen classification	Dfb	Dwa	Dfa	Af	BSk
Avg. annual temp, °C	2.2	12.1	10	26.8	17.4
Warmest aver. Temp, °C	20.7 (Jul)	26.3 (Jul)	23.5 (Jul)	27.4 (May)	24.9 (Aug)
Coldest average temp, °C	−21.4 (Feb)	−9.3 (Jan)	−9.1 (Jan)	22.3 (Jan)	6.3 (Jan)
Annual precipitation, mm	308	610	918	2378	445
Elev. above sea level, m	347	43.5	179	16	15

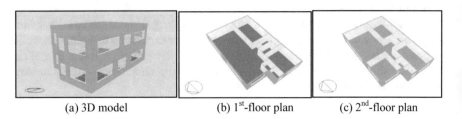

(a) 3D model   (b) 1st-floor plan   (c) 2nd-floor plan

Fig. 1. Reference building

For all regions, heating is provided by natural gas while cooling is due to the air-conditioning system powered by the electricity. Domestic hot water (DHW) is supplied using a gas boiler. Two types of external wall envelope were analyzed. One component wall that consists of 20 mm wood was chosen as the reference wall as it has the most basic configuration. The other wall used insulated polyurethane (IPUF) foam as the primary insulating material. Roof, ground floor, internal wall, and partition wall envelopes were kept unchanged throughout the project. Details of building envelopes are provided in Table 2.

Effect of window types was investigated using three types of windows with different composition and thermal transmittance values. These windows are double glazed reference window, double glazed window, and triple glazed window. They are analyzed for four different window-to-wall ratios that are 16%, 25%, 34%, and 41%. Details of windows are provided in Table 3.

It is assumed that the occupancy of the building will be 0.02 people/m^2 which is a typical value for residential houses. Cooling setpoint is at 24 °C while heating setpoint is at 16 °C. Regular schedules for occupancy, heating, cooling, and lighting are provided in Table 4.

Table 2. Building envelopes

	Reference wall (Type I)	IPUF
External walls	20 mm wood	20 mm wood, 50 mm polyurethane foam, 100 mm glass fiber batt insulation, 10 mm gypsum board
Roof	20 mm wood siding + 100 mm expanded polystyrene + 50 mm polyurethane foam + 15 mm gypsum board	
Slab on grade floor	150 mm concrete slab + 50 mm extruded polystyrene + 20 mm wooden flooring	
Internal floor	20 mm wooden flooring	
Partition walls	15 mm gypsum board + 20 mm air gap + 15 mm gypsum board	

Table 3. Window types and physical properties

Window type	Composition, mm	Thermal transmittance, W $m^{-2}K^{-1}$	Solar transmittance, g
Reference window	3/13/3	1.960	0.690
Double glazed	6/13/6	2.665	0.497
Triple glazed	3/13/3/13/3	0.982	0.474

Table 4. Occupancy, heating, cooling, and lighting schedules

Type	Schedule
Occupancy	0.00–7.00 *100%*, 7–9.00 *50%*, 9–16.00 *0%*, 16–19.00 *80%*, 19–00.00 *100%*
Heating	0–8.00 *100%*, 8–16.00 *0%*, 16–0.00 *100%*
Cooling	0–8.00 *100%*, 8–16.00 *0%*, 16–0.00 *100%*
Lighting	0–7.00 *0%*, 7–9.00 *50%*, 9–17.00 *0%*, 17–20.00 *50%*, 20–0.00 *100%*

EnergyPlus is a widely used simulation tool for whole building energy performance with an extensive database of weather files (climate conditions) and building materials. It provides a broad range of energy performance modeling, evaluation capabilities, and heat transfer computations, advanced HVAC system configurations, algorithms for calculation of thermal comfort of occupants, environmental effect, and cost evaluation [4].

3   Results and Discussions

3.1   Effect of Location and Wall Envelope on Annual Energy Consumption

In this section, the results of annual energy consumption simulations for two wall envelopes are presented for each city. Table 5 shows annual heating, cooling, and total energy use for wall envelope in the selected cities. It can be noticed that obtained values of energy demand conform to climate conditions of locations. For instance, for the reference envelope in cold regions with cold winters, energy utilized on heating purpose dominated. From Table 5, it can also be noticed that the insulated wall envelope leads to a reduction in heating energy consumption.

Table 5. Annual energy consumption with selected external wall envelopes

Cities	Heating energy		Cooling energy		Total energy	
	Ref. wall, kWh	IPUF, kWh	Ref. wall, kWh	IPUF, kWh	Ref. wall, kWh	IPUF, kWh (ESP %)
Nur-Sultan	200,355	43,286	2,960	4,805	203,315	48,092 (76.4%)
Beijing	67,179	7,242	6,297	7,494	73,476	14,736 (79.9%)
Chicago	97,010	12,052	4,278	6,311	101,288	18,363 (81.9%)
Valencia	14,108	32	7,446	9,993	21,555	10,026 (53.5%)
Singapore	0	0	18,892	19,306	18,892	19,306 (−2.2%)

Annual energy saving potentials (ESP) was calculated for IPUF wall envelope and shown in the last column of Table 5. It was estimated by ESP (%) = (X − Y) * 100/X, where X is the total energy consumption of the reference envelope, and Y is the total energy consumption of the wall envelope. Overall, the values of ESP varied slightly among the chosen locations. There is a tangible difference in ESP values between wall without insulation and wall with insulation. It is to be noted that insulated envelope negatively influenced the energy consumption in Singapore. Besides, in all regions, the effect of wall envelope modification was found varying throughout the year, namely, during the winter period, ESP showed better performance than the summer period.

3.2   Effect of Window Characteristics on Annual Energy Consumption

Simulations were carried out for three types of window with varying window to wall ratios (WWR) for reference wall and IPUF wall. When the reference wall envelope was used, increment in WWR led to an increase in cooling energy consumption and decrease in heating and total energy consumption. Figure 2a depicts changes in annual energy consumption due to the rise in WWR and variation of window type in Nur-Sultan. When IPUF wall was used, heating, cooling, and total energy consumptions escalated upwards due to increase in WWR. Figure 2b illustrates changes in total energy consumption due to the rise in WWR and alteration of window type in Nur-Sultan. The results show that the increase in WWR tends to raise annual energy consumption in well-insulated buildings, whereas the same drops in uninsulated buildings. For both wall envelopes, Nur-Sultan, Beijing, and Chicago have the lowest annual energy consumption when a triple glazed window was used. In Valencia and Singapore energy performance of double glazed and triple glazed windows were nearly the same. Nevertheless, for both cities, the use of double-glazed window resulted in slightly lower annual energy demand.

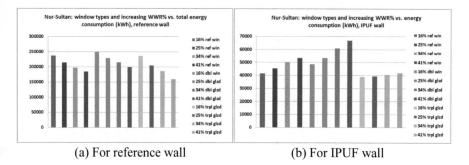

(a) For reference wall   (b) For IPUF wall

Fig. 2.   Window type vs. annual energy consumption for Nur-Sultan

Table 6. Amount of CO_2 emissions in Nur-Sultan for window types and WWR

	Nur-Sultan CO_2 emissions, kg (Ref. wall)				Nur-Sultan CO_2 emissions, kg (IPUF)			
	16%	25%	34%	41%	16%	25%	34%	41%
Ref	64,713	62,551	60,390	58,716	29,926	31,033	32,220	33,169
Double	65,443	63,702	61,955	60,590	30,662	32,091	33,547	34,692
Triple	64,037	61,422	58,774	56,691	29,005	29,508	30,063	30,528

3.3   Effect of Window Characteristics on the Amount of CO_2 Emissions

Table 6 shows the amount of CO_2 emissions in Nur-Sultan for all three window types and WWR for both reference wall and IPUF wall. For reference wall envelope, increase in WWR leads to a decrease in CO_2 emissions. In the case of IPUF wall envelope, the amount of CO_2 emissions increased as WWR increased. Moreover, the use of triple glazed window resulted in the lower CO_2 emissions in Nur-Sultan, Beijing, and Chicago for both wall envelopes whereas in Valencia and Singapore, results for double glazed and triple glazed windows were nearly the same. Use of insulating material significantly decreased CO_2 emissions. However, insulation did not have a considerable effect in cities with warm climate as Valencia and Singapore (Table 7). These results comply with general climate conditions of cities.

Table 7. Amount of CO_2 emissions in all cities for both wall types with reference window with WWR = 16%

	Reference wall, kg	IPUF wall, kg
Nur-Sultan	64,713	29,926
Beijing	39,545	25,542
Chicago	41,382	22,506
Valencia	28,888	25,381
Singapore	32,803	31,051

4   Conclusions and Recommendations

The study has shown that taking one component wall that the wall insulation can considerably enhance the energy performance of the building. However, results are not the same for different locations and wall envelopes. For instance, more energy can be saved by using IPUF in Nur-Sultan, Beijing, and Chicago. However, this may not be true for other locations. Use of insulated wall envelopes significantly enhanced the energy performance of buildings in cold regions by substantially decreasing heating loads. There are no significant energy saving outcomes for Singapore with an equatorial hot and humid climate. During summer, the effect of insulation tends to decrease, and ESP approaches zero for some locations. Hence, the efficiency of wall envelopes is highly dependent on materials used as well as climate conditions. Insulated walls tended to increase CDD and significantly decrease HDD in all regions except Singapore, where CDD slightly dropped. As insulation reduced the HDD, it can be stated that such envelopes are best suitable for areas with cold climates.

For reference wall, increase in WWR led to the reduction of annual total energy consumption. However, when the wall envelope with an insulation layer was used, annual energy demand increased with increase in WWR. Thus, it can be concluded that the rise in WWR leads to an increase in annual energy consumption in well-insulated buildings and decreases yearly energy use in uninsulated structures. For reference wall envelope and IPUF envelope increasing WWR led to the increment of CDD and decline of HDD. For reference wall, the increase in WWR associated with the reduction of CO_2 emissions while for IPUF wall it was vice versa. Use of insulating materials considerably reduces the amount of CO_2 emissions in regions with cold winters.

Acknowledgement. This research was supported by the Project: Academic Cooperation in Postgraduate Engineering Education (ACE) - CPEA-ST-2019/10029 funded by the Norwegian Agency for International Cooperation and Quality Enhancement in Higher Education (Diku) under Eurasia Programme 2019.

References

1. Lei, J., Yang, J., Yang, E.H.: Energy performance of building envelopes integrated with phase change materials for cooling load reduction in tropical Singapore. Appl. Energy **162**, 207–217 (2016)
2. Marin, P., Saffari, M., de Gracia, A., Zhu, X., Farid, M.M., Cabeza, L.F., Ushak, S.: Energy savings due to the use of PCM for relocatable lightweight buildings passive heating and cooling in different weather conditions. Energy Build. **129**, 274–283 (2016)
3. Sadineni, S.B., Madala, S., Boehm, R.F.: Passive building energy savings: a review of building envelope components. Renew. Sustain. Energy Rev. **15**(8), 3617–3631 (2011)
4. Saffari, M., de Gracia, A., Fernández, C., Cabeza, L.F.: Simulation-based optimization of PCM melting temperature to improve the energy performance in buildings. Appl. Energy **202**, 420–434 (2017)
5. Simona, P.L., Spiru, P., Ion, I.V.: Increasing the energy efficiency of buildings by thermal insulation. Energy Procedia **128**, 393–399 (2017)
6. Fang, Z., Li, N., Li, B., Luo, G., Huang, Y.: The effect of building envelope insulation on cooling energy consumption in summer. Energy Build. **77**, 197–205 (2014)
7. Nadeem, A.: Building energy simulations for different climate zones and building envelopes. In: 1st International Conference on High Performance Energy Efficient Buildings and Homes (HPEEBH 2018), August 1–2, 2018, Lahore, Pakistan, pp. 77–84 (2018)
8. Aldawi, F., Alam, F., Khan, I., Alghamdi, M.: Effect of climates and building materials on house wall thermal performance. Procedia Eng. **56**, 661–666 (2013). 5th BSME International Conference on Thermal Engineering
9. Amaral, A.R., Rodrigues, E., Gaspar, A.R., Gomes, A.: A thermal performance parametric study of window type, orientation, size, and shadowing effect. Sustain. Cities Soc. **26**, 456–465 (2016)
10. Gasparella, A., Pernigotto, G., Cappelletti, F., Romagnoni, P., Baggio, P.: Analysis and modelling of window and glazing systems energy performance for a well-insulated residential building. Energy Build. **43**, 1030–1037 (2011)

Re-imagining the Architecture of the City in the Autonomous Vehicles Era

Dalia O. Hafiz[1(✉)] and Ismail H. Zohdy[2]

[1] School of Architecture and Design, AL-Ghurair University,
Dubai, United Arab Emirates
d.hafiz@agu.ac.ae
[2] Expert, Road and Transportation Authority, Dubai, United Arab Emirates
ismail.zohdy@gmail.com

Abstract. The adoption of self-driving technology will likely birth new cities typologies with unique buildings and needs. Starting from centralized hubs where the cars park themselves to fewer congestions and empty parking lots and Autonomous Vehicles (AVs) repair shops. Although such technology can be a great tool and facilitate the passengers' movement and minimize travel time, careful implementation to city configuration and urban planning is needed when applied. Priority should be given to people and places to minimize all possible undesirable effects such technologies might cause. Consideration should be given to pedestrians, cyclist, and walkability in the city.

This paper aims at examining possible design configurations in the new and smart cities with Autonomous and connected vehicles to maintain safety, sustainability and walkability in the city while implementing such technologies.

Keywords: Autonomous vehicles · Smart cities · Architecture of the city · Smart urban planning

1 Introduction

One of the main transportation technology of the future is "Autonomous Vehicles" (or as they are usually called: Self-driving, Unmanned vehicles)-AVs. The implication of AVs can be considered one the most profound shifts in urban land use any architect has seen, arguably since the transition from horses to cars a century ago. The autonomous vehicle is set to be as life-changing as the invention of the motor vehicle itself [1]. Today, our vehicles are parked close to 95% of the time. There are millions of parking spaces worldwide, which may not be needed with the autonomous connected vehicles. There will be far fewer vehicles on the road. *"We could take 90% of the cars off the road without any reduction in transit time, while also reclaiming 20% more space inside cities currently occupied by parking and roads."* By Michael Schmidt.

In addition, autonomous vehicles can travel very close together at high speeds without endangering humans, which means roads themselves can be much narrower (road diet). This will free up a huge amount of land that is currently congested with vehicles. All these aspects have tremendous implications for how cities and buildings will be designed.

© The Author(s), under exclusive license to Springer Nature Switzerland AG 2021
I. El Dimeery et al. (Eds.): JIC Smart Cities 2019, SUCI, pp. 278–286, 2021.
https://doi.org/10.1007/978-3-030-64217-4_32

There has been several arguments regarding the implementation of Autonomous Vehicles in our cities and everyday experience. One of these arguments looks at the walkability in the city; With AVs, walking from parking spaces to the work location is no-more needed. Instead, AVs will drop off the commuters at the workplace and then drive themselves to the parking spaces. On the other hand, the departure/arrival profile will change when compared to the literature with non-autonomous vehicles (non-AVs). Furthermore, recent research started to examine a monocentric (a single center city) model with a possibility that AVs might be parked at home during the day to avoid parking fees [2]. An altered approach proposed that it is unnecessary for AVs to park at all. Instead, they can cruise (circle around at lower speed) or transport other passengers [3].

Modern cities have been shaped largely by the mobility that is achieved through motor vehicles, providing transport services to people and goods alike and supported by major road networks worldwide (Kulmala, Jääskeläinen and Pakarinen 2019). When examining the effect of AVs on the city, architects and planners can examine two key aspects of the city -people and places- to strategize how each can be affected.

While Autonomous Vehicles are considered an important step towards the future of transportation, such technology application may repeat previous urban planning mistake of prioritizing traffic efficiency over walkability and community vitality.

In a research conducted by MIT, the simulations demonstrated cities without traffic lights with cars which are easily traveling through an intersection when compared traditional streets. However, there is a key missing component in such simulations, humans; the simulations did not show pedestrians, bus riders or cyclists. While intersections represent the most walkable locations in the entire United States, and worldwide [4].

Recent research on connected and Autonomous vehicles (CAVs) mainly focus of the vehicles design, industry, and manufacturing process. Studies on the impact of CAVs is focused on the highway. Very little research examined the impact on people, buildings and urban design and how the city and buildings can adapt to accommodate CVs, while in the near future there will be a transition period where CVs will be on the roads with human-driven cards [5].

1.1 People

Pedestrians need to feel safe and comfortable, but they also need to be entertained. This represents a very important role that architects and urban planners can take. According to Steve Mouzon theory of "walk appeal," holding that how far we will walk is all about what we encounter along the way. Stores and businesses with street-level windows and architectural details were found to be the most encouraging walkability features in cities, while lined parking and vertical building lines discourage it [15].

Transportation that prioritizes active transportation in walkable, bikeable communities oriented around mass transit. Several Utopian urban planning projects failed not because of their architecture, but because they forgot the most important aspect of any community: the people. A very famous example of such failure is the radiant city (City Radieux) by the famous architect and urban planner -Le Corbusier- where he imagined the city planning with a Cartesian grid separating cars, pedestrians and commercial spaces. He saw the Radiant City's geometric layout, standardization and repetitive towers as "the perfect form" of urbanism. His designs were based on his theory "A city made for speed is made for success". His theories were adopted in several neighborhood and resulted in centers for crimes, poverty and social decay, as they fail to predict the people behavior and enhance their comfort [6, 7].

Spaces
To achieve successful spaces, architects need to make friendly and unique buildings, while eliminating the empty spaces between them. Such process can be done through transitional spaces, or passages. Such spaces, are essential links for a sustainable mobility. They can be represented in tunnels, footbridges, escalators, urban funiculars, corridors that facilitate transition between various transportation modes, or between urban ambiences. They can incorporate landmarks and remarkable elements that enhance pedestrians experience and wayfinding [8].

Consequently, the purpose of this paper is to apply modern techniques of future cities on a use case (Dubai) taking in consideration the new mobility solutions (Autonomous vehicles) with capturing the main aspects of successful cities and elements of livability for citizens.

2   Dubai: Re-immagining The Future City

Dubai, like several cities, is considered with futuristic inventions. The city plans to become one of the world's smart and sustainable cities. With one of its key goals is to convert 25% of all trips in the city of Dubai to be smart and driverless by 2030 [9, 10]. The selected case study is located on Sheikh Zayed road, the longest road in the Emirates, stretching along the Persian Gulf.

The case study is re-imagination of a block on Sheikh Zayed road. It focuses on the human interaction, space-making, and community in the city of Dubai, which may be more challenging with the hot summer weather of Dubai.

The key elements in the examined block are the museum of the Future, the Metro station, and the existing commercial activities. From the literature there are six main pillars for the design of future cities: safety, comfort, interaction and space making, walkability, sustainability and healthy environment.

There are sets of guidelines for each pillar that need to be maintained when applying AVs technology as presented in the following sections.

Pillar 1: Safety
According to the United Nations' International Crime Victim Survey' (ICVS): "*urbanization is the strongest predictor for crime and victimization*" [11]. Although several new research conducted that the United Arab Emirates is one of the safest places to live, especially the city of Abu Dhabi and Dubai, it is important to consider the city urban safety with the country growth. Previous research based on real cities showed that enhancing environmental and urban design provides the best opportunities to safety through: a) decreased unoccupied spaces (parking lots and garages); b) Narrower streets surrounded by public spaces, and c) enhanced views to public transportation stops, lighting, and vegetation, which are proposed in the case studies.

Pillar 2: Comfort

Comfort is defined as *"the condition of mind which expresses satisfaction with the environment"* [12] and *"when individuals have the psychological, social and physical resources they need to meet a particular psychological, social and/or physical challenge"* [13]. Although Comfort is very subjective and it can vary from one person to another. Architects, designers and urban planners aim at reaching higher comfort levels. There are several factors that can affect the level of comfort which are mainly related to human senses, levels of activities and ambient. Generally the key aspects of comfort can be subdivided into:

a) Thermal comfort: People satisfaction with their thermal environment can affect their ability to function effectively. There are several parameters other than air temperature that can affect thermal comfort (perceived temperature) including direct/diffused sun, wind, and humidity.

In the proposed case study thermal comfort was enhanced in outdoor spaces using radiant cooling that can be used in the warm season (winter). In addition to thermally controlled shopping and recreational spaces (where energy is achieved from PV panels), which can be used in the hot summer season, Fig. 1.

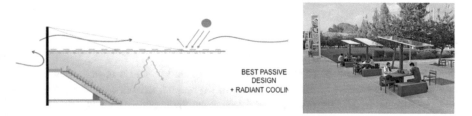

Fig. 1. Passive and active cooling

b) Visual Comfort: is to be able to perform the tasks safely and comfortably, in a visually pleasant environment. To achieve visual comfort there must be an efficient quality and quantity of light (Dodge, et al. n.d.). In the selected case study visual comfort was achieved through daylighting in the whole urban fabric, while mainly introduced in the underground- thermally controlled- level. Innovative systems are needed to bring natural daylight in underground spaces using light tubes and fiber optics, which can reach down to of 40 meters underground to maintain daylighting to the deepest zones, Fig. 2.

Acoustical Comfort: Acoustical comfort aims at reducing noise and disturbance that might affect health of quality of life. The proposed design shift to cut off traffic noise; it was developed in order to offer some solutions to noise problems generated by the main road. A continuous walkway approach, showing success in several European cities was used. It involves a combination of water features, greenery, smart noise barrier, and businesses to help in the process by holding public discussions [14]. The design of the continuous walkway has also included more attractive playgrounds, sports areas and green spaces which have increased the level of lively sounds from human activities, Fig. 3.

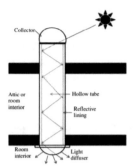

Fig. 2. Daylighting penetration for underground public and commercial spaces

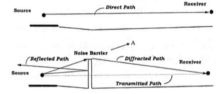

Fig. 3. Smart noise barriers

Pillar 3: Interaction and Space Making

Since unsuccessful city plans were based on speed and isolation of activities. continuous visual and physical interaction needs to be maintained between the pedestrians and the streets with AVs and motorcycles. Such interaction aims at minimize the "bubble" effect where activities are clearly differentiated into motors and people. Space making is possible with AVs road dieting in different ways including lanes size and number reduction, in addition to removal of the unwanted off street parking. Such diet gives a vast opportunity in the proposed case study for other urban and social activities; Ex. protected and improved bike lanes, wider sidewalks, street furniture, street vending, public art, fountains, streetscape and landscape, outdoor dining, jogging paths, and playgrounds [8].

Pillar 4: Walkability

The transportation profession has long called for more walkable communities for many reasons: address traffic congestion, minimize sources of pollution, and enhance the residents health with the big increase in obesity and decreased level of humans physical activities amoung the cities residents [15]. Walkability was achieved in the case study through the connection with public transportation (Dubai Metro), cultural experience (The museum of the Future), commercial experience (stores and restaurants), in addition to the underground thermally controlled however - daylit spaces.

Pillar 5: Sustainability

A sustainable city can be defined as; "Improving the quality of life in a city, including ecological, cultural, political, institutional, social and economic components without leaving a burden on the future generations" [16]. Recently several features of sustainability especially sustainable urban planning where climate change, clean air and water, renewable energy and land use are making large impact on the global sustainability; where Architecture and transportation are two key ones. In the proposed case study sustainable urban planning involves several disciplines including 1) recycled irrigation grey water (from nearby buildings), 2) shaded pedestrian paths/walkways covered with Photovoltaic (PV) panels which produce enough energy to ventilate the walkways, 3) daylighing in the underground surface using light tubes. In addition mixed-use developments, walkability, greenways and open spaces are found to be key features of healthy city and residents well-being. Such solutions can help improving the urban city sustainability, and create more efficient spaces.

Pillar 6: Healthy Environment
In times of rapid urbanization, health and well-being of citizens are increasingly recognized as a challenge and that a remarkable amount of research on the potential associations between urban areas and health or well-being has been conducted [14].

Higher percentages of green-spaces are generally associated with better health-related quality of life. Fewer vacant land, Architectural features that facilitate visual and social contacts are found to be associated with healthier environments. Attractive cities were often characterized by neighborhood safety, aesthetics, walkability, increased services, and access to cultural, shopping or sport amenities [17].

2.1 Dubai: Sheikh Zayed Road Re-immagination Summary

While re-imagining one arterial UAE road (Sheikh Zayed road) and applying all the aspects of smart livable cities, to achieve the optimum levels of comfort, safety, sustainability, and well-being as shown in Table 1 and Fig. 4.

Table 1. Existing vs imagined features of Sheikh Zayed road with autonomous vehicles

1. Safety		
	Existing	Re-imagined
Streets		
Lane width (Avg)	3.6	2.7
Lane no (Avg - 2 ways)	12	8
Motorcycle lanes	N/A	1 each way
Bike lanes	N/A	2 each way (Isolated from vehicles and motorcycles)

(continued)

Table 1. (*continued*)

Public/commercial spaces		
	Side walks	Wider sidewalks - Continuous commercial spaces - Underground commercial spaces - Seating areas
Views		
Unblocked and enhanced views	Buildings	Greenery – Walkways - commercial spaces - transparent smart noise barriers - water surfaces
2. Comfort		
Thermal	Shaded/thermally controlled metro stations	Passive Cooling - radiant cooling with summer heat - Passive cooling (Power from PV panels) - Also PV panels for shading - Water features
Visual		
	N/A	Shaded walkways - Daylighting underground (light tubes and fiber optics)
Acoustical		
	Solid noise barriers (in residential zones) smart noise barriers for the metro	Smart noise barriers (Streets) - water features (fountains) - Commercial spaces (Cafes and shops) - Greenery (trees) - Public spaces (walkways, bike lanes, and public seating)
3. Interaction and space-making		
	N/A	Commercial and public spaces - water features - AVs drop off/pick up zones - Greenery
4. Walkability		
	Unprotected sidewalks	Shaded and cooled sidewalks - continuous commercial/social and cultural activities - shaded bike lanes
5. Sustainability		
	Partial grey water irrigation	PV panels for cooled walkways - Daylighting - Grey water recycling for irrigation - bike lanes - no off road parking
6. Healthy environment		
	Some commercial activities	Enhanced walkability - less vehicles = less emissions-continuous access to activitie - thermally controlled outdoor spaces

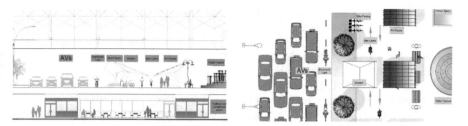

Fig. 4. Re-imagined block plan and section of Sheikh Zayed road

3 Conclusion and Future Research

It is very important to understand how to integrate the AV technology in the proposed city planning and transportation goal rather than completely embracing AVs or rejecting them entirely. Such step is a key phase to fully benefit from the opportunities such technology can offer, and also avoid fully embracing it without considering its consequences and effect on the urban environment and the life of the city residences while turning it into an automobile-dominated city with AVs.

AV technologies should be examined in the livable city where public transit, walkability, public space and most importantly-people can thrive.

References

1. Preparing for autonomous vehicles in Canada (2015)
2. Zakharenko, R.: Self-driving cars will change cities. Reg. Sci. Urban Econ. **61**, 26–37 (2016)
3. Millard-Ball, A.: The autonomous vehicle parking problem. Transp. Policy **75**, 99–108 (2019)
4. Fridman, L.: Human-centered autonomous vehicle systems: principles of effective shared autonomy (2018)
5. Autonomous Vehicles: Hype and Potential | CNU. https://www.cnu.org/publicsquare/2016/09/06/autonomous-vehicles-hype-and-potential
6. Tungare, A.: Le Corbusier's principles of city planning and their application in virtual environments (2001)
7. Le Corbusier: How a utopic vision became pathological in practice | Orange Ticker. https://orangeticker.wordpress.com/2013/03/05/le-corbusier-how-a-utopic-vision-became-pathological-in-practic/
8. Sahlqvist, S., Goodman, A., Cooper, A.R., Ogilvie, D.: Change in active travel and changes in recreational and total physical activity in adults: longitudinal findings from the iConnect study. Int. J. Behav. Nutr. Phys. Activity **10**(1), 28 (2013)
9. Mohammed BIN RASHID APPROVES the Traffic and Transport Plan 2030 (2016)
10. 2021–2030 - The Official Portal of the UAE Government. https://government.ae/en/more/uae-future/2021-2030
11. A Set of European CPTED Standards for Secure Cities Safe & Secure cities through urban design and planning: standardizing the process (2013)
12. Jakubiec, J.A.: Comfort and Perception in Architecture. Springer, Heidelberg (2016)

13. Dodge, R., Daly, A., Huyton, J., Sanders, L.D.: The challenge of defining wellbeing. Int. J. Wellbeing **2**, 222–235 (2012)
14. Hosking, J., Macmillan, A., Connor, J., Bullen, C., Ameratunga, S.: Organisational travel plans for improving health. In: Macmillan, A. (ed.) Cochrane Database of Systematic Reviews. Wiley, Chichester (2010)
15. Ogilvie, D., Egan, M., Hamilton, V., Petticrew, M.: Promoting walking and cycling as an alternative to using cars: systematic review. BMJ **329**(7469), 763 (2004)
16. Austin, J.L., Jütersonke, O.: Understanding the grammar of the city: urban safety and peacebuilding practice through a semiotic lens. Paper Series of the Technical Working Group on the Confluence of Urban Safety and Peacebuilding Practice (2016)
17. Smart City Challenge | US Department of Transportation. https://www.transportation.gov/smartcity

Key Challenges of Smart Railway Station

R. E. Shaltout[1,2(✉)]

[1] Mechanical Power Engineering Department, Faculty of Engineering, Zagazig
University, Zagazig, Egypt
Reamer@eng.zu.edu.eg
[2] Civil Engineering Program, Faculty of Engineering and Material Science,
German University in Cairo, Cairo, Egypt

Abstract. This paper introduces the main concepts of establishing a smart railway station under the context of smart city. In the recent years, cities are digitally developing to enhance the all aspects of the urban life including economic, social and environmental aspects. These three aspects are considered to be the main pillars of sustainable development of a smart city. A core element in the smart city development is the mobility. This might include the physical mobility of people or the economic mobility. Introducing the smartness concept in the transportation infrastructure will influence the population growth and business needs. Railway points of contact such as the stations that can offer excellent economics development, environmental performance, and punctuality for passengers are recognized as a fundamental component of urban development. Stations are always considered the interconnection between different transportation modes which had to be adapted to the urbanisation growth. The proposed work presents a state of the art review for rail smart station design and illustrates the key challenges that can be faced when thinking in establishment of smart railway station.

Keywords: Rail station · Smart station · Smart mobility · Intelligent transportation · Smart city

1 Introduction

Population growth, enormous level of industrialization and urbanization are constantly pushing in the infrastructure and creating a jostling situation for the consumption of resources. Cities have an important role in the economic and environmental aspects all over the world [1]. The concept of smart cities became popular in the scientific literature in the last two decades [2]. Currently, all cities should find all innovative solutions and ways in order to face and manage the new challenges they exposed to. These challenges may include urban planning, mobility, security, energy, transportation, healthcare, utility usage and governance. Various researchers have studied the smart city definition [3–5].

The concept of smart cities was discussed in different ways in academic articles. The main common aspects between these studies is that, smart cities considers the basic elements of sustainable urban communities including: environmental, social, economic and cultural aspects. A smart city is sustainable city as it takes into account all the

© The Author(s), under exclusive license to Springer Nature Switzerland AG 2021
I. El Dimeery et al. (Eds.): JIC Smart Cities 2019, SUCI, pp. 287–294, 2021.
https://doi.org/10.1007/978-3-030-64217-4_33

aforementioned aspects in addition to enhancing the quality of life and fairness for all citizens [6].

A smart city uses digital technologies or information and communication technologies to enhance the quality and performance of urban facilities by carrying out the control over the delivered services to the urban residents [7, 8]. In a smart city, researchers highlighted that all various system should work in an integrated form and no system works in isolation of the others, especially in dens environments. In the literature, many researchers defined various dimensions of the smart cities [9–12]. By investigating all dimensions of smart cities, it can be noticed that transportation and infrastructure present a crucial axes of a smart city. Rail stations form a connection between these two axes.

Rail stations are considered to be the point of trust between the passengers and the service providers. As it can be defined the point of interest in the end-to-end journey, stations should be designed to better accommodate all modes of transportation to increase the passenger comfort and decrease the time wasted during the journey. Congestion is a major problem for all exiting railway stations especially at peak hours. Recent studies by the department for transport in UK predicted that the rail demand will increase by around 85 billion passenger kilometers in 2033. It was also found that 40% of trains are required during the peak period which can be enabled by the digitalization of the provided railway services.

The present paper discuss the main challenges facing the implementation of smart rail station with in the context of smart cities. The paper introduces an overview analysis on one of the important axes of the design of smart cities, that is, smart rail transportation system that supports the economic growth and urban planning of smart cities.

2 Smart Transportation Systems

Smart transportation systems form a major element in the smart city model. Fig. 1, indicates the major elements of a smart city model. As it can be noticed, the transportation module within the model is subsequently based on smart mobility, smart connectivity and smart energy consumption. Smart transportation is needed for the proper and efficient mobility of people, freight and services. The future of transportation systems lies not only on the construction of new stations, roads, tunnels and bridges, but also on the use of intelligent transportation system (ITS) [13].

The main idea of using ITS, is to provide the traveller or the transport service user with all necessary information to increase the service reliability as well as sustainability. This can be done by the application of computer, microchips, electronics, information and communication technologies (ICT) and management strategies in an integrated manner. Implementing ITS solutions will definitely develop the way that people commute which directly influence the quality of life in modern hi-tech cities. These solutions can also provide consumers an access to a smarter, safer, and faster way to travelling [13, 14].

Smart mobility, energy and connectivity are considered to be the triangle of the smart transportation systems with in a smart city Fig. 1. To solve part of the traffic

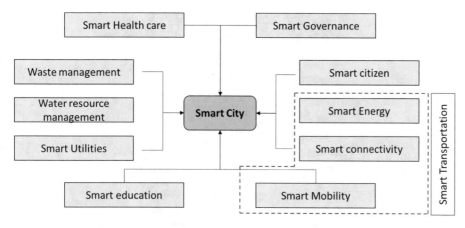

Fig. 1. Smart transportation system in the smart city model

congestion problems, efficient public transportation system may be applied. But, the smart mobility concept searches for innovative solutions for people mobility in smart cities. For example, smart mobility solutions implements the advances and technology and proactive passenger behaviour. Mobility has many aspects rather than the people behaviour, it also includes economic and environmental aspects. Smart connectivity supports both mobility and energy concept in the transportation system by the application of ICTs in optimizing the traffic flow and efficient energy consumption. This will have a significant environmental impact of implementing smart energy solutions in reducing CO_2 emissions.

3   Smart Station Concept

Railway transport is always in continuous growth and development. Rail stations form also part of this revolution in the rail industry. Since the rail stations already exist since the beginning of 19th century, they have influenced by the recent advances in the transportation sector by introducing the concept of intelligent transportation systems. The importance of rail stations comes from its role and services provided to the station users and the whole environment surrounding the station. There is no unique criteria for defining the concept of smart stations, however it should be defined by the same criteria used for the definition of smart city.

Smart stations are designed in a revolutionary way to make use of available data and technologies to increase the rail transportation's reliability and sustainability. All infrastructure managers and political organizations devote for the importance of renovating rail stations to maximize the role they play beyond being only a transportation hub (Fig. 2).

Modern design of smart stations should built on the basis of smart infrastructure, smart mobility, smart management and sustainability. Therefore, rail station can be smart by promoting its important place in smart city. The main three pillars of sustainability targeted by the smart station design are: *1)* Environmental targets-including

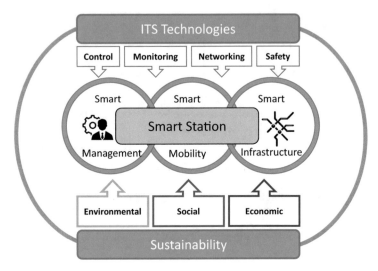

Fig. 2. Smart station concept

the use of a low carbon power source (i.e., a renewable energy source based on solar capturing technologies as well as noise reduction by improved preventive maintenance system through monitoring of the railway system parameters; *2)* Social targets–including the improvement of safety inside the rail stations and regaining the mode share concept for increased urban mobility objectives; *3)* Economic targets–including the reduction of track maintenance and operation costs as well as reduction of overall life cycle cost (LCC) of the rail system.

4   Smart Station Challenges

Stations present a key role in the rail transportation as they are considered the point of interaction between the transportation authority, operators and passengers. Smart rail stations have a direct impact on the smart city planning and design. In the following subsections, the key challenges that might face the design and commissioning of rail station will be stated.

4.1   Safety and security

People safety is the number one priority in rail transportation. The safety and security challenges in the rail stations may include: terrorism act prevention; suicide attacks; fire and explosion risks; abandoned bags identification, control of access points and bomb threats. When these challenges are managed successfully by the smart security and monitoring systems in rail stations, all rail users and employees will have a confidence in the rail stations' safety and the overall railway system safety. Smart Railway Stations transport systems have benefited from innovative technologies for railway infrastructure managers and train operating organizations to help them make more effective decisions and improve railway station security and safety.

Embedded computing systems have been used for rail station surveillance and monitoring [15]. Those systems are made for checking system status from a physical-security perspective, so as to distinguish interruptions and other natural occasions. Embedded systems (ES) also are utilized in cyber-physical checking and control applications highlight progressively unpredictable architectures and strict requirements about security, privacy and dependability.

4.2   Intermodality

Stations acts as a key point in the intermodal concept. A recent UK study including 10000 respondents showed that the pain points for passengers are at the points at which they shift their mode of transport. It was found that 75% of UK journeys are affected by pain points. This percent increase to 86% when rail transport is involved. Smart rail stations should be designed in such form that evolves the following aspect: Improving the speed of passenger interchange; Reducing the complications associated to multi-modal journeys planning and execution; Improving the reliability and connectivity offered to modal shift users.

4.3   Connectivity

Connectivity is more than just a pillar in smart cities, connectivity is the foundation of smart planning. One of the main items in connectivity is Internet of Things (IOT) sensors which rely inertly on their ability to transmit data in real time, which help in smart planning decision making. Smart rail stations implement sensors that can be added to the vehicle and station infrastructure. The data transmission is done by using efficient connection and communication system design that implement low-power, short-range wireless network (such as Zigbee) with a local data collection unit (or 'gateway') for condition monitoring of the rail system [16]. The collected data can then be processed and transmitted in real time to the driver, operator, control room, entities in charge of maintenance and even to the station users. The challenge is to define possible solutions for data transmission and communication using either wayside devices (density of wayside readers for effective coverage is not known). This will be based on the density of data transferred.

4.4   Energy Consumption

In the recent years the energy efficiency gained a lot of attention due to its significant impact on the reduction of CO_2 emissions. It was found that, in 2010 transportation is emitting almost 14% of the total amount of greenhouse gas emissions and this percent is forecasted to be increased by the double in 2050 [17]. Electric railway systems (ERSs) is one of the biggest consumers of energy in the transportation sector. It is also considered the first targeted candidate in the reduction of CO_2 emissions by using alternative energy resources and maximizing the energy efficiency. Smart rail stations present a key role in such initiatives.

Energy storage systems can be fitted in rail stations to store the amount of energy produced from alternative resources such as Photo Voltaic panels (PV) and Regenerative Braking (RB) which can reduce the total energy consumption of ERSs by 1 - to 40% [18]. In addition to the use of smart and energy efficient systems inside the stations like (efficient escalators and lightening systems), smart stations also should use the smart grid concept which provide a new prospective of energy management strategies in ERSs [19, 20].

4.5   Governance

Under the context of smart cities, the concept of smart governance has an utmost importance. Smart governance meaning is the use of Internet and Computer Technologies (ICTs) in crafting a progressive public-government partnership [21, 22].

By investigating the different models of smart governance which include: Government to Citizen (G2C); Government to Business (G2B); Government to Government (G2G) and Government to Employee (G2E), smart stations are most likely to be affected by the C2C and C2B models. The smart governance opens a direct communication channels between the government and individuals. It will allow station users to directly report their feedback and can easily get a response from the infrastructure managers about the provided service in the stations. This might include safety and security problems, data protection issues and regulation of the relationship between rail customers, rail operators and authorities that manage the stations like the infrastructure managers [7, 22, 23].

Reducing the problems faced by entrepreneurs and business leaders in the interaction with the government will always help in the economic growth. It will encourage private sector in investing in smart stations and will foster the implementation of new technologies in the transportation sector in general towards a better mobility and provide real time monitoring, save cost and time. This can be also beneficial to companies that needs a direct knowledge about the latest policies regulations and even get an access to data collected by the government which can be used for economic forecasting and planning.

4.6   Integrity and Social Recognition

At the governmental level, a clear transportation strategy should be adopted throughout an innovative techniques and measures that provide decreasing all levels of barriers for investors and innovators to implement the new technologies in the design, construction and operation of futuristic rail stations. A pilot station model can be adopted by the transportation authorities with the help of other transportation modes operators to act as a measure for the advances in the smart rail station design and it compatibility with the other modes. This can be considered as a framework towards the analysis of the impact of new technological advances implemented in the stations.

5 Conclusions

The present paper highlights the challenges that might face the smart rail stations within the context of smart city concept. Rail stations became a point of interest as it considered the point of interactions between passengers and transportation authorities all over the world. It is necessary to increase the volume of investments in smart stations to decrease the level of congestions and overcrowding.

All rail resources must be aligned to encourage innovators and endorsement of industrial partners to provide a proof of concept of the technologies and innovations that can increase the performance of rail station as well as the rail competition.

Smart stations should be designed in away that maximizes the role it plays in a smart city rather than being a simple transportation hub. By making stations smart, its influence will be translated to a successful business model.

References

1. Mori, K., Christodoulou, A.: Review of sustainability indices and indicators: towards a new city sustainability index (CSI). Environ. Impact Assess. Rev. **32**(1), 94–106 (2012)
2. Albino, V., Berardi, U., Dangelico, R.M.: Smart cities: definitions, dimensions, performance, and initiatives. J. Urban Technol. **22**(1), 3–21 (2015)
3. Chen, T.M.: Smart grids, smart cities need better networks [Editor's Note]. IEEE Netw. **24** (2), 2–3 (2010)
4. Bakıcı, T., Almirall, E., Wareham, J.: A smart city initiative: the case of Barcelona. J. Knowl. Econ. **4**(2), 135–148 (2013)
5. Barrionuevo, J.M., Berrone, P., Ricart, J.E.: Smart cities, sustainable progress. IESE Insight **14**(14), 50–57 (2012)
6. Hollands, R.G.: Will the real smart city please stand up? Intelligent, progressive or entrepreneurial? City **12**(3), 303–320 (2008)
7. Nam, T., Pardo, T.A.: Conceptualizing smart city with dimensions of technology, people, and institutions. In: Proceedings of the 12th Annual International Digital Government Research Conference: Digital Government Innovation in Challenging Times, pp. 282–291. ACM (June 2011).
8. Shapiro, J.M.: Smart cities: quality of life, productivity, and the growth effects of human capital. Rev. Econ. Stat. **88**(2), 324–335 (2006)
9. Mahizhnan, A.: Smart cities: the Singapore case. Cities **16**(1), 13–18 (1999)
10. Giffinger, R., Pichler-Milanović, N.: Smart cities: ranking of European medium-sized cities. Centre of Regional Science, Vienna University of Technology (2007)
11. Alawadhi, S., Aldama-Nalda, A., Chourabi, H., Gil-Garcia, J.R., Leung, S., Mellouli, S., Nam, T., Pardo, T.A., Scholl, H.J., Walker, S.: Building understanding of smart city initiatives. In: International Conference on Electronic Government, pp. 40–53. Springer, Heidelberg (September 2012)
12. Lombardi, P., Giordano, S., Farouh, H., Yousef, W.: Modelling the smart city performance. Innov.: Eur. J. Soc. Sci. Res. **25**(2), 137–149 (2012)
13. Chowdhary, M.A., Sadek, A.: Fundamentals of Intelligent Transportation Systems Planning. Artech House Inc., US (2003)
14. Williams, B.: Intelligent transportation systems standards. Artech House, London (2008)

15. Bocchetti, G., Flammini, F., Pragliola, C., Pappalardo, A.: Dependable integrated surveillance systems for the physical security of metro railways. In: 2009 Third ACM/IEEE International Conference on Distributed Smart Cameras (ICDSC), pp. 1–7. IEEE (August 2009)
16. Ulianov, C., Hyde, P., Shaltout, R.: Railway applications for monitoring and tracking systems. In: Sustainable Rail Transport, pp. 77–91. Springer, Cham (2018)
17. Pachauri, R.K., Allen, M.R., Barros, V.R., Broome, J., Cramer, W., Christ, R., Church, J.A., Clarke, L., Dahe, Q., Dasgupta, P., Dubash, N.K.: Climate change 2014: synthesis report. Contribution of Working Groups I, II and III to the Fifth Assessment Report of the Intergovernmental Panel on Climate Change, p. 151. IPCC (2014)
18. González-Gil, A., Palacin, R., Batty, P.: Sustainable urban rail systems: strategies and technologies for optimal management of regenerative braking energy. Energy Convers. Manag. **75**, 374–388 (2013)
19. Şengör, İ, Kılıçkıran, H.C., Akdemir, H., Kekezoğlu, B., Erdinc, O., Catalao, J.P.: Energy management of a smart railway station considering regenerative braking and stochastic behaviour of ESS and PV generation. IEEE Trans. Sustain. Energy **9**(3), 1041–1050 (2017)
20. Collotta, M., Pau, G.: An innovative approach for forecasting of energy requirements to improve a smart home management system based on BLE. IEEE Trans. Green Commun. Netw. **1**(1), 112–120 (2017)
21. Meijer, A., Bolívar, M.P.R.: Governing the smart city: a review of the literature on smart urban governance. Int. Rev. Adm. Sci. **82**(2), 392–408 (2016)
22. Chourabi, H., Nam, T., Walker, S., Gil-Garcia, J.R., Mellouli, S., Nahon, K., Pardo, T.A., Scholl, H.J.: Understanding smart cities: an integrative framework. In: 2012 45th Hawaii International Conference on System Sciences, pp. 2289–2297. IEEE (January 2012)
23. Bătăgan, L.: Smart cities and sustainability models. Informatica Economică **15**(3), 80–87 (2011)
24. Pereira, G.V., Parycek, P., Falco, E., Kleinhans, R.: Smart governance in the context of smart cities: a literature review. Inf. Polity **23**(2), 143–162 (2018)

A Smart-Left Decision Support System for Flashing Yellow Arrow Traffic Signals

Hatem Abou-Senna$^{(\boxtimes)}$, Essam Radwan, and Hesham Eldeeb

Center for Advanced Transportation Systems Simulations (CATSS), Department
of Civil, Environmental and Construction Engineering (CECE), University of
Central Florida, 4000 Central Florida Blvd., Eng II-Room 301E,
P.O. Box 162450, Orlando, FL 32816-2450, USA
{habousenna, Ahmed.Radwan}@ucf.edu,
Hesham_eldeeb@yahoo.com

Abstract. The flashing yellow arrow (FYA) signal display creates an opportunity to enhance the left-turn phase with a variable mode that can be changed on demand. This paper presents phase II of the research. Phase I developed a decision support system (DSS) to select the FYA left-turn mode, and changing by time of day at intersections. There was a need to continue to refine the interactive framework to improve its service. However, the ultimate objective of the continued research of phase II was to demonstrate the ability to execute the automation of the process. Phase II of the FYA project provided additional intersection data that refined the model. Virtual testing of the DSS was first conducted using VISSIM application programming interface (API) before the field testing environment. A Custom communications software was developed to retrieve instantaneous channel input data, synchronize opposing thru green phase, analyze traffic information, provide the algorithm decision, and generate a real-time log recording the events to determine whether it would be optimal to switch the red arrow to a flashing yellow arrow. The algorithm determines the time interval between the successive arrivals of vehicles and computes the corresponding headway for each lane by cycle on a second-by-second basis. The DSS was ultimately tested at two different intersections in Seminole County. The FYA 4-section configuration provides the opportunity for a fully adjustable system and provides the TMCs with more tools to operate the intersections as efficiently as possible at peak and off-peak times.

1 Introduction

The all-new four-arrow configuration with the flashing yellow arrow (FYA) signal display creates an opportunity to enhance the left-turn signal with a variable mode that can be changed by time of day on demand. Phase I of this research project provided the framework and detailed process of developing a decision support system (DSS) with the use of an interactive model [1]. The DSS facilitates the selection of the flashing yellow arrow left-turn mode and changing by time of day at intersections. There was a need to continue to refine the interactive framework to improve its service as a decision support system. The framework already allowed for an interactive evaluation of the permissive left-turn phase and was able to recommend phasing mode by time of day.

I. El Dimeery et al. (Eds.): JIC Smart Cities 2019, SUCI, pp. 295–315, 2021.
https://doi.org/10.1007/978-3-030-64217-4_34

However, the ultimate objective of the continued research of phase II was to demonstrate the ability to execute the automation of the process in a field testing environment through the use of an active controller.

The phase II portion of the flashing yellow arrow project provided additional video data that was extracted on a second-by-second basis [2]. The master database was increased to 38 intersections with locations across the State of Florida. The data extraction process in phase II was completed to match the basic prioritized parameters that were used to refine the developed model in phase I. With an expanded database, the model's coefficient of correlation was improved because of the increased model domain. The final total hours used in the statistical analysis were 1,058 h. Based on the analysis, the neural networks model provided the highest correlation between the independent variables, with a coefficient of correlation reaching 90%. The final refined neural network model, along with the decision support system criteria, was first tested in a simulated environment before moving on to the field testing environment. Virtual testing or Software-in-the-Loop-Simulation (SILS) is used to prove or test the software. Virtual testing was conducted using the latest version of the microscopic traffic simulation model VISSIM 7.13 along with its application programming interface modules, which included the use of COM (Component Object Module) server as well as the VISVAP (VISSIM Vehicle Actuated Programming) module [3]. These components, unified under the Windows operating environment and integrated with VISSIM, provide the ability to simulate one or more intersections with a unifying controller management interface and the ability to model both standard and custom saturated timing strategies. Virtual testing of the decision support system using VISSIM application programming interface (API) confirmed the applicability and validity of the procedure and logic.

The main objective of this research is to test the final refined decision support system (DSS) and the algorithm criteria based on the cycle by cycle data in a field testing environment as a "proof of concept" before actual implementation in the field. The testing was conducted at Seminole County Traffic Engineering Lab in Orlando, Florida where actual intersection field data was obtained through loop detector mapping to the controller in the lab in real-time mode. This process is called HILS (Hardware-in-the-Loop-Simulation) testing where an actual traffic controller is needed along with a controller interface device (CID) such as the data logger DI-161. The term HIL is used to describe a test methodology where executable code such as algorithms or even an entire controller strategy, usually written for a particular system, is tested within a field environment that can help prove a concept or test a software package.

Seminole County Traffic Engineering Staff were very helpful in setting up the testing environment and mapping the intersection loop detectors from the field to the cabinet in the lab. The CCTV cameras were also setup to monitor both the study approach as well as the traffic signal indication. The intersection vehicle detection system through the loop occupancy and the CCTV cameras were connected to the data logger and the communication software to receive data signaling the traffic flow on a second by second basis. The permissive green times and the opposing thru traffic were determined on a cycle-by-cycle basis from the field by the data logger software. The logic was based on modeling the inter-arrival time of vehicles and calculating the minimum headway and gap time per lane for the opposing traffic from the loop detectors data for the first two to three cycles before recommending a decision for the left-turn signal head, either flashing or not for the next cycle. This iterative process is repeated throughout the day on a cycle by cycle basis as will be explained in greater detail in the following sections.

2 Literature Review

Several references were examined in the process of determining the background information and motivation for this research. Numerous past studies have been conducted to develop guidelines or warrants for determining left turn signal control at signalized intersections which are often presented in a sequence format such as flowcharts or a step-by-step process using a ranking score. These studies examined various traffic parameters that have an effect on the signal operation such as traffic volume, delay, geometry, crash data, speed and many other related factors. Furthermore, studies related to FYA focused mainly either on driver's comprehension of the FYA indication or its safety performance.

Agent developed guidelines for the use of protected/permissive left turn phasing based on accident analysis and concluded that it is the preferred approach of left-turn phasing because of savings in time compared with protected only phasing [4]. However, it creates an increased accident potential and it should not be used if certain conditions related to speed limits, number of crossing lanes and historical crash data exist. In a follow up to Agent's study by Stamatiadis et al., a two-step approach is considered in the decision whether to install left-turn phasing [5]. The recommendations made take into account many variables, including left-turn volumes, accident rates, product of opposing and left-turn volumes, and left-turn delays. Al-Kaisy and Stewart developed a volume based warrant approach for protected left-turn phase at signalized intersections using simulation [6]. They indicated that the volume of opposing through traffic may have little impact on when a protected left-turn phase is warranted and concluded that the mode was not to be determined by traffic volumes alone and is a combination of multiple traffic variables as alluded to previously in the parameters deemed by Agent [4]. Zhang et al. tried to combine both existing empirical warrants and an optimization-based volume warrant similar to that proposed by Al-Kaisy and Stewart [6] to develop a comprehensive decision flowchart for the selection of left-turn control modes [7]. Chang et al. developed a hybrid model for left turns based on saturation flow rates using simulation [8].

Hu et al. introduced a new approach based on the analytical hierarchy process to determine left-turn control types using a ranking score considering several traffic factors in a case study of 14 intersections in the Reno-Sparks, Nevada area [9]. Chen et al. evaluated the safety impacts of changing left-turn signal phasing from permissive to protected/permissive or protected-only at 68 intersections in New York City using a rigorous quasi-experimental design accompanied with regression modeling [10]. The results suggested that left-turn phasing should not be treated as a universal solution, considering the trade-offs between safety and delay, and many other factors such as geometry, traffic flows and operations. Gal-Tzur et al. developed a decision support system for controlling traffic signals using a systematic scanning of a wide range of alternatives and includes a special algorithm that recommends the most promising strategies in a statistical decision tree format [11]. Ozmen et al. introduced a new guideline for determining left turn control type based on the principles of multi-criterion decision analysis and provided an index based recommendation using weights and scores; a numerical scale was used to compare each type of left-turn control with the others instead of an absolute type [12].

Yu et al. [13] developed guidelines for recommending the most appropriate left-turn phasing treatments at signalized intersections utilizing both operational and safety impacts to construct a flowchart similar to NCHRP 457 report by Bonneson and Fontaine [14]. However, the chart indicated some additional parameters for consideration.

As can be concluded from the literature, a majority of studies tried to develop warrants and guidelines for the permissive left turn phase. Although the developed guidelines are applicable, they are not considered practical to be implemented in the field. Prior information is needed for before and after study conditions. For left-turn volume warrants, almost all studies were consistent in applying the cross product methodology of left-turn and opposing through volumes as the main warrant. A cross product is generally accepted as a signal warrant but lacks the ability to be inclusive of all intersections especially at different time scales. Furthermore, Cycle-by-cycle data have been shown to be effective in the analysis of a signalized intersection with measures of effectiveness such as volume-to-capacity ratios, arrival type, headway calculation and average vehicular delay. This research develops an integrated general purpose data collection module that time stamps detector and phase state changes within a National Electrical Manufacturers Association actuated traffic signal controller and uses those data to provide recommendations for the flashing yellow arrow left-turn mode on a cycle-by-cycle basis.

3   Hardware/Software Description

The board used in this project is a DATAQ Instruments™ event logger model DI-161 as shown in Fig. 1. The board is local area network (LAN) based and connects to a computer using an Ethernet cable. It has eight input channels, each of which can operate in one of three modes: **Count**, **Event** or **State**. The Count mode sums the total number of events during each reporting interval. The Event mode reports a single occurrence during each interval even if multiple events may occur. The State mode reports how long an event lasts.

The pilot study is designed to monitor up to four lanes in each direction as well as the start/stop state of the thru green phase. The four lanes are monitored in the field either via loop or video detectors; each is connected to an input channel, channels F0–F3. These channels are configured to operate in **Count** mode. The start/stop state of the thru green phase is monitored by Channel F4 which is configured to operate in **State** mode. Figure 1 shows the wires, channel connections and the light bulb on channel F4 indicating that the opposing thru green phase is ON.

The basic communications software that accompanied the DI-161 board was limited compared to what was required in this project. It essentially establishes connection

Fig. 1. DATAQ model DI-161 board with wires and connections

with the board and generates a text file with the data received through the input channels. However, in order to access the text file, data logging has to stop. What was needed, however, was real-time access to the channel data as it is received by the board so that the algorithm can analyze traffic information in real-time and make accurate decisions. A custom communications software was needed on top of the basic software which has three main functions; control the hardware, display real-time status and execute the proposed FYA algorithm. The UCF research team developed a specific code to retrieve instantaneous channel input data, synchronize opposing thru green phase, analyze traffic information, provide the algorithm decision, and generate a real-time log recording the events. The software was developed using the C# language under Microsoft's™ Visual Studio 2013 development environment. The main screen of the developed software and its different components are shown in Fig. 2.

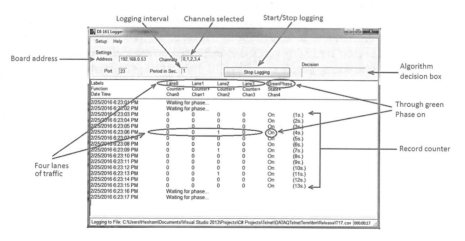

Fig. 2. UCF custom data logger software.

4 Input Data

The custom software monitors up to five channels simultaneously; up to four channels for the traffic lanes in **Count Mode** and one channel for the thru green phase in **State Mode**. The algorithm analyzes the traffic flow data received during the thru green phase which is synchronized by the input on the phase channel. There is also a configuration file for specifying different parameters needed for each intersection. The configuration file specifies the opposing number of lanes, analysis period to determine the number of cycles to be analyzed before providing a decision, application period which specifies the frequency to provide a decision after the analysis period whether after each cycle or more and lastly, the actuated cycle length in seconds.

5 Flashing Yellow Arrow Algorithm

5.1 Headway Modeling

Modeling the arrival of vehicles was an essential step in the algorithm logic. The vehicle arrival is obviously a random process. Hence, vehicle arrival needs to be characterized statistically. Vehicle arrivals can be modeled in two inter-related ways; modeling the time interval between the successive arrivals of vehicles or modeling how many vehicles arrive in a given interval of time. In the former approach, the random variables represent the time denoting interval between successive arrivals of vehicles and hence some suitable continuous distribution can be used to model the vehicle arrival. In the later approach, the random variables represent the number of vehicles arrived in a given interval of time and hence takes some integer values. In this case, a discrete distribution can be used to model the process.

The developed algorithm utilizes the former approach and uses continuous distributions to model the vehicle arrival process. However, the inter-arrival time or the time headway is not constant due to the stochastic nature of vehicle arrival and also the behavior of vehicle arrival is different at different flow conditions. Therefore, it may be possible that different distributions may work better at different flow conditions.

The negative exponential distribution is used when the traffic is low and is the simplest of the distributions in terms of computation effort. The normal distribution on the other hand is used for highly congested traffic and its evaluation requires standard normal distribution tables. The Pearson Type III distribution is the most general case of negative exponential distribution and can be used for intermediate or normal traffic conditions. Unlike many other distributions, one of the key advantages of the negative exponential distribution is the existence of a closed form solution to the probability density function. The negative exponential distribution is closely related to the Poisson distribution which is a discrete distribution. The probability density function of Poisson distribution is given as:

$$p(x) = \frac{\lambda^x e^{-\lambda}}{x!} \tag{1}$$

Where, p(x) is the probability of x events (vehicle arrivals) in some time interval (t), and λ is the expected (mean) arrival rate in that interval. If the mean flow rate is q vehicles per hour, then $\lambda = \frac{q}{3600}$ vehicles per second. Here, λ is defined as the average number of vehicles arriving in time t. If the flow rate is q vehicles per hour, then,

$$\lambda = \frac{q \times t}{3600} = \frac{t}{\mu} \tag{2}$$

Since mean flow rate is the inverse of mean headway, an alternate way of representing the probability density function of negative exponential distribution is given as

$$f(t) = \frac{1}{\mu} e^{\frac{-t}{\mu}} \tag{3}$$

Where $\mu = \frac{1}{\lambda}$ or $\lambda = \frac{1}{\mu}$. Here, μ is the mean headway in seconds which is again the inverse of flow rate. Using the above equations, the observed headway frequency distribution between any interval and flow rate can be computed. Statistical analysis was carried out for the field data to determine the minimum acceptable gap time based on the observed headways. Figure 3 shows the distribution of the observed headways for each case of crossing lanes one to four. The analysis shows that the observed headways follow a negative exponential distribution.

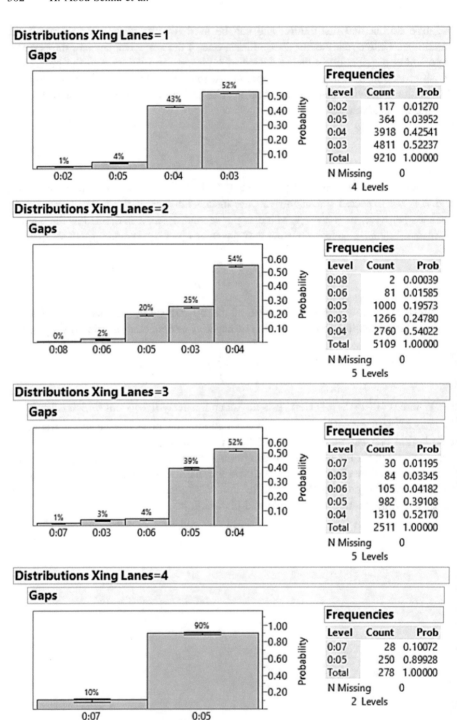

Fig. 3. Distribution of gaps by number of crossed lanes.

5.2   Algorithm Logic

The proposed algorithm is implemented with the goal of safely optimizing traffic operations. In the case of a red arrow signaled for a left-turn, the opposing thru traffic during the green phase is constantly analyzed in real-time to determine whether it would be optimal to switch the red arrow to a flashing yellow arrow. The decision is made based on a number of parameters which include the minimum headway of vehicles in the opposing traffic, the number of lanes to cross, and the number of cycles to be analyzed prior to making the decision.

The algorithm determines the time interval between the successive arrivals of vehicles for each lane independently and computes the corresponding headway for each lane by cycle on a second-by-second basis. It then determines the minimum gap duration by dividing the headway by the flow per lane.

$$\text{Gap per Lane} = \text{Headway/Flow} \qquad (4)$$

The algorithm then picks the minimum headway and compares it to the minimum acceptable gap in seconds needed for a vehicle to safely cross the given number of lanes. The thresholds used for different crossing number of lanes were obtained from the database of 30,000 cycles collected from the field. If the minimum headway for the corresponding number of lanes is achieved and repeated for a certain number of times, for example, at least five times during the analysis period (whether one or two cycles) which is also an input to the algorithm, the decision is made to switch to a flashing yellow mode. Otherwise, a red arrow is decided upon. The following durations in seconds, shown in Table 1, are the minimum acceptable thresholds used to determine the minimum headways for different number of lanes crossed which are used in the decision making process. These thresholds were developed based on the statistical analysis of the cycle by cycle data collected from the field. Tables 2 and 3 are excerpts from the field data for crossing three and four lanes, respectively.

Table 1. FYA algorithm criteria

No. of Opposing Lanes Crossed	Min acceptable Gap Time	Comments
1 Lane	3.0 s.	1 Thru lane
2 Lanes	3.5 s.	2 Thru lanes or 1 Thru + 1 RT
3 Lanes	4.0 s.	3 Thru lanes or 2 Thru + 1 RT
4 Lanes	4.5 s.	4 Thru lanes or 3 Thru + 1 RT

Table 2. Minimum acceptable gap time for crossing three lanes by cycle

Collection Period			Speed	Crossing Lanes	Start Clock Time	End Clock Time	Permitted Green Time	Permitted Left Turns	Opposing Volumes			Left Turns	
Day	Hour	Peak							Through	Right	Total	Gap	Follow-Up
Wed	7:00	Peak	45	3	0:00	2:17	2:17	1	13	0	13	0:04	0:02
Wed	7:00	Peak	45	3	2:31	3:07	0:36	0	8	0	8	0:04	0:02
Wed	7:00	Peak	45	3	3:21	4:35	1:14	0	13	0	13	0:04	0:02
Wed	7:00	Peak	45	3	4:51	5:18	0:27	0	3	0	3	0:04	0:02
Wed	7:00	Peak	45	3	5:31	6:52	1:21	2	15	0	15	0:04	0:02
Wed	7:00	Peak	45	3	7:04	7:29	0:25	0	5	0	5	0:04	0:02
Wed	7:00	Peak	45	3	7:42	8:24	0:42	0	9	0	9	0:04	0:02
Wed	7:00	Peak	45	3	8:34	9:21	0:47	0	8	0	8	0:04	0:02
Wed	7:00	Peak	45	3	9:39	9:57	0:18	0	2	0	2	0:04	0:02
Wed	7:00	Peak	45	3	10:14	11:38	1:24	0	19	0	19	0:04	0:02
Wed	7:00	Peak	45	3	11:53	12:30	0:37	1	4	0	4	0:04	0:02
Wed	7:00	Peak	45	3	12:43	13:21	0:38	0	12	2	14	0:04	0:02
Wed	7:00	Peak	45	3	13:35	16:07	2:32	0	19	1	20	0:04	0:02
Wed	7:00	Peak	45	3	16:21	17:48	1:27	2	13	3	16	0:04	0:02
Wed	7:00	Peak	45	3	18:02	19:25	1:23	1	19	0	19	0:04	0:02
Wed	7:00	Peak	45	3	19:40	21:04	1:24	0	18	0	18	0:04	0:02
Wed	7:00	Peak	45	3	21:20	22:45	1:25	3	22	1	23	0:04	0:02
Wed	7:00	Peak	45	3	23:13	26:04	2:51	3	30	1	31	0:04	0:02
Wed	7:00	Peak	45	3	26:53	27:45	0:52	0	16	0	16	0:04	0:02
Wed	7:00	Peak	45	3	28:00	29:24	1:24	0	25	1	26	0:04	0:02
Wed	7:00	Peak	45	3	29:54	31:06	1:12	2	15	1	16	0:04	0:02
Wed	7:00	Peak	45	3	31:20	32:43	1:23	3	25	0	25	0:04	0:02
Wed	7:00	Peak	45	3	33:14	34:25	1:11	1	21	1	22	0:04	0:02
Wed	7:00	Peak	45	3	34:42	36:05	1:23	1	23	0	23	0:04	0:02
Wed	7:00	Peak	45	3	36:34	37:45	1:11	1	31	0	31	0:04	0:02
Wed	7:00	Peak	45	3	38:15	39:27	1:12	2	21	0	21	0:04	0:02
Wed	7:00	Peak	45	3	39:57	41:05	1:08	0	21	0	21	0:04	0:02
Wed	7:00	Peak	45	3	41:21	42:45	1:24	1	30	2	32	0:04	0:02
Wed	7:00	Peak	45	3	43:00	44:24	1:24	0	25	1	26	0:04	0:02
Wed	7:00	Peak	45	3	44:38	46:08	1:30	3	40	0	40	0:04	0:02
Wed	7:00	Peak	45	3	46:33	47:45	1:12	1	20	1	21	0:04	0:02
Wed	7:00	Peak	45	3	48:00	49:24	1:24	2	33	2	35	0:04	0:02
Wed	7:00	Peak	45	3	49:55	51:05	1:10	2	41	0	41	0:04	0:02
Wed	7:00	Peak	45	3	51:20	52:45	1:25	2	26	1	27	0:04	0:02
Wed	7:00	Peak	45	3	53:13	54:24	1:11	0	44	0	44	0:04	0:02
Wed	7:00	Peak	45	3	54:40	56:07	1:27	2	34	0	34	0:04	0:02
Wed	7:00	Peak	45	3	56:20	57:44	1:24	1	37	4	41	0:04	0:02
Wed	7:00	Peak	45	3	58:00	59:25	1:25	0	34	0	34	0:04	0:02
Wed	7:00	Peak	45	3	59:41	59:59	0:18	0	6	0	6	0:04	0:02
Wed	9:00	Non	45	3	0:32	0:58	0:26	1	1	0	1	0:04	0:03
Wed	9:00	Non	45	3	1:10	1:44	0:34	1	12	1	13	0:04	0:03
Wed	9:00	Non	45	3	2:01	2:36	0:35	0	9	1	10	0:04	0:03
Wed	9:00	Non	45	3	3:01	3:25	0:24	1	8	1	9	0:04	0:03
Wed	9:00	Non	45	3	3:46	4:17	0:31	0	2	0	2	0:04	0:03
Wed	9:00	Non	45	3	4:48	5:14	0:26	0	5	0	5	0:04	0:03
Wed	9:00	Non	45	3	5:48	6:17	0:29	1	13	2	15	0:04	0:03
Wed	9:00	Non	45	3	6:37	7:11	0:34	1	3	0	3	0:04	0:03
Wed	9:00	Non	45	3	8:13	8:39	0:26	1	6	0	6	0:04	0:03

Table 3. Minimum acceptable gap time for crossing four lanes by cycle

Day	Hour	Peak	Speed	Crossing Lanes	Start Clock Time	End Clock Time	Permitted Green Time	Permitted Left Turns	Through	Right	Total	Gap	Follow-Up
Mon	6:53	Peak	45	4	0:25	0:58	0:33	1	6	0	6	0:05	0:02
Mon	6:53	Peak	45	4	2:46	3:31	0:45	3	2	2	4	0:05	0:02
Mon	6:53	Peak	45	4	4:37	5:12	0:35	0	13	0	13	0:05	0:02
Mon	6:53	Peak	45	4	6:34	7:23	0:49	3	15	1	16	0:05	0:02
Mon	6:53	Peak	45	4	8:57	9:39	0:42	2	15	0	15	0:05	0:02
Mon	6:53	Peak	45	4	11:21	11:49	0:28	2	2	0	2	0:05	0:02
Mon	6:53	Peak	45	4	12:50	13:21	0:31	1	10	0	10	0:05	0:02
Mon	6:53	Peak	45	4	14:41	15:09	0:28	0	4	2	6	0:05	0:02
Mon	6:53	Peak	45	4	16:34	17:07	0:33	1	5	1	6	0:05	0:02
Mon	6:53	Peak	45	4	18:41	19:20	0:39	0	11	1	12	0:05	0:02
Mon	6:53	Peak	45	4	21:17	22:06	0:49	3	25	0	25	0:05	0:02
Mon	6:53	Peak	45	4	23:44	24:33	0:49	1	23	0	23	0:05	0:02
Mon	6:53	Peak	45	4	26:42	27:15	0:33	0	26	1	27	0:05	0:02
Mon	6:53	Peak	45	4	29:34	30:22	0:48	0	36	1	37	0:05	0:02
Mon	6:53	Peak	45	4	32:52	33:20	0:28	1	20	3	23	0:05	0:02
Mon	6:53	Peak	45	4	35:35	36:14	0:39	1	26	1	27	0:05	0:02
Mon	6:53	Peak	45	4	38:21	39:06	0:45	2	25	0	25	0:05	0:02
Mon	6:53	Peak	45	4	41:34	42:17	0:43	3	18	0	18	0:05	0:02
Mon	6:53	Peak	45	4	44:40	45:20	0:40	0	35	2	37	0:05	0:02
Mon	6:53	Peak	45	4	47:44	48:32	0:48	1	43	0	43	0:05	0:02
Mon	6:53	Peak	45	4	50:53	51:41	0:48	1	39	1	40	0:05	0:02
Mon	6:53	Peak	45	4	53:56	54:40	0:44	0	36	5	41	0:05	0:02
Mon	6:53	Peak	45	4	57:07	57:51	0:44	1	34	1	35	0:05	0:02
Mon	7:53	Peak	45	4	0:00	0:47	0:47	0	50	1	51	0:05	0:02
Mon	7:53	Peak	45	4	2:57	3:45	0:48	1	51	3	54	0:05	0:02
Mon	7:53	Peak	45	4	6:15	7:04	0:49	3	38	0	38	0:05	0:02
Mon	7:53	Peak	45	4	9:15	9:43	0:28	0	14	1	15	0:05	0:02
Mon	7:53	Peak	45	4	12:08	12:57	0:49	1	36	0	36	0:05	0:02
Mon	7:53	Peak	45	4	15:12	15:58	0:46	1	30	1	31	0:05	0:02
Mon	7:53	Peak	45	4	18:16	19:03	0:47	0	30	1	31	0:05	0:02
Mon	7:53	Peak	45	4	21:15	22:04	0:49	0	27	2	29	0:05	0:02
Mon	7:53	Peak	45	4	24:21	24:52	0:31	1	17	1	18	0:05	0:02
Mon	7:53	Peak	45	4	27:20	27:50	0:30	0	13	0	13	0:05	0:02
Mon	7:53	Peak	45	4	29:44	30:17	0:33	0	9	0	9	0:05	0:02
Mon	7:53	Peak	45	4	32:16	32:58	0:42	0	16	1	17	0:05	0:02
Mon	7:53	Peak	45	4	34:57	35:41	0:44	1	16	2	18	0:05	0:02
Mon	7:53	Peak	45	4	38:07	38:43	0:36	1	31	2	33	0:05	0:02
Mon	7:53	Peak	45	4	40:41	41:15	0:34	0	18	0	18	0:05	0:02
Mon	7:53	Peak	45	4	43:11	44:01	0:50	0	23	1	24	0:05	0:02
Mon	7:53	Peak	45	4	46:23	47:11	0:48	2	34	1	35	0:05	0:02
Mon	7:53	Peak	45	4	49:15	49:37	0:22	2	6	1	7	0:05	0:02
Mon	7:53	Peak	45	4	50:59	51:38	0:39	1	21	1	22	0:05	0:02
Mon	7:53	Peak	45	4	52:55	53:42	0:47	3	4	0	4	0:05	0:02
Mon	7:53	Peak	45	4	55:57	56:45	0:48	1	25	0	25	0:05	0:02
Mon	7:53	Peak	45	4	59:03	59:51	0:48	1	22	3	25	0:05	0:02

5.3   Decision

The decision of the algorithm is displayed in a text box on the screen. If the decision is to switch to a flashing yellow arrow mode, the message "Flashing Yellow Arrow" is displayed in Yellow. If the decision is to switch to a red arrow mode, the message "Red Arrow" is displayed in Red.

5.4   Quality Assurance and Verification

The software outputs and stores all the input data and decisions performed by the algorithm to a log file in real-time during the algorithm operation. This log file is intended for algorithm verification and future improvement as well as to help better understand the decision process during various traffic situations. The following section provides the results of one of the two case studies for the DSS lab testing.

6   DSS Lab Testing Procedure and Results

As mentioned earlier, the testing was conducted at Seminole County Traffic Engineering Lab through the Staff help. They ran a peer-to-peer logic to map the controller data from the field to the lab controller as well as the loop detectors. Vehicle detection was in real-time mode and monitored by CCTV cameras through the Bosch Video Management Software (BVMS). The DSS was tested on two intersections within Seminole County; US 17-92 at Church Avenue and SR 436 (Semoran Blvd) at CR 427 (Ronald Reagan Blvd).

6.1   US 17-92 at Church Avenue

At the vicinity of the intersection, US 17-92 is a six-lane divided arterial which runs in the north-south direction with a posted speed limit of 45 mph. Church Avenue is a two-lane undivided local road on one side and a parking lot on the other side as shown in Fig. 4. Commercial land uses exist on both sides of the road such as McDonald's, Burger King and Long John Silver's. The area gets busy during the lunch hour. The intersection has exclusive northbound and southbound left-turn lanes. The NB and SB left-turn lanes have a four-section head display which operates in a protected permissive mode. The intersection is monitored by CCTV cameras as shown in Fig. 5, which feed into the County's Traffic Management Center (TMC). In order to test the DSS algorithm, the DI-161 data logger channels were connected to the loop detectors in the cabinet to receive real-time traffic data. Figure 6 shows the DI-161 light bulbs for channels F0 and F2 which indicates that lanes 1 and 3 detected two vehicles at the same time. The intersection's cycle length varies according to the demand but was approximately 200 s.

Fig. 4. US 17-92 and Church Avenue geometry.

Fig. 5. US 17-92 at Church Avenue CCTV camera feeds.

Fig. 6. DI-161 Data logger detection with channels F0 and F2 bulbs lit.

6.2   DSS Results

The intersection was monitored for approximately one hour during lunch time between 12:00 and 1:00 pm. Table 4 displays the DSS log file and outputs for the intersection of US 17-92 and Church Avenue on a second by second basis for one cycle along with the algorithm decision.

As can be seen on Table 4, the customized data logger software displays the date and time step in real time mode on a second-by-second basis. The channels 0–2 represent the opposing three thru lanes and detects the arrivals of the vehicles in each lane while Channel 3 detects the start and end times of the opposing thru phase during which the flashing yellow arrow phase should be working. The developed software also includes the FYA algorithm, which specifies the minimum acceptable gap time for the corresponding number of lanes crossed and also the frequency of this minimum gap time in each cycle. For example, the study intersection has four opposing lanes to be crossed which correspond with a minimum acceptable gap time of 4.5 s as defined in Table 1. However, this minimum gap needs to be repeated at least five times, as specified in the algorithm, before deciding on a flashing yellow arrow mode. The algorithm kept receiving data for the first two cycles to calculate the minimum acceptable gap. Then the decision is provided in the third cycle and each cycle afterwards. The red boxes shown on Table 4 display the gap pattern and its frequency showing the five times specified in the algorithm to be able to decide on FYA mode. It should be noted that a minimum of five gaps repeated in each cycle is found to be reasonable especially for cycle lengths of 120 s or more. This criterion is updated in the algorithm based on the cycle length of the intersection.

Table 4. DSS output log file for US 17-92 and Church Avenue (one cycle)

Date Time	Ch0	Ch1	Ch2	Ch3	Interval
2/26/2016 12:19:24 PM	0	0	0	On	103 s
2/26/2016 12:19:25 PM	Waiting for phase...				
2/26/2016 12:19:26 PM	Waiting for phase...				
2/26/2016 12:19:27 PM	Waiting for phase...				
2/26/2016 12:19:28 PM	Waiting for phase...				
2/26/2016 12:19:29 PM	Waiting for phase...				
2/26/2016 12:19:30 PM	Waiting for phase...				
2/26/2016 12:19:31 PM	Waiting for phase...				
2/26/2016 12:19:32 PM	Waiting for phase...				
2/26/2016 12:19:33 PM	Waiting for phase...				
2/26/2016 12:19:34 PM	Waiting for phase...				
2/26/2016 12:19:35 PM	Waiting for phase...				
2/26/2016 12:19:36 PM	Waiting for phase...				
2/26/2016 12:19:37 PM	Waiting for phase...				
2/26/2016 12:19:38 PM	Waiting for phase...				
2/26/2016 12:19:39 PM	Waiting for phase...				
2/26/2016 12:19:40 PM	Waiting for phase...				
2/26/2016 12:19:41 PM	Waiting for phase...				
2/26/2016 12:19:42 PM	Waiting for phase...				
2/26/2016 12:19:43 PM	Waiting for phase...				
2/26/2016 12:19:44 PM	Waiting for phase...				
2/26/2016 12:19:45 PM	Waiting for phase...				
2/26/2016 12:19:46 PM	Waiting for phase...				
2/26/2016 12:19:47 PM	Waiting for phase...				
2/26/2016 12:19:48 PM	Waiting for phase...				
2/26/2016 12:19:49 PM	Waiting for phase...				
2/26/2016 12:19:50 PM	Waiting for phase...				
2/26/2016 12:19:51 PM	Waiting for phase...				
2/26/2016 12:19:52 PM	0	0	0	On	104 s
2/26/2016 12:19:53 PM	0	0	0	On	105 s
2/26/2016 12:19:54 PM	0	0	0	On	106 s
2/26/2016 12:19:55 PM	0	1	0	On	107 s
2/26/2016 12:19:56 PM	0	1	0	On	108 s
2/26/2016 12:19:57 PM	1	0	0	On	109 s
2/26/2016 12:19:58 PM	0	0	1	On	110 s
2/26/2016 12:19:59 PM	0	0	1	On	111 s

Date	Time	Ch0	Ch1	Ch2	Ch3	Interval
2/26/2016	12:20:00 PM	0	0	0	On	112 s
2/26/2016	12:20:01 PM	0	1	0	On	113 s
2/26/2016	12:20:02 PM	1	1	0	On	114 s
2/26/2016	12:20:03 PM	0	0	1	On	115 s
2/26/2016	12:20:04 PM	1	1	0	On	116 s
2/26/2016	12:20:05 PM	1	0	0	On	117 s
2/26/2016	12:20:06 PM	0	0	0	On	118 s
2/26/2016	12:20:07 PM	1	0	0	On	119 s
2/26/2016	12:20:08 PM	0	0	0	On	120 s
2/26/2016	12:20:09 PM	0	0	1	On	121 s
2/26/2016	12:20:10 PM	1	1	1	On	122 s
2/26/2016	12:20:11 PM	1	0	0	On	123 s
2/26/2016	12:20:12 PM	0	0	0	On	124 s
2/26/2016	12:20:13 PM	0	1	0	On	125 s
2/26/2016	12:20:14 PM	0	0	0	On	126 s
2/26/2016	12:20:15 PM	0	0	0	On	127 s
2/26/2016	12:20:16 PM	0	0	0	On	128 s
2/26/2016	12:20:17 PM	0	0	0	On	129 s
2/26/2016	12:20:18 PM	0	0	0	On	130 s
2/26/2016	12:20:19 PM	0	0	0	On	131 s
2/26/2016	12:20:20 PM	0	0	0	On	132 s
2/26/2016	12:20:21 PM	1	1	0	On	133 s
2/26/2016	12:20:22 PM	0	0	0	On	134 s
2/26/2016	12:20:23 PM	1	1	0	On	135 s
2/26/2016	12:20:24 PM	0	1	0	On	136 s
2/26/2016	12:20:25 PM	0	0	0	On	137 s
2/26/2016	12:20:26 PM	0	0	1	On	138 s
2/26/2016	12:20:27 PM	0	0	0	On	139 s
2/26/2016	12:20:28 PM	0	0	0	On	140 s
2/26/2016	12:20:29 PM	0	1	0	On	141 s
2/26/2016	12:20:30 PM	1	0	1	On	142 s
2/26/2016	12:20:31 PM	0	1	1	On	143 s
2/26/2016	12:20:32 PM	0	1	0	On	144 s
2/26/2016	12:20:33 PM	0	0	0	On	145 s
2/26/2016	12:20:34 PM	0	0	0	On	146 s
2/26/2016	12:20:35 PM	0	0	0	On	147 s
2/26/2016	12:20:36 PM	0	0	0	On	148 s
2/26/2016	12:20:37 PM	0	0	0	On	149 s
2/26/2016	12:20:38 PM	0	0	1	On	150 s
2/26/2016	12:20:39 PM	0	1	0	On	151 s
2/26/2016	12:20:40 PM	0	0	0	On	152 s
2/26/2016	12:20:41 PM	0	0	0	On	153 s
2/26/2016	12:20:42 PM	0	1	1	On	154 s
2/26/2016	12:20:43 PM	0	0	0	On	155 s

Date	Time	Ch0	Ch1	Ch2	Ch3	Interval
2/26/2016	12:20:44 PM	0	0	0	On	156 s
2/26/2016	12:20:45 PM	0	0	0	On	157 s
2/26/2016	12:20:46 PM	0	0	0	On	158 s
2/26/2016	12:20:47 PM	0	0	0	On	159 s
2/26/2016	12:20:48 PM	0	0	0	On	160 s
2/26/2016	12:20:49 PM	0	0	0	On	161 s
2/26/2016	12:20:50 PM	0	0	0	On	162 s
2/26/2016	12:20:51 PM	0	0	0	On	163 s
2/26/2016	12:20:52 PM	0	0	0	On	164 s
2/26/2016	12:20:53 PM	0	0	0	On	165 s
2/26/2016	12:20:54 PM	0	0	0	On	166 s
2/26/2016	12:20:55 PM	0	0	0	On	167 s
2/26/2016	12:20:56 PM	0	0	0	On	168 s
2/26/2016	12:20:57 PM	0	0	0	On	169 s
2/26/2016	12:20:58 PM	0	0	0	On	170 s
2/26/2016	12:20:59 PM	0	0	0	On	171 s
2/26/2016	12:21:00 PM	0	0	0	On	172 s
2/26/2016	12:21:01 PM	1	0	0	On	173 s
2/26/2016	12:21:02 PM	0	0	0	On	174 s
2/26/2016	12:21:03 PM	0	1	0	On	175 s
2/26/2016	12:21:04 PM	1	0	0	On	176 s
2/26/2016	12:21:05 PM	1	0	0	On	177 s
2/26/2016	12:21:06 PM	0	0	0	On	178 s
2/26/2016	12:21:07 PM	1	0	0	On	179 s
2/26/2016	12:21:08 PM	0	1	0	On	180 s
2/26/2016	12:21:09 PM	0	0	0	On	181 s
2/26/2016	12:21:10 PM	0	0	0	On	182 s
2/26/2016	12:21:11 PM	1	0	0	On	183 s
2/26/2016	12:21:12 PM	0	0	0	On	184 s
2/26/2016	12:21:13 PM	0	0	0	On	185 s
2/26/2016	12:21:14 PM	0	0	0	On	186 s
2/26/2016	12:21:15 PM	0	0	0	On	187 s
2/26/2016	12:21:16 PM	0	0	0	On	188 s
2/26/2016	12:21:17 PM	1	0	0	On	189 s
2/26/2016	12:21:18 PM	0	1	0	On	190 s
2/26/2016	12:21:19 PM	0	0	0	On	191 s
2/26/2016	12:21:20 PM	0	0	0	On	192 s
2/26/2016	12:21:21 PM	1	0	0	On	193 s
2/26/2016	12:21:22 PM	0	0	0	On	194 s
2/26/2016	12:21:23 PM	0	0	0	On	195 s
2/26/2016	12:21:24 PM	0	0	0	On	196 s
2/26/2016	12:21:25 PM	0	1	0	On	197 s
2/26/2016	12:21:26 PM	1	0	0	On	198 s
2/26/2016	12:21:27 PM	0	0	0	On	199 s

Date	Time	Ch0	Ch1	Ch2	Ch3	Interval
2/26/2016 12:21:28 PM		1	0	1	On	200 s
2/26/2016 12:21:29 PM		1	1	0	On	201 s
2/26/2016 12:21:30 PM		0	0	0	On	202 s
2/26/2016 12:21:31 PM		0	0	0	On	203 s
2/26/2016 12:21:32 PM		0	0	0	On	204 s
2/26/2016 12:21:33 PM		0	0	0	On	205 s
2/26/2016 12:21:34 PM		0	0	0	On	206 s
2/26/2016 12:21:35 PM		0	0	0	On	207 s
2/26/2016 12:21:36 PM		1	1	0	On	208 s
2/26/2016 12:21:37 PM		1	0	0	On	209 s
2/26/2016 12:21:38 PM		0	1	0	On	210 s
2/26/2016 12:21:39 PM		0	0	0	On	211 s
2/26/2016 12:21:40 PM		0	1	1	On	212 s
2/26/2016 12:21:41 PM		0	0	0	On	213 s
2/26/2016 12:21:42 PM		0	0	1	On	214 s
2/26/2016 12:21:43 PM		0	0	0	On	215 s
2/26/2016 12:21:44 PM		0	0	0	On	216 s
2/26/2016 12:21:45 PM		0	0	0	On	217 s
2/26/2016 12:21:46 PM		0	0	1	On	218 s
2/26/2016 12:21:47 PM		0	1	0	On	219 s
2/26/2016 12:21:48 PM		1	0	0	On	220 s
2/26/2016 12:21:49 PM		0	0	0	On	221 s
2/26/2016 12:21:50 PM		0	0	0	On	222 s
2/26/2016 12:21:51 PM		1	0	0	On	223 s
2/26/2016 12:21:52 PM		0	0	0	On	224 s
2/26/2016 12:21:53 PM		0	0	0	On	225 s
2/26/2016 12:21:54 PM		0	0	0	On	226 s
2/26/2016 12:21:55 PM		0	0	0	On	227 s
2/26/2016 12:21:56 PM		0	0	1	On	228 s
2/26/2016 12:21:57 PM		0	0	0	On	229 s
2/26/2016 12:21:58 PM		0	1	0	On	230 s
2/26/2016 12:21:59 PM		0	1	0	On	231 s
2/26/2016 12:22:00 PM		0	0	0	On	232 s
2/26/2016 12:22:01 PM		0	0	0	On	233 s
2/26/2016 12:22:02 PM		0	0	0	On	234 s
2/26/2016 12:22:03 PM		1	0	0	On	235 s
2/26/2016 12:22:04 PM		0	0	0	On	236 s
2/26/2016 12:22:05 PM	Applying decision	Flashing Yellow Arrow				
2/26/2016 12:22:05 PM	Waiting for phase...					
2/26/2016 12:22:06 PM	Waiting for phase...					
2/26/2016 12:22:07 PM	Waiting for phase...					
2/26/2016 12:22:08 PM	Waiting for phase...					

6.3 DSS Validation

It should be noted that the intersection was video recorded during the DSS testing for validation purposes. The validation procedure involved matching the same time step from the video file with the DSS log file. The intersection was recorded for 15 min which corresponded to five cycles. During the 15 min period, 12 left-turn vehicles arrived during the permissive phase and were waiting for an acceptable gap. It was worth mentioning that the 12 vehicles were able to find an acceptable gap during the recorded 15 min and cleared the intersection. For the reported cycle data in Table 4, five vehicles arrived and cleared the intersection. Two consecutive vehicles made the left-turn during the first gap from 12:20:14 to 12:20:20; a total of 7 s which included the min gap time of 4.5 s and a follow up time of 2.5 s. Another truck arrived at 12:20:44 and cleared the intersection during the big gap of 18 s. Another two vehicles utilized the remaining two gaps at 12:21:12 and 12:21:30.

7 Conclusions

The flashing yellow arrow phase II project provided additional intersection video data that was extracted and utilized in order to refine the model developed in phase I. The additional videos, garnished by FDOT representatives, were an asset to the project and contributed to its success. The usable data of the master database was increased to 38 intersections with locations across the State of Florida. The data extraction process in phase II was completed to match the basic prioritized parameters that were used to refine the developed model in phase I. Additional parameters such as the left-turn timing, left-turn gap, opposing lane utilization and left-turn stop delay were extracted as necessary, broadening the data analysis.

Model refinement required the expansion of the database to increase the domain and improve reliability of the developed model. Several modeling techniques were investigated which included stepwise regression, time series analysis and neural networks. Based on the analysis, neural networks model provided the highest correlation between the independent variables with coefficient of correlation reaching 90%. Virtual testing of the decision support system using VISSIM application programming interface (API) confirmed the applicability and validity of the procedure and logic. This was a critical juncture before running a field test.

A custom communications software was developed which has three main functions; control the hardware, display real-time status and execute the proposed FYA algorithm. The UCF research team developed a specific code to retrieve instantaneous channel input data, synchronize opposing thru green phase, analyze traffic information, provide the algorithm decision, and generate a real-time log recording the events. The software was developed using the C# language under Microsoft's™ Visual Studio 2013 development environment.

The proposed algorithm is implemented with the goal of safely optimizing traffic operations. In the case of a red arrow signaled for a left-turn, the opposing thru traffic during the green phase is constantly analyzed in real-time to determine whether it would be optimal to switch the red arrow to a flashing yellow arrow. The decision is made based on a number of parameters which includes: the minimum headway of vehicles in the opposing traffic, the number of lanes to cross, and the number of cycles to be analyzed prior to making the decision. The algorithm determines the time interval between the successive arrivals of vehicles for each lane independently and computes the corresponding headway for each lane by cycle on a second by second. The thresholds used for different crossing number of lanes were obtained from the database of 30,000 cycles collected from the field. If the minimum headway for the corresponding number of lanes is achieved and repeated for certain number of times, for example, at least five times during the analysis period (whether one or two cycles) which is also an input to the algorithm, the decision is made to switch to a flashing yellow mode. Otherwise, a red arrow is decided upon.

The decision support system was ultimately tested at two different intersections in Seminole County. Video data was collected at the same time period as the algorithm was tested in order to validate the algorithm decisions. The DSS testing confirmed the applicability and validity of the developed DSS as well as the aforementioned procedure, criteria and logic. The value of the DSS in making real-time traffic decisions is crucial to improving the performance of the left-turning traffic and can be applied at any flashing yellow arrow system.

Acknowledgement. The work reported in this paper is part of a research project under contract number BDV 24-977-10, which is sponsored by the Florida Department of Transportation (FDOT). The views expressed in this paper do not necessarily reflect those of the sponsors. The authors would like to thank FDOT for their support.

References

1. Abou-Senna, H., Radwan, E., Harb, R., Navarro, A., Chalise, S.: Interactive decision support system for predicting flashing yellow arrow left-turn mode by time of day. Transp. Res. Rec. **2463**(1), 16–25 (2014)
2. Miovision Technologies Scout Video Collection Unit Specification. http://www.miovision.com/products/scout-vcu/
3. AG, PTV Planung Transport Verkehr. VISSIM Manual–VAP Version 2.14 APPENDIX B: ADD-ON VisVAP. PTV Corporation (2003)
4. Agent, K.R.: Guidelines for the use of protected/permissive left-turn phasing. ITE J. **57**(7), 37–43 (1987)
5. Stamatiadis, N., Agent, K., Bizakis, A.: Guidelines for left-turn phasing treatment. Transp. Res. Rec. **1605**(1), 1–7 (1997)
6. Al-Kaisy, A.F., Stewart, J.A.: New approach for developing warrants of protected left-turn phase at signalized intersections. Transp. Res. Part A: Policy Pract. **35**(6), 561–574 (2001)
7. Zhang, L., Prevedouros, P.D., Li, H.: Warrants for protected left-turn phasing. ASCE J. Transp. Eng. **1073**, 28–37 (2005)
8. Chang, G.-L., Chen, C.-Y., Perez, C.: Hybrid model for estimating permitted left-turn saturation flow rate. Transp. Res. Rec. **1566**(1), 54–63 (1996)
9. Hu, P., Tian, Z.Z., Zang, L.: Methods for selecting left-turn control types based on analytical-hierarchy-process approaches at signalized intersections. ITE J. **82**(11), 34 (2012)
10. Chen, L., Chen, C., Ewing, R.: Left-turn phase: permissive, protected, or both?. Paper presented at the Transportation Research Board 91st Annual Meeting (2012)
11. Gal-Tzur, A., Mahalel, D., Prashker, J.N.: Decision support system for controlling traffic signals. Transp. Res. Rec. **1421**, 69–75 (1993)
12. Ozmen, O., Tian, Z.Z., Gibby, R.: Guidelines for multicriterion decision-based left-turn signal control. Transp. Res. Rec. **2128**(1), 96–104 (2009)

13. Yu, L., Qi, Y., Yu, H., Guo, L.: Development of left-turn operations guidelines at signalized intersections. Report No. TxDOT 0-5840-1 for Texas Department of Transportation (TxDOT) (2008)
14. Bonneson, J.A., Fontaine, M.D.: NCHRP Report 457: engineering study guide for evaluating intersection improvements, pp. 22–25 (2001)

Using Computational Social Science Techniques to Identify Coordinated Cyber Threats to Smart City Networks

Mustafa Alassad[1], Billy Spann[1], Samer Al-khateeb[2], and Nitin Agarwal[1(✉)]

[1] University of Arkansas, Little Rock, AR, USA
{mmalassad, bxspann, nxagarwal}@ualr.edu
[2] Creighton University, Omaha, NE, USA
SamerAl-Khateebl@creighton.edu

Abstract. Smart cities are increasingly facing cyber-attacks due to the endeavors they have made in technological advancements. The challenge for smart cities, that utilize complex digital networks to manage city systems and services, is that any device that relies on internet connectivity to function is a potential cyber-attack victim. Smart cities use smart sensors. Online Social Networks (OSNs) act as human sensors offering significant contributions to the amount of data used in smart cities. OSNs can also be used as a coordination and amplification platform for attacks. For instance, aggressors can increase the impact of an attack by causing panic in an area by promoting attacks using OSNs. Public data can help aggressors to determine the best timing for attacks, scheduling attacks, and then using OSNs to coordinate attacks on smart city infrastructure. This convergence of the cyber and physical worlds is known as cybernetics. Quantitative socio-technical methods such as deviant cyber flash mob detection (DCFM) and focal structure analysis (FSA) can provide reconnaissance capabilities that enable cities to look beyond internal data and identify threats based on active events. Assessment of powerful actors using DCFM detection methods can help to identify and prevent attacks. Groups of powerful hackers can be identified through FSA which is a model that uses a degree centrality method at the *node-level* and spectral modularity at *group-level* to measure the power of a focal structure (a subset of the network). DCFM and FSA models can help cyber-security experts by providing a better picture of the threat which will help to plan a better response.

Keywords: Online social networks · Focal structure analysis · Flash mob detection · Spectral modularity · Degree centrality · Network clustering · Smart cities · Cyber security

1 Introduction

Transformation to smarter cities presents many challenges for researchers and engineers as they face new procedures, data management platforms, and operations. Building new and efficient smart systems open the door for many new issues such as

© The Author(s), under exclusive license to Springer Nature Switzerland AG 2021
I. El Dimeery et al. (Eds.): JIC Smart Cities 2019, SUCI, pp. 316–326, 2021.
https://doi.org/10.1007/978-3-030-64217-4_35

privacy, security, big data coming from various sensors, public and private services, and social systems. Although these systems are being transformed into smart systems by connecting them to the internet A.K.A. the Internet of Things (IoT), it still needs more cybersecurity enhancements.

The critical infrastructure of smart cities should have monitoring capabilities for optimizing security methods, reducing vulnerability, increasing reliability. This will enhance the transportation systems, the security of smart power grids and various energy systems such as: petroleum refineries, health, and food systems. Such monitoring systems require data collection, real-time processing, analysis, and decision-making capabilities.

Today, various industries have services that monitor many malicious activities and threats, such as hackers who try to access crucial department databases to steal information or damage a provided service. In recent years, many malicious cyber activities across the world have been reported to cause enormous damage to various critical smart systems [1].

In this research, we are considering the massive growth of social media platforms in the recent years such as Twitter, Facebook, YouTube, and WeChat, and how many social applications must work with significant amounts of personal and public data. People use these tools to share information, opinions, and activities with their relatives, friends and other cultural organizations. However, in the last few years, the use of these platforms was changed by a few radical organizations. Malicious actors misused social media to amplify and share terrorist activities and malicious threats, information dissemination, propagating radical behaviors, spreading fake news, and conducting cyberattacks on public and private online smart infrastructure networks [2].

Quantitative methods have been applied to help to analyze the complex social networks in recent years. Some of the most common approaches to quantitative network analysis use measures such centrality and modularity to help define network structure and model the networks. Node-based community detection algorithms using the degree centrality method [3, 4] and group-based community detection algorithms using the modularity method [5, 6], are considered in our research and presented in Sect. 2. However, merely considering these two community detection categories alone, lacks the depth and insight into the most influential aggressors and network links that would maximize the damage to a smart city infrastructure grid. Therefore, we propose a mixed model, developing the node-level measure which considers the individual's centrality value, and then spectral modularity (group-level) is employed to measure the groups' influence at the Network-Level. The resultant model is a Bi-Level centrality-modularity maximization model called Focal Structure Analysis (FSA). These focal structures (sub network or sub graphs) are the hidden intensive groups that can influence maximum number of users in the network.

The contributions from the model in this research considers the shortcomings in the regular community detection algorithms, where the node-based methods cannot identify these groups, and the group-based method cannot cluster intensive small groups. We are proposing a mix of the node and group-based community detection algorithms, whereby we create a model consisting of two major sections: the bi-level optimization section, and the deviant cyber flash mob detection method. Other supplementary sections are also used to help in clustering the network. Finally, the model utilizes small real-world metrics to identify FSA sets and then evaluate them using the deviant cyber flash mob detection (DCFM) method to determine if the aggressors' and sets can influence the entire network.

Multiple case studies leveraged the two aforementioned approaches independently, such as Sen et al. [8], utilized a greedy model on a Facebook network and concluded that Facebook was used to mobilize crowds in 2007 during the Egyptian Revolution [9]. The authors in [9] studied a Twitter network, where they identified a small influential set of users who are responsible for the 2011 Saudi Arabia women's right to drive campaign [9]. Alassad et al. [10] studied a network of commenters on YouTube that disseminated disinformation. He used a decomposition optimization model to identify small influential sets of commenters responsible for commenting on various videos.

For our practical implementation of this mixed-mode model that would extend to the smart city domain, we consider an ISIS dataset provided in a study conducted by the International Centre for the Study of Radicalization and Political Violence (ICSR) which shows a group of individuals who helped ISIS recruiters to disseminate their propaganda on Twitter and other social media platforms [7]. This mixed-mode model was also applied on a YouTube channel that was spreading fake news in the South China Sea [11].

1.1 Problem Statement

The aim of this study is to apply a *non-traditional cybersecurity network approach* to cluster and analyze influential sets of social media users. These users are highly central disseminators who can amplify information spread to a maximum number of individuals in the network. One of the big challenges that are facing network scientists is to identify and suspend such hidden coordinating groups of malicious users in complex social network. These focal structures of malicious users in the network can be influential and can disseminate their radical or terrorist propaganda to threaten smart cities' intelligent systems very effectively.

These sets of aggressors (focal structures) can coordinate attacks on various smart city infrastructures by utilizing well-known social media platforms, for example, they can post directions, locations, and other coordination activities on social media informing their followers. Since smart cities rely on internet services, the government would not want to shut down internet service across the entire smart city network and risk financial, economic, or security lose. The success of a deviant cyber flash mob targeting a smart city infrastructure would likely have a crippling effect on the smart city. Identifying hidden influential groups and suspending them without impacting the total infrastructure network is essential. In this research we use a network of commenters who are posting radical directions on Twitter to paralyze infrastructure in smart cities. These FSAs could be responsible for organizing multi-cyber-attacks to maximize the damages to the network, spread fake news and convince other nodes in the network to participate in or create their own cyber-attacks. In this paper, we identify these malicious set of users, and then suspend them from their locations in the network to stop their influence without taking down the remaining network.

The rest of the paper is organized as follows. Section 2 summarizes the data set and the research methodology. Section 3, we apply the proposed model to the dataset collected and demonstrate the model efficiency. Finally, we conclude with intended future work in Sect. 4.

2 Methodology

The proposed model is designed to (1) overcome the shortcomings in regular community detection methods [4, 10], (2) advance the FSA model proposed by Sen et al. [9, 12], and (3) use the DCFM model developed by Al-khateeb et al. [13] to identify a sets of powerful actors in complex social networks aiming to conduct deviant acts that can damage smart cities' infrastructures.

2.1 Data Set

In this research, we collected data of a Twitter network consisting of 1,453 nodes and 1,487 edges. An initial set of Twitter usernames were provided in a report published by the International Centre for the Study of Radicalization and Political Violence (ICSR) in which they provided a list of individuals who help ISIS disseminate their propaganda on Twitter and other social media platforms [16]. We crawled these usernames' friends and followers then cross-intersected them with another dataset collected during three beheading events conducted by ISIS in Egypt, Libya, and Palestine [17]. For the users in the resultant dataset, we calculated control, interest, and power to estimate the power of each node (user) in the network. We built the communication (retweets and mentions) network for these users then ran our model to determine the focal structures within the network. These FSAs are ranked based on the sum of power for all users within that focal structure.

2.2 Node-level and Group Level Measures

The first step in collecting the necessary measurements for the model after identifying the user network is to calculate node-level power, degree centrality (node-level measurements) and the clustering coefficient (group-level measurements). The power of each node is calculated using a collective action-based model developed by Al-khateeb et al. [18, 19]. The degree centrality method is utilized to measure a node's sphere of influence [4]. Figure 1 shows the average degree centrality for all 53 FSA sets. In addition to degree centrality, the model needs to consider the node's neighbors' friendship as well, to determine if the friends-of-friends are also his/her friend. Hence, we used the clustering coefficient as shown in Fig. 2 to determine if a node exhibits this behavior or not [3, 4]. The final analysis uses not only the node's degree centrality, but also the network power calculated by the DCFM method.

The result of the two methods combined, i.e., the degree centrality and clustering coefficient, are ranked as sets of active local communities consisting of highly central nodes that have active neighbors (can communicate with each other). The measurements from these two methods will be exported to the Network-Level to measure their ability to maximize the network's sparsity or their communication to other aggressors' groups.

2.3   Network-Level Analysis

The spectral modularity method [6], is used to measure the graph's sparsity inheriting the nodes' sets from the Node-Level. The objective function as shown in the Network-Level in Fig. 3, is to import sets from the previous level and then find sets that can maximize the graph's modularity value [10, 11]. The model is searching for sets of groups that can produce the maximum number of aggressors in the network [5, 10, 11, 14].

These focal structures include the maximum number of influential nodes in the network who have the power to convince other nodes in the network to participate in deviant actions, such as multi-cyber-attacks. Also, these nodes can be part of other groups (other focal structures, and can supervise other nodes, control information dissemination, and amplify their radical actions to other parts of the network.

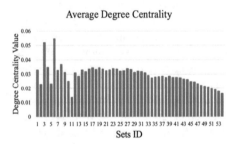

Fig. 1. FSA Sets' average degree centrality values

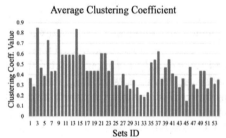

Fig. 2. FSA Sets' average clustering coefficient values

2.4   FSA Evaluation and DCFM

Decomposition of the identified focal structures help to measure the sets characteristics from different points of views as follows:

2.4.1   Small Real-World Network Metrics

Our model used two measures, namely degree centrality and clustering coefficient to determine its output. The goal of the model is to find subsets in the network (called focal structures) that can maximize the average degree centrality of each node and the average clustering coefficient of these central nodes (to measure the members' connectivity within the sets). Figure 1 shows the identified focal structures (influential sets) average degree centrality values while Fig. 2 shows the interaction between these subsets nodes by utilizing the clustering coefficient of the group.

2.4.2   DCFM Metrics

The DCFM phenomenon can be considered a form of a cyber-collective action that is defined as an action aiming to improve a group's conditions (such as, status or power). If we can identify those strong influential groups organizing DCFM, we can design counter measures to stop the aggressors from attacking smart city infrastructure. Previous work by Al-khateeb and Agarwal [13] developed a collective action based theoretical model which identified factors to predict success or failure of a Deviant Cyber Flash Mob (DCFM).

In their model, the identified factors are – Utility (U) (the benefits an individual gain if the DCFM success or fail), Interest (I) (how much interest an aggressor has based on the utility gained), Control (C) (how much control the aggressor has on the outcome of the DCFM), and Power (P) (how powerful an aggressor is in the group). In this study, we calculate the structural characteristics of our sample DCFM network and assess the impact of these collective action measurements (i.e., I, C, and P) using our Focal Structure Analysis (FSA) model.

3   Experimental Results

We applied our model to the ISIS Twitter network shown in Fig. 4. The model identified the highly influential sets of aggressors in the dataset that maximizes the graph sparsity, influences maximum number of individuals, and includes members acting in different group as shown in Fig. 5. Also, the interconnection between pairwise focal structure reveals a spoke and hub communication structure, where a set conveys information to other groups who then carry out operations as shown in Fig. 5.

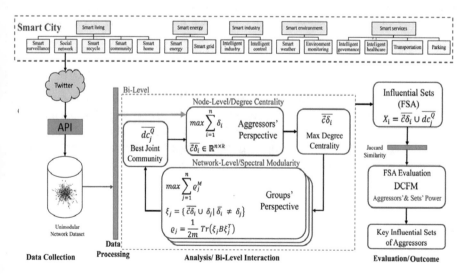

Fig. 3. Focal Structure Analysis (FSA) structure in Smart City, where social media is part of the smart city structure. The model will import the data constructed from the social media platforms such as Twitter.

The DCFM method calculated the sets' power (influence), whereby the more power they have the darker the sets' color as shown in Fig. 5.

To understand the focal structures' impact inside the network as shown in Fig. 6 , we employed two methods as basis to make the evaluation. First, the Girvan-Neman modularity method [15], which returned a modularity value of 0.645 and clustered 40 communities as shown in Fig. 4. Second, Trajan et al. [16] found only one weakly connected user in the network. These are the baseline network measurements for this collection of users and their corresponding network structure.

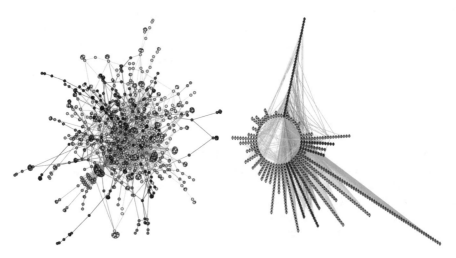

Fig. 4. ISIS Twitter network clustered into 40 groups via modularity method.

Fig. 5. Commenters are clustered into 54 influential sets. The darker the color, higher the influence.

Moreover, Fig. 7, shows the top twenty influential aggressors and the count of FSA sets containing each aggressor. Since we identified that very powerful actors appear in multiple network sets, it enables the authorities to measure, predict, and allocate the influential aggressors' active strategies, possible spots for information dissemination, and cyber-attacks' locations.

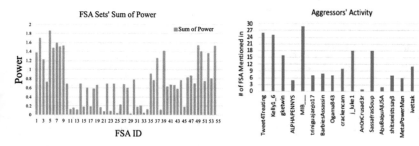

Fig. 6. Sets' power measured by DCFM method.

Fig. 7. Top 20 Influential aggressors measured by DCFM.

Table 1 shows the top twenty influential sets of aggressors, where the impact of each set on the modularity and connectivity were measured accordingly. We found that, each of these sets can maximize the graph modularity value into the interval of [0.7–0.83], and they can maximize the graph sparsity from 40 groups into [103–426] groups. This means that by removing the most influential users from the network, the clusters or communities within the network lose their connection to other parts of the network. That is, by removing the aggressors in any given FSA ID in Table 1, the overall network of users becomes isolated to other communities.

Table 1. Top 20 influential sets.

FSA ID	Sum of power	Count of users	# of Weakly conn. Users	Count of comm	Max modularity value	FSA allocation
5	1.86	42	408	426	0.83	19
2	1.7	41	401	423	0.82	13
7	1.58	21	287	313	0.78	9
49	1.53	47	256	280	0.81	15
9	1.52	21	268	290	0.78	9
54	1.52	130	278	300	0.82	20
8	1.5	21	330	352	0.8	13
6	1.47	11	273	298	0.76	9
39	1.41	13	241	268	0.75	12
50	1.4	49	261	286	0.78	15
1	1.37	19	270	292	0.77	7
52	1.36	54	230	254	0.8	19
37	1.25	13	145	175	0.72	5
3	1.22	7	165	189	0.7	5
35	0.91	12	161	187	0.74	8

(*continued*)

Table 1. (*continued*)

FSA ID	Sum of power	Count of users	# of Weakly conn. Users	Count of comm	Max modularity value	FSA allocation
47	0.87	32	73	103	0.72	10
46	0.83	25	161	190	0.73	8
53	0.81	76	115	142	0.76	20
30	0.78	9	131	191	0.71	4
36	0.77	12	174	198	0.72	5

In addition, based on Trajan et al. [16] the min-max numbers of weakly connected users caused by these sets increase from one to an interval between [73–408] weakly connected users. Most importantly, we were able to identify each FSA set's attack locations by identifying how many FSA sets each aggressor appeared in, (as shown in Fig. 7). Proposing that each set of aggressors can attack multiple places at the same time. For example, FSA (5), the top influential set consisted of 42 aggressors, they can influence 408 individuals, are able to attack 19 different locations in the network and can divide the network into 426 other groups. Therefore, by removing the users within this FSA (5), the network is divided into 426 communities vs 40 communities with FSA (5) included, so aggressors would have to work harder to disseminate information across the network.

4 Conclusion and Discussion

In this research, we have studied social media cybersecurity risks at a network level using computational social science techniques, where aggressors utilize Twitter platforms to perform cyber attacks. Considering the shortcomings of regular community detection algorithms and taking a non-traditional cybersecurity approach, the proposed bi-level model was able to identify hidden influential sets of aggressors in the network. The proposed model was able to identify a spoke and hub communication structure, where a single influential FSA set conveys information to other sets who can carry out deviant behaviors. Such focal structures are more prevalent in terrorist networks.

Throughout this research, we were able to allocate the aggressors' activities, track all their possible cyber-attacks locations, provide an overview of the influential sets, and estimate the aggressors' influence in the network. Using this model could enhance and harden smart cities' strategies against cyber-attacks when they originate from social media platforms by suspending any of those focal aggressors' structures to prevent the massive damages that can be caused to a smart cities critical foundation.

Acknowledgment. This research is funded in part by the U.S. National Science Foundation (OIA-1920920, IIS-1636933, ACI-1429160, and IIS-1110868), U.S. Office of Naval Research (N00014–10-1–0091, N00014–14-1–0489,N00014–15-P-1187, N00014–16-1–2016, N00014–16-1–2412, N00014–17-1–2605, N00014–17-1–2675,N00014–19-1–2336), U.S. Air Force Research Lab, U.S. Army Research Office (W911NF-16–1-0189), U.S.Defense Advanced

Research Projects Agency (W31P4Q-17-C-0059), Arkansas Research Alliance, and the Jerry L. Maulden/Entergy Endowment at the University of Arkansas at Little Rock. Any opinions, findings, and conclusions or recommendations expressed in this material are those of the authors and do not necessarily reflect the views of the funding organizations. The researchers gratefully acknowledge the support.

References

1. Hall, R.E., Braverman, J., Taylor, J., Todosow, H.: The Vision of A Smart City, Ins, Fr. Sept. 28, Paris, Fr. 2nd Int. Present. Life Ext. Technol. Work. Paris, pp. 1–6 (2000)
2. Abouzakhar, N.: Critical Infrastructure Cybersecurity: A Review of Recent Threats and Violations (2013)
3. Freeman, L.C.: Centrality in social networks. Soc. Netw. **1**, 215–239 (1978)
4. Zafarani, R., Abbasi, M.A., Liu, H.: Social Media Mining: An Introduction. Cambridge University Press, Cambridge (2014)
5. Girvan, M., Newman, M.E.J.: Community structure in social and biological networks. Proc. Natl. Acad. Sci. **99**(12), 7821–7826 (2002)
6. Tsung, C.K., Ho, H., Chou, S., Lin, J., Lee, S.: A spectral clustering approach based on modularity maximization for community detection problem. In: Proceedings of 2016 International Computer Symposium ICS 2016, pp. 12–17 (2017)
7. Mohammad Yasin, N.A.: Impact of ISIS Online Campaign in Southeast Asia. International Centre for Political Violence and Terrorism Research, vol. 7, no. 4, pp. 26–32 (2015)
8. Romano, V., Duboscq, J., Sarabian, C., Thomas, E., Sueur, C., MacIntosh, A.J.J.: Modeling infection transmission in primate networks to predict centrality-based risk. Am. J. Primatol. **78**(7), 767–779 (2016)
9. Şen, F., Wigand, R., Agarwal, N., Tokdemir, S., Kasprzyk, R.: Focal structures analysis: identifying influential sets of individuals in a social network. Soc. Netw. Anal. Min. **6**(1), 17 (2016)
10. Alassad, M., Agarwal, N., Hussain, M.N.: Examining intensive groups in YouTube commenter networks. In: Proceedings of 12th International Conference, SBP-BRiMS 2019, no. 12, pp. 224–233 (2019)
11. Alassad, M., Hussain, M.N., Agarwal, N.: Finding fake news key spreaders in complex social networks by using bi-level decomposition optimization method. In: International Conference on Modelling and Simulation of Social-Behavioural Phenomena in Creative Societies, pp. 41–54 (2019)
12. Sen, F., Wigand, R.T., Agarwal, N., Mahata, D., Bisgin, H.: Identifying focal patterns in social networks. In: Proceedings of the 2012 4th International Conference on Computational Aspects of Social Networks, CASoN 2012, no. November 2012, pp. 105–108 (2012)
13. Al-Khateeb, S., Agarwal, N.: Modeling flash mobs in cybernetic space: Evaluating threats of emerging socio-technical behaviors to human security. In: Proceedings of the 2014 IEEE Joint Intelligence and Security Informatics Conference JISIC 2014, vol. 7, no. 1, p. 328 (2014)
14. Newman, M.E.J.: Modularity and community structure in networks. Proc. Natl. Acad. Sci. **103**(23), 8577–8582 (2006)
15. Girvan, M., Newman, M.: Community structure in social and biological networks. PNAS **99** (12), 7821–7826 (2002)
16. Tarjan, R.: Depth-first search and linear graph algorithms. SIAM J. Comput. **1**(2), 146–160 (1972)

17. Samer, A., Agarwal, N.: Examining botnet behaviors for propaganda dissemination: a case study of isil's beheading videos-based propaganda. In: 2015 IEEE International Conference on Data Mining Workshop (ICDMW). IEEE, pp. 51–57 (2015)
18. Al-khateeb, S., Agarwal, N.: Developing a conceptual framework for modeling deviant cyber flash mob: a socio-computational approach leveraging hypergraph constructs. J. Digit. Forensics Secur. Law **9**(2), 10 (2014)
19. Samer, A., Agarwal, N.: Analyzing deviant cyber flash mobs of ISIL on Twitter. In: International Conference on Social Computing, Behavioral-Cultural Modeling, and Prediction, pp. 251–257. Springer, Cham (2015)

LoRa Wide Area Network Pragmatic Heterogeneous IoT Applications, Deployment Using Different Spreading Factors

Minar El-Aasser[(✉)], Mohamed Ashour, and Tallal Elshabrawy

Information and Engineering Technology Department, German University in Cairo, New Cairo, Egypt
{minar.elaasser,mohamed.ashour,
tallal.el-shabrawy}@guc.edu.eg

Abstract. Long Range Low Power Wide Area Network (LoRa LPWAN) offers ubiquitous connectivity in IoT applications, while keeping network structure and management simple. The technology is offering developers an attractive route to build IoT installations involving sensors located at kilometers rather than meters from the nearest gateway. LoRa uniquely relies on 6 spreading factors (SFs) ranging from 7 to 12. Higher distances are covered by the use of higher SFs at the expense of reduced data rate. On the other hand, higher SF consumes longer transmission time on air. The technology had been widely deployed world- wide and in order to support scientific research several network simulators have been conducted to answer the what-if analysis. This paper presents an Omnet++ simulation model developing the two conventional SF allocation schemes; modeling LoRaWAN with SF tiers representing Adaptive Data Rate activation and parallel quasi-orthogonal SF network. Moreover, the simulation model allows for heterogeneous IoT application deployment that shall act as a pragmatic network planning tool. The idea is to utilize fast transmissions of smaller SF networks by configuring end-devices with applications of large packet sized to use the small SF networks and utilize the developed simulator to inspect and analyze network performance.

Keywords: LoRaWAN · Spreading factor · Heterogeneity · IoT applications · Network simulation

1 Introduction and Related Work

Long Range Low Power Wide Area Network (LoRa LP- WAN) is becoming a vital solution that supports major portion of the rapidly growing Internet of Things (IoT) applications. While there are many LPWAN technologies, e.g.: Sigfox, NB- IoT, etc., and standards, LoRa LPWAN technology is under active development and deployment and is arguably the most adopted [1]. The key to its wide-spread is the two components on which the technology is built upon.

First, LoRa Physical layer which designates the Chirp Spreading Spectrum (CSS) modulation utilized to create the long range communication link. LoRa uniquely relies on 6 spreading factors (SFs) ranging from 7 to 12. Each end-device adapts it's SF locally to the number of re-transmissions or based on information from the gateway embedded in down- link messages. Maximum service radius of each SF is defined by a Signal-to-Noise Ratio (SNR) threshold for a correct signal demodulation at the intended gateway maintained by receiver sensitivity. Higher distances are covered by the use of higher SFs at the expense of reduced data rate. On the other hand, higher SF consumes longer transmission time on air.

Second, LoRaWAN MAC protocol enables end-devices to communicate using LoRa modulation with a gateway in an Aloha-like medium access method. A LoRaWAN cell is typically configured as a star topology. In principle, an ED with a SF can be located in a cell anywhere ranging from the gateway to the maximum SF service radius. Packets transmitted on the same frequency with different SFs can be decoded simultaneously. Thus a subset of EDs with the same SF represents a logical SF network and thus LoRa supports multiple quasi-orthogonal simultaneous logical networks; all overlapping in the same area as demonstrated in Fig. 1a. However, for power and data rate optimization; conventional adaptive data rate (ADR) strategy is currently used in most LoRaWAN deployments where each end-device opts to communicate by selecting the minimum SF which permits correct reception at the intended GW. This strategy, thus, arranges the network into multiple SF coverage tiers as shown in Fig. 1b. In addition to the simplified network architecture LoRaWAN adopts to simplify network management and reduce cost, LoRa further operates in the reasonable worldwide available license exempt frequency bands. Given that LoRaWAN utilizes random access, thus this imposes a per sub-band duty cycle regulations enforced by the European Telecommunication Standard Institute (ETSI).

Despite its widespread, the technology has limitations that need to be clearly understood to avoid inflated expectations and disillusionment. In order to support scientific research and technological development, numerous endeavored efforts assess the network performance by simulation; [2, 3, 4 and 5]. A simulation model would greatly simplify a conceptual deployment decision and quickly provide performance indicators and potential scalability bottleneck.

Nonetheless, all simulation models considered only homogeneous IoT applications deployment whilst a realistic network planning should include heterogeneous deployment of IoT applications on end-devices. An IoT application can be modeled by a pair of attributes: data payload and inter-arrival time of packet generation. This defines the total traffic aggregated at a gateway and hence impacts the network performance of a LoRa-LPWAN. Therefore, the basis of the conducted work is to attempt to allocate SF of an ED locally based on the installed application deployed on the end-device. The key concept is to utilize fast transmissions of smaller SF networks by configuring end-devices with applications of large packet sized to use the small SF networks.

There is no much research done in the domain of modeling LoRa-LPWAN with heterogeneous IoT applications deployed. Some authors have utilized multi-technology networks to improve the performance of heterogeneous IoT applications [6, 7] and [8]. Different technologies have their pros and cons in terms of offered throughput, power consumption, covered distances, traffic latency, etc. Different heterogeneous applications can benefit by on-the-fly smart selection of communication technology. Consequently, LoRa LPWAN and other technologies were tested to co-exist in order to

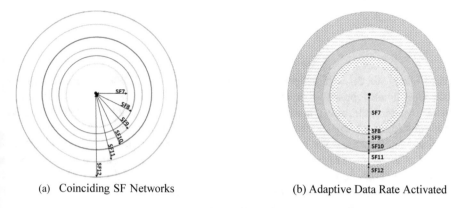

(a) Coinciding SF Networks   (b) Adaptive Data Rate Activated

Fig. 1. LoRaWAN conventional SF allocation strategies

improve flexibility and reliability of the communication. Nonetheless, there are several challenges for multi-model network implementation, e.g.: protocol stack design and management, application security over different technologies, applications detection mechanisms might have different performance in terms of detection latency. To the best of our knowledge, this is the first attempt in literature to comprehensively study the realistic network planning deployment of heterogeneous IoT applications utilizing LoRa LPWAN exclusively. In addition, it is a first attempt to assign SFs to end-devices based on the application load. This paper is structured as follows describing the contributions of our work. In Sect. 2, the needed principles of the LoRa LPWAN are briefly described. In Sect. 2, we display features of IoT applications and better categorize them with respect to data payload and packet generation inter-arrival times. Section 3 explains the implementation of adopted simulation models using Omnet++. In Sect. 4, we display the results and analysis conducted. All potential future work is listed in Sect. 5. Finally, in Sect. 5 the conclusions of this work are drawn.

2 LORA LPWAN

A. LoRa

It is a proprietary radio technology owned by Semtech [9]. Modulation in LoRa network is based on a derivative of CSS achieving low receiver sensitivities. LoRa has several configurable parameters; e.g.: carrier frequency (CF), SF, coding rate (CR), transmission power (TP) and bandwidth (BW), from which an end-device can choose. It utilizes quasi-orthogonal SF and thus implements variable data rate transmissions. Accordingly the system is designed to trade data rate for range or power, depending on field of deployment, in order to optimize network performance in a constant BW. A typical LoRa network operates on a channel with BW of either 125 kHz, 250 kHz or 500 kHz. At a given SF, both the symbol rate and the bit rate are proportional to the BW. Accordingly, doubling the BW will effectively double the transmission rate. The relation between the symbol bit rate R_s, BW, SF and nominal bit rate R_b, is given by (1) and (2).

$$R_s = BW/2^{SF} \tag{1}$$

$$R_b = BW * R_s * CR \tag{2}$$

where CR is the forward error correction rate used by the LoRa modem that offers protection against bursts of interference. The time-on-air TOA is the duration of a packet transmission time, where it is a function of R_b and application payload size. The configurable SF parameter ranging from 7 to 12 are the pillar for LoRa LPWAN six simultaneous transmissions on the same channel. Therefore, the use of SF could be defined as the enabler of simultaneous receptions of packets, in addition to the enabler of long range communication in LoRa. On the other hand, the higher the SF, the lower the data rate and thus the more prone to collisions due to its long interval of transmission time. The received packet allocated a specific SF will be successfully decoded at a GW if its received power exceeds SNR value derived from receiver sensitivities from Semtech transceiver's data-sheet. Consequently, the corresponding distance of a SF network radius can be computed representing the area of deployment.

The LoRa radio usage in sub-GHz exempt free spectrum worldwide has been standardized by the LoRaWAN alliance. For Europe the unlicensed free spectrum, known as Short Range Devices (SRD) [10], ranges from 863–870 MHz. Since LoRaWAN follows an Aloha-like scheme in the unlicensed spectrum and dont adopt a Listen-Before-Talk (LBT) policy, therefore occupying a channel is a key restriction and so it is subject to radio duty cycle regulation imposed by the ETSI. As per LoRaWAN specifications [11], the maximum duty-cycle is defined as the maximum percentage of TOA in which an ED can occupy a channel. Each time a frame is transmitted in a given sub-band, the same sub-band cannot be used again during the next $Toff_{subband}$ seconds calculated by (3). Each ED is allowed to transmit on channels belonging to different sub-bands in order to increase the aggregate TOA as long as the duty cycle limit in each sub-band is respected.

$$Toff_{subband} = \frac{TOA}{DutyCycle} - TOA \tag{3}$$

There are three different categories of SRD sub-bands distinguished in Europe. First, g1.0 (867-868 MHz), g1.1 (868-868.6 MHz), g1.4 (869.7870 MHz), with 1% duty cycle to be shared between all sub-channels in each sub-band, and an ERP limit of 14 dBm. Second, g1.2 (868.7 869.2 MHz), with 0.1% duty cycle, and a ERP limit of 14 dBm. Third, g1.3: (869.4869.65 MHz) with 10% duty cycle, and 27 dBm ERP limit.

B. LoRaWAN

LoRaWAN is an open-standard MAC and networking layer governed by the LoRa Alliance [11]. Its topology comprises of three main components; end-device (ED), gateway (GW) and network server (NS). EDs communicates with a single or multiple GWs using LoRa modulation. A message transmitted by an ED can be received by multiple GWs, which in turn forward the collected messages to an NS that interacts with the different application servers. EDs are battery powered and intended to last several years; 10 to 20 years. GWs forward packets received from ED to NS over a high throughput IP backhaul interface, such as Ethernet or 3G. An ED is not associated with a certain GW and thus a data packet can be processed by several GWs. The NS de-duplicates the packets sent by the EDs, generates packets that should be sent back to the EDs, if any, and chooses a GW to finally forward those replies to the intended ED. LoRaWAN network is a star topology which helps the technology maintain a simple network architecture that is easily managed and at low cost. Bidirectional communication is supported although uplink transmission from EDs to NS is strongly dominant. LoRaWAN defines three different types of EDs for the various needs of IoT sensors applications; Class A, B and C. The LoRaWAN classes operate in an Aloha like fashion and differ only in their downlink transmission. Class A is the default class that must be implemented by all EDs and other classes need to stay compatible with it. Class A downlink transmission is only allowed after a successful uplink transmission. For which two downlink receive windows are opened listening for a response. It is the class with the least power consumption. The three classes can coexist in the same network and EDs can switch from one class to another based on the deployed application.

Assuming perfect orthogonality between different SFs [12]; each SF network represents a parallel logical ALOHA network with a maximum network radius, all overlapping in the same area. The ADR functionality will be used if requested by the EDs by setting the ADR flag in the uplink message. If the ADR flag is set he NS can control the EDs transmission parameters (SF, CF and TP). ADR should be used in stable radio frequency situations where ED are stationary or quasi stationary during which ADR is enabled throughout their motionless periods. ADR is an important feature of LoRaWAN that attempts to optimize packet transmission power, time and used data rate. Actual implementation of ADR is explained thoroughly in LoRaWAN specification [11].

C. **IoT Applications**

IoT applications are conventionally modeled as a pair of at- tributes in any simulation model. We define those attributes as data payload and inter-arrival time of packet generation. These attributes define the total traffic aggregated at a GW and hence impact the network performance of a LoRa-LPWAN. Rationally, this classifies the applications by a two dimensional criteria as demonstrated in Fig. 2; based on end-device's reporting intervals and based on the report payload size. The in-text of this paper is trying to utilize the different parameters of an application and allocate the SF to an ED accordingly. This led to the urge of further classifying the different IoT applications according to the defined attributes and consequently a relation between the installed application and allocated SF can be realized. Figure 2 also lists several IoT applications under each category, referenced from M2M communication use cases derived from IEEE 802.16 Broadband Wireless Access Working Group [13]. As displayed, an application will be categorized based on the ED reporting interval (frequent, average or infrequent) and the reporting data payload size (long or short messages). Frequent transmissions range in the order of seconds up to 1 min, average transmissions defined as such transmitting with intervals in the order of minutes and infrequent shall be defined for application that report their data in the order of days.

According to [11], maximum MAC payload size varies from 51 to 222 Bytes derived from limitation of the physical layer depending on the effective modulation rate used taking into account. Packets of large payload are fragmented according to maximum frame payload supported during transmission. Therefore as to simplify the classification, we define short messages as reports with less than 51 Bytes and otherwise defined as long messages. So as not to lose generality, only applications with short messages will be analyzed in what follows. This is done in order to avoid fragmentation which will increase the complexity of the developed simulation model.

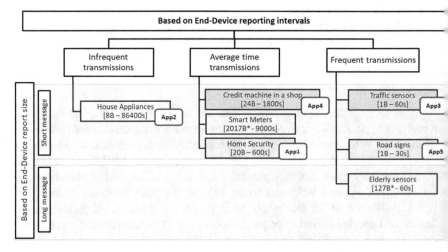

Fig. 2. IoT applications classification

3   LoRa LPWAN Simulation Model

Omnet++ is used for LoRa LPWAN model simulation; where EDs are distributed around a single centered GW. Network density is defined as the total number of EDs randomly distributed within total cell area. The network cell radius is equivalent to that of SF = 12 service radius. We use LoRa receiver sensitivity provided by [9] and follow most findings in literature [14] to represent LoRa LPWAN area of deployment by a log-distance path-loss model. Table 1 lists the receiver sensitivities Γ used and the corresponding maximum service radius R_C for each SF using (4).

$$\Gamma = P_{TX} + G_{ANT} - K - 10nlog(R) - P_{Noise} \tag{4}$$

where P_{TX} = 14 dBm is set to the maximum transmit power, G_{ANT} = 2 dB is the antenna LoRa GW gain, K= 128.95 dB is the intercept of path-loss model and P_{Noise}= 117 dBm is the background noise calculated from $P_{Noise} = 174 - 10nlog(B) - NF$ with B = 125 kHz and NF = 6 dB is the noise figure. All parameters values used are in-line with the findings of several LoRa channel measurement campaigns as referenced by [15]. We only change the path-loss exponent $n = 3$ representing an obstructed home environment. At the beginning of the simulation, EDs are assigned a CR = 4/5, an SF ranging from 7 to 12 and IoT application to define packet payload and the frequency of transmission. The packet transmission time TOA is calculated provided application payload and LoRa nominal bit rate for different SFs provided by (2). We consider that all EDs and the GW have only the three fundamental channels of the same sub-band installed of 1% duty cycle. Each ED selects a CF available to transmit, and then it enforces a silent period provided by (3) before being able to transmit again. This silent period is applied on channel frequencies of the same sub-band as per LoRaWAN specifications. Inter-system interference for non-LoRa signals operating in ISM frequencies have not been considered. Thus, we only model interference caused by collisions that are due to LoRa signals. We consider all EDs use 125 kHz bandwidth for their transmission, and thus collision is defined as the overlap of packet in time with same SF and CF. A packet is successfully decoded by a GW if it's received power exceeds a the announced receiver sensitivities. In all following simulation tests, we will analyze only the short messages classes to avoid fragmentation. The simulation model has 2 modes of operation; parallel Aloha-like networks model and ADR activated model. In parallel Aloha-like networks model, each SF network is setup with a total number of $N_{SF-total}$ ED that are uniformly distributed in their network area ranging from the GW till it's SF network radius. All SF networks residing in the same area are assumed quasi- orthogonal and support collision free communication between the different SF networks. When ADR mode is activated, N_{total} total number of EDs are uniformly distributed within the total network area of radius corresponding to that of SF = 12 network. Thereafter, EDs are organized in multiple SF tiers. See Fig. 1 for an illustration of the 2 modes supported by the developed simulation model.

Table 1. LoRa SNR thresholds for correct signal demodulation

SF	$Gamma_{symbol}(dB)$	R(km)
7	−7.5	2.42661
8	−10	2.93991
9	−12.5	3.56178
10	−15	4.31519
11	−17.5	5.22797
12	−20	6.33383

4   Results and Analysis

Several simulation campaigns are conducted to test the impact of IoT heterogeneous applications' deployments on network performance. The used IoT applications' indexing and parameters are displayed in Fig. 2. Any simulation run time is equivalent to 10 days. The goal of these simulations is to analyze network success probability and network success rate as a function of network density. Network success probability is the ratio between number of successful packets and total packets received at a GW. Network success rate defined as successful packet per second is the number of successfully received packets within total simulation time.

A. Uniform Deployment of Heterogenous IoT Applications

In this subsection, ADR mode is activated. We start by displaying the network success probability as a function of network density for conventionally adopted homogeneous applications deployment as in all LoRaWAN literature. In each simulation run, all EDs are homogeneous with respect to the deployed IoT application. The 5 IoT applications used are listed in the first 2 rows of Table 2 with their parameters. Figure 3 shows that the Home Appliance application has the highest network success probability since it has very small payload size of 8 Bytes that is reported once a day.

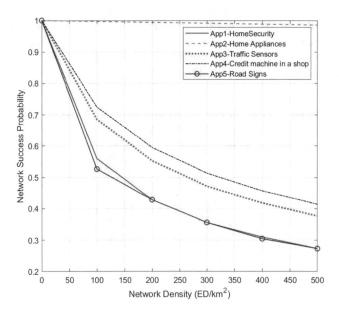

Fig. 3. Network success probability for different homogeneous IoT applications' deployments

From Fig. 3 it is clear that both Home Security and Road Signs application have the same network success probability for all network densities above 200ED/km2. For Traffic Sensors and Credit Machine in a Shop application, both have analogous network success probability for different network densities with an average of 0.0399 success probability difference. This figure proves that the intuitive classification of IoT application based on their attributes is not adequate for the propose of this paper; where we opt to allocate SF based on the installed IoT application. Instead we will classify IoT applications based on their network performance behavior; which is a joint function of a payload size and frequency of transmission. Accordingly, we'll contemplate both App.1 and App.5 to be the same and App.3 and 4 to be analogous as well. In what follow, we'll vary the percentage of IoT deployed applications on the uniformly distributed EDs. This deployment resembles a more realistically representative deployment than conventionally adopted homogeneous arrangements in all LoRaWAN literature of network performance analysis. We use different scenarios of varied percentage of the deployed applications installed on EDs. Each scenario consists of two applications marked in Table 2; where scenario (a) holds 50% of each application, scenario (b) hold 15% of the first marked application and 85% of the second application and finally scenario (c) holds 80% of the first application and 20% of the second one.

Table 2. Heterogeneous IoT applications' parameters and deployment scenarios used

	-App1-	-App2-	-App3-	-App4-	-App5-
	Home security	House appliances	Traffic sensors	Machine in shop	Road signs
	20 Bytes 600 s	8 Bytes 86400 s	1 Byte 60 s	24 Bytes 2400 s	1 Byte 30 s
Scenario 1	x	x			
Scenario 2	x		x		
Scenario 3		x	x		

Figure 4 displays the network success probability for each scenario exhibiting different percentage of deployed IoT applications as described above. It is evident from the three figures that the performance of homogeneous applications involved in the scenario acts as lower and upper bounds to the intermediary scenarios. The more percentage of an application deployed in the network, the more it's performance curve is pulled towards the homogeneous application network performance. This simulation campaign emphasizes the need to investigate more into heterogeneous IoT deployments for enhanced LoRa LPWAN network planning and capacity analysis.

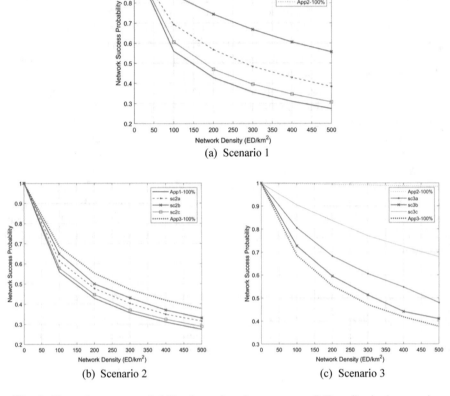

(a) Scenario 1

(b) Scenario 2   (c) Scenario 3

Fig. 4. Network success probability for various heterogeneous IoT application's scenario

B. Proposed Non-Uniform Heterogenous IoT Application Deployment

Our hypothesis is that an IoT application with degraded network probability shall use the fast transmitting SF networks. Based on the fact that lower SF has shorter transmission times, we shall utilize this to assign ED with deployed IoT applications of long transmission time or large packet payload. Given this rational, we start with the simplest allocation scheme where SF = 7 network is locked for the use of homogeneous application resulting in lower network success probability. For this simulation, we investigate scenario 2 where Home security and Traffic sensors applications are used. ADR mode is activated. All EDs located within the SF = 7 network radius have Application 1 installed on it, otherwise Application 2 will be installed. For a uniform distribution of ED with the total network radius, the number of ED located with SF = 7 network region represents 14.8% which is the ratio between number of ED in this region over the total number of devices in the network cell. Since these devices have been assigned to application 1, leaving the rest of the EDs to operate for application 3; thus this can be comparable to Scenario 2(b) with 15%-85% applications deployment. Figure 5 displays the network success probability as a function of network density for the proposed IoT application deployment and the comparable deployment of Scenario 2(b). It is clear that for our simple proposed application deployment is superior than the uniform heterogeneous deployment; specially at network density \(500 ED/km^2\); which exhibits a 19.5\% increase in the total network success probability. However, this leaves the space to address more issues that need to be explored in order to make use of SF attributes and pragmatically plan the network with different IoT applications. What if it is required to deploy an application elsewhere, which SF will it choose that still ensures correct demodulation at the intended gateway. The solution could be answered either by investigating more gateway deployment in the vicinity of such end-devices or simply forcing the end-devices to transmit using the higher order SF and assessing the heterogeneous IoT application deployment or further deactivating the ADR mode and utilizing the whole SF network region that the parallel quasi-orthogonal SF networks model offer.

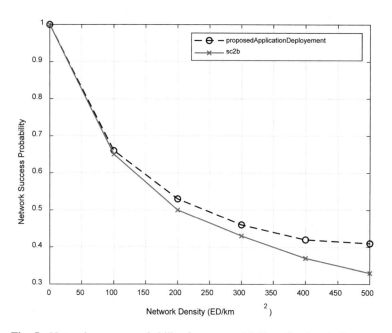

Fig. 5. Network success probability for proposed IoT application deployment

5 Future Work and Conclusion

LoRa LPWAN has recently received a lot of attention from network operators and solution providers to deliver major portion of IoT applications. This paper first devised a simulator using Omnet++ to model LoRaWAN in 2 modes; with ADR activated and modelling the network as parallel quasi-orthogonal SF networks. Then, it investigates the deployment of heterogeneous IoT applications on EDs to represent a more pragmatic deployment than the currently used homogenous representations in most

LoRaWAN analysis in literature. The conducted work is an initial assessment of how network planning based on the realistic heterogeneous deployment could impact network performance and analysis. However, in order to have concrete findings; further work should be conducted to find a mathematical relation by placing an optimization problem to attempt to allocate the SF based on deployed application on an ED and aggregated traffic at the GW.

References

1. Centenaro, M., Vangelista, L., Zanella, A., Zorzi, M.: Long-range communications in unlicensed bands: the rising stars in the iot and smart city scenarios. In: IEEE Wireless Communications, vol. 23 (2016)
2. Bor, M.C., Roedig, U., Voigt, T., Alonso, J.M.: Do lora low-power wide-area networks scale? In: Proceedings of the 19th ACM International Conference on Modeling, Analysis and Simulation of Wireless and Mobile Systems, ser. MSWiM 2016. New York, NY, pp. 59–67. ACM (2016)
3. Jun, A.D., Hong, S., Lee, W., Lee, K., Joe, I., Lee, K., Park, T.-J.: Modeling and simulation of lora in opnet. In: Park, J.J.J.H., Chen, S.-C., Raymond Choo, K.-K. (eds.) Advanced Multimedia and Ubiquitous Engineering, pp. 551–559. Springer, Singapore (2017)
4. Magrin, D., Centenaro, M., Vangelista, L.: Performance evaluation of lora networks in a smart city scenario. In: 2017 IEEE International Conference on Communications (ICC), May 2017, pp. 1–7 (2017)
5. Van den Abeele, F., Haxhibeqiri, J., Moerman, I., Hoebeke, J.: Scalability analysis of large-scale lorawan networks in ns-3. IEEE Internet Things J. **49**(6), 2186–2198 (2017)
6. Mikhaylov, K., Stusek, M., Masek, P., Petrov, V., Petajajarvi, J., Andreev, S., Pokorny, J., Hosek, J., Pouttu, A., Koucheryavy, Y.: Multi-rat lpwan in smart cities: Trial of lorawan and nb-iot integration. In: 2018 IEEE International Conference on Communications (ICC), May 2018, pp. 1– 6 (2018)
7. Almeida, R., Oliveira, R., Sousa, D., Luis, M., Senna, C., Sargento, S.: A multi-technology opportunistic platform for environmental data gathering on smart cities. In: 2017 IEEE Globecom Workshops (GC Wkshps), December 2017, pp. 1–7 (2017)
8. R Almeida, R Oliveira, M Luis, C Senna, S Sargento A multi-technology communication platform for urban mobile sensing. In: IEEE Globecom Workshops (GC Wkshps) vol. 18, p. 1184 (2018)
9. Semtech, Sx1272/73 860 MHz-1020 MHz LP long range transceiver
10. ETSI, Short range devices (srd) operating in the frequency range 25 mhz to 1000 mhz; part 2: Harmonised standard for access to radio spectrum for non-specific radio equipment (2017) Patent V3.2.0 Draft EN 300 22, September 2017
11. Alliance, L.: Lorawan 1.1 specification (2017). https://loraalliance.org/resource-hub/lorawantm-specification-v11
12. Georgiou, O., Raza, U.: Low Power Wide Area Network Analysis: Can LoRa Scale? IEEE (2017)
13. Srinivasan, L.J.R.N.R., Zhuang, J., Park, J.: IEEE 802.16 m Evaluation Methodology Document (EMD), IEEE 802.16 Broadband Wireless Access Working Group
14. Petajajarvi, J., Mikhaylov, K., Roivainen, A., Hanninen, T., Pettissalo, M.: On the coverage of lpwans: range evaluation and channel attenuation model for lora technology. In: 2015 14th International Conference on ITS Telecommunications (ITST), December 2015, pp. 55–59 (2015)
15. Elshabrawy, T., Robert, J.: Capacity planning of lora networks with joint noise-limited and interference-limited coverage considerations. IEEE Sens. J. **19**(11), 4340–4348 (2019)

Autonomous Vehicle Prototype for Closed-Campuses

Abdelrahman Anwar$^{(\boxtimes)}$, Amr Elmougy, Mohamed Sabry,
Ahmed Morsy, Omar Rifky, and Slim Abdennadher

German University in Cairo, Cairo, Egypt
Abdel-rahman.abbas@guc.edu.eg

Abstract. Autonomous driving technology is an invention that mainly aims at giving a vehicle the ability to control itself and to successfully manage situations without human intervention. In this paper, we propose a prototype autonomous system that can manage moving from a starting point to a destination point on in a closed campus. Here, a unique set of challenges emerge. In particular, the interaction with an area with a high density of pedestrians that do not have a fixed crossing point. These closed campus challenges have not received sufficient attention in the literature. Our prototype is comprehensive and includes multiple modules such as Mapping, Perception, Planning and Control. Furthermore, we test the performance of these modules and identify open challenges that should be addressed in closed campus self-driving environments.

Keywords: In-vehicle technology · Navigation systems · Autonomous driving modules · Closed-campus environment · Sensing · Perception · Planning

1 Introduction

The current development in the autonomous driving technologies has yielded outstanding performance results. This development showed that multiple approaches can be taken into consideration to tackle each module in the autonomous technology. A clear robust performance can be traced back to the DARPA Urban Challenge [1]. The use of different sensors in these systems contributed significantly to the robustness of the performance. This is mainly due to the fact that each sensor provides certain advantages. Moreover, the accuracy of each sensor can differ significantly depending on the situation at hand. For example, using the LIDAR sensor to detect pavement edges is more robust than using the camera due to brightness changes and shadows. However, using the LIDAR is computationally more expensive than using the camera. Hence, if there is a situation where proper algorithms were used to solve a task while eliminating the problems of brightness, the camera can be the better candidate in this case. Given that conclusion, the setup used on the prototype in this work includes multiple sensor types such as LIDAR, camera and GPS.

To take advantage of the mentioned sensors, the autonomous system needs modules that can benefit significantly from the data provided from the sensors and use them efficiently. Meaning that the data from the sensors need to be utilized correctly in order

© The Author(s), under exclusive license to Springer Nature Switzerland AG 2021
I. El Dimeery et al. (Eds.): JIC Smart Cities 2019, SUCI, pp. 339–348, 2021.
https://doi.org/10.1007/978-3-030-64217-4_37

to receive the desired output that ensures the safe navigation of the vehicle in dynamic environments. The main modules used are Mapping, Perception, Planning and Control.

The navigation of an autonomous vehicle can be either in open world interacting with all the obstacles such as pedestrians, cars and road rules etc. Or the navigation can be in closed areas such as universities or compounds where the challenges are different and the environment is semi-controlled. Given that the closed areas are generally small in size, a map of the area can be established and later used in multiple modules. This procedure is called the Mapping stage which produces a 3D Point cloud map of a certain area.

After the mapping stage the vehicle can use the produced map mainly for localization. Moreover, the vehicle needs to percept the environment. Perception is getting information about the environment. This leads to the need for the detection of moving obstacles, drivable areas and detection of road obstacles such as speed bumps. In addition, data can be also be received to have information about the location of the vehicle from the GPS.

Given that the perception is done, path planning has to be made in order to establish the route that the vehicle should take to arrive to the desired destination. Finally, given the path that the vehicle should pursue, a controller is needed to give the orders to the microcontroller in charge of moving the motors. That is mainly to ensure the motion of the vehicle is as smooth as possible and the desired destination is reached.

We discuss environment related perception challenges such as detecting drivable areas without road markings (no lanes or pavement markings), detecting marked and unmarked road bumps and finally detecting vehicles and pedestrians.

The main contribution in this paper is implementing a prototype self-driving car architecture on a golf cart. The prototype was tested on several paths within a closed campus using multiple modules related to localization, drivable area detection, obstacle detection and classification.

2 Related Work

Our work builds on top of a combination of previous implementations related to localization, planning and perception. To be able to navigate through the world, the vehicle has to know its location. This localization process is responsible for determining the precise location of the vehicle in the world. A GPS can help locating a car but with an error range of 1–2 m in the best case. This range is unacceptable as a car might think it's located in the center of its lane while it could be off by 1–2 m and really be on the sidewalk or running into things. Thus, there is a need for an accuracy that is a lot better than 1–2 m which the GPS provides. Fortunately, there are several approaches for tackling this problem. [2] proposed a technique called NDT Mapping and Matching. They proposed mapping the environment using laser sensors such as LiDARs, then calculating a normal distribution transform (NDT) for each cluster of points resulting in having a grid of probabilistic functions from the laser points map. Biber et al. used this NDT-based map in the localization process where they take the live data from the laser sensor and apply the same NDT algorithm then starts matching the live data with grid parts of the map. The approach described by Biber *et al.* is actually described on a 2D space, but the same principles will apply on a 3D space. Using this technique increases the accuracy of the localization as its error range is about 2–10 cm and decreases the computational power needed for the localization process since we are trying to match NDTs instead of matching laser points. Another approach to tackle the localization problem is the Simultaneous Localization and Mapping (SLAM) [3]. SLAM has been a topic of research in robotics and artificial intelligence fields for several years, only recently it started to attract attention from the autonomous driving community. The core idea behind SLAM algorithm is the estimation of the motion of a moving vehicle solely based on the data collected from its perception sensors, unlike the map-based approaches that needs a high definition 3D map as input to be able to localize the vehicle.

For our work in localization we decided to go for a map-based localization approach as it aligns with our main objective which is making the vehicle able to navigate autonomously between a set of previously marked locations on the map. SLAM was proven to not fit this objective as it cannot locate the vehicle on a previously created map.

The second problem that comes after localization is planning the route that the vehicle needs to take in order to reach its goal destination. To be able to plan a route for the vehicle we need to define our world in some way. Although the point cloud map used in the localization is giving a very good perception for the environment, it does not define any road related information such as street and lane information.

[1] proposed a technique to solve this problem, they described the road information in terms of *Waypoints*. A *Waypoint* is an object that consists of *(x,y,z and velocity)*. A lane is defined in terms of a path of connected *Waypoints* as shown in Fig. 1.

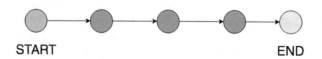

START   **END**

Fig. 1. A *Waypoint*-based lane representation

Consider having two lanes *a* and *b* that we want to connect together, we will do this by connecting the ending node of lane *a* to the starting node of lane *b*. Using this technique, we will be representing our map in terms of a set of connected nodes *G(N, E)*, where N corresponds to the set of nodes (*Waypoints*), and E is the set of edges. Consequently, our planning problem can now be defined as a classical path planning problem where we can use existing path planning algorithms such as Dijkstra's shortest path algorithm to plan a route from the vehicle's current location and its desired destination. In our work, we decided to use this technique to tackle our route planning problem as it fits our goal in mapping and defining all road rules for a closed-campus environment. The waypoint-based definition of the road rules merged with the point cloud map provides us with the needed information. This information assists the car to navigate from its initial location to its desired destination.

After a path of *Waypoints* is generated for the vehicle to reach its destination. The vehicle needs a way to follow this path. One way to do this is to use a classical pure pursuit controller [5]. Ohta *et al.*proposed an implementation for pure pursuit controller that is suitable for autonomous vehicles. Their implementation uses the *Waypoints* to calculate the angular velocity needed to follow the path. We followed their implementation as it showed to be reliable as it was used in other prototypes such as Boss; the winner vehicle in DARPA Urban Challenge.

Given that our implementation is targeting a closed campus environment. Our biggest challenge is to detect, track and interact with the surrounding dynamic obstacles such as pedestrians and other vehicles. Given that pedestrians do not have defined crosswalks or crossing points. A robust pedestrian and obstacle detection solution had to be implemented to overcome this challenge.

As camera-based detection and classification using DNNs showed a huge success lately, we decided to work upon the state-of-the-art detection and classification deep learning model YOLOv3 [4]. YOLOv3 has proven to have the best accuracy to speed ratio in detecting and classifying objects, so we used it in our camera-based detection module to ensure robustness.

3   Autonomous Vehicle Architecture

Figure 2 shows an overview of the pipeline of our vehicle system architecture. The vehicle system architecture is based on *ROS* (Robot Operating System) [6], so each module is represented as a ROS Node and it broadcasts its data through publishing it to one or more topics and receives data from other nodes by subscribing to their respective topics.

Our sensors module is responsible for gathering information from the following sensors: Grasshopper 3 Camera, Velodyne VLP-16 LiDAR, Skylabs SKM-33 GPS module, 9 turn potentiometer for the steering feedback and the embedded vehicle speed encoder. For each sensor, we do the needed preprocessing tasks on its raw data then we publish the processed data to their relevant topic in the ROS network. In this paper we focus on the implementation of the perception module, route planning and trajectory planning submodules.

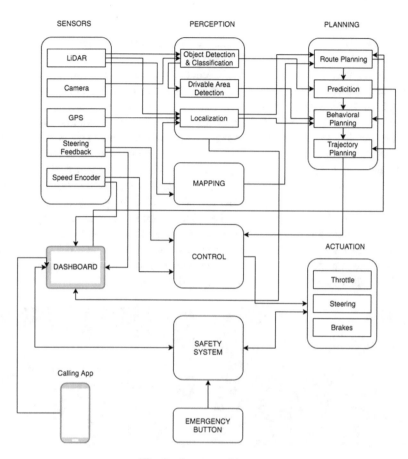

Fig. 2. System architecture

3.1 Object Detection and Classification

Given that pedestrians do not have defined crosswalks or crossing points. A robust pedestrian detection and tracking solution had to be implemented to overcome this challenge.

The aim of this module is to give insights about the objects that are within the vicinity of the vehicle. This module is divided into two submodules. The first sub-module is responsible for detecting and classifying objects by processing images from the camera input. Images are processed through YOLOv3 [4] convolutional neural network (CNN), which detects and classifies the objects within each frame received from the camera. However, cameras are not a reliable source of data in low lighting conditions. Moreover, some obstacles are not easily detected with cameras such as pavement edges as they are unmarked within our campus. To compensate for these limitations from the camera, we had to process point cloud data from LiDAR sensor to detect obstacles in the vicinity of the vehicle. This was done by the Euclidean point cloud clustering algorithm like [7].

3.1.1   Camera-Based Detection

To be able to detect and classify obstacles such as pedestrians, vehicles, speed bumps and cones in real-time YOLOv3 CNN was used.

A dataset was collected from the German University in Cairo campus from a Grasshopper 3 camera mounted on the vehicle. In order to collect a balanced dataset to deal with variable lighting conditions, multiple datasets were collected at different times of the day. The full dataset had approximately 4500 training images. After collecting the dataset, each training sample was manually labeled by cones and speed bumps as shown in Fig. 4. A separate predefined module is used to detect cars and pedestrians as shown in Fig. 3. In order to achieve a high performance, PyTorch-based YOLOv3 implementation was used as in our vehicle, as it showed to be faster than the original darknet framework implementation. To be able to specify obstacles that are within the region of interest in front of the vehicle, values of pixels of the right and left sides of the bounding box were mapped to correspond to the vehicle collision range. If the pixel values are within a certain range this means that a possible collision may occur with the vehicle. Thus, a signal is published to the respective topic in the ROS network so other modules can take the required action.

Fig. 3. Vehicle and Pedestrian detection   **Fig. 4.** Speed bump detection

3.1.2   LiDAR-Based Detection

In order to ensure the reliability of the perception module on the vehicle, a LiDAR sensor is used to detect obstacles using the Euclidean clustering method. The advantage of this method is clustering the obstacles present around the vehicle regardless of its class and assigning each obstacle with a bounding box. In addition, this method reliably gets the distance between the objects surrounding the vehicle and the vehicle itself. If an object is detected within the danger zone in front of the vehicle, a signal is sent to the safety system to act upon.

3.2   Drivable Area Detection

In order to define the safe drivable area for the vehicle, we need to track the objects within the vicinity of the vehicle. The perception module is able to detect the vehicle surroundings. However, in an environment with moving objects, we have to take into consideration other factors such as the heading of our vehicle. If the vehicle decides to go from a lane to the other, factors like moving objects behind or in front of the vehicle can affect its drivable area. The drivable area detection module uses the output bounding boxes from the LiDAR-based detection submodule to know the locations of the surrounding objects. It also takes the steering wheel encoder data as input to track the heading direction of the vehicle and update the safe drivable area accordingly.

3.3 Localization

Localization is one of the most crucial modules in the system as it is responsible for getting a precise location of the vehicle in the world. A GPS can locate the vehicle but only with an accuracy of 1-2m which is not enough as it will not provide important information such as the vehicles' current lane. So, we need something that can get us a more precise position. That is where the localization module comes in action. For this problem, we used the map-based NDT matching approach proposed by [2].

The implemented localization module is divided into two phases:

1. The matching phase, were a matching algorithm tries to compare the live lidar scans to a point cloud map retrieved from the mapping module till there is a match. A match means that the vehicle's position is successfully identified on the map.
2. The tracking phase, where after the vehicle is accurately localized within its environment, the location is updated on the map according to the vehicle actions and movements.

A point cloud map was created for the German University in Cairo campus as shown in Fig. 5 using the NDT Mapping technique and data received from our LiDAR sensor. Fig. 6 shows live LiDAR data correctly matched with the point cloud map.

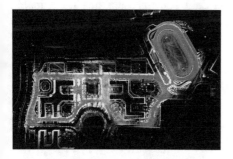

Fig. 5. GUC Point Cloud Map

Fig. 6. NDT Matching

3.4 Planning

To achieve the task of navigating the vehicle from the current location to its desired destination. A vector map that contains the *Waypoints* describing the road information inside our campus was created. Using the vector map and the point cloud map created previously for the mapping module, we can calculate the shortest path between any two

Waypoints within our campus. Finally, in order to drive autonomously through the campus, we used the pure pursuit controller to follow the generated path within our map.

4 Results

4.1 Perception

Our vehicle was able to detect and classify objects in front of the vehicle using the camera input as shown in Figs. 7 and 8. In addition, it was able to detect all the objects within its vicinity using the LiDAR detection submodule as shown in Fig. 9. Each submodule in the perception module publishes its output to its defined topic in real-time so leverage modules can use this data.

  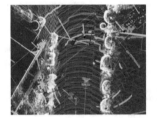

Fig. 7. Vehicle and Pedes- **Fig. 8.** Speed bump detection **Fig. 9.** LiDAR-based detection
trian detection

4.2 Mapping and Localization

We were able to map the German University in Cairo campus using the NDT mapping technique and the input data from the LiDAR sensor as shown in Fig. 10. We were able to use this map to localize the vehicle within the campus as seen in Fig. 11. In order for the vehicle to localize itself, current GPS location is given as an input for the localization module, then the matching algorithm starts searching the area around the GPS point until it finds a matching frame as seen in Fig. 11.

Fig. 10. NDT-Map for GUC Campus   **Fig. 11.** Localization on NDT Map

4.3   Path Planning

In order to test our planning module, we generated a path for the vehicle to drive around the football track shown in Fig. 11. The vehicle was able to navigate and drive autonomously completing a full lap around the football track using the pure pursuit controller. Figure 12 shows the *Waypoints* path assigned to the vehicle in blue, and the curve generated by the pure pursuit to follow these *Waypoints*.

Fig. 12. Vehicle driving on specified Waypoints

5   Conclusion and the Way Forward

In this paper, an autonomous vehicle architecture prototype was implemented on a golf cart including all its core modules. The prototype vehicle was tested and demonstrated in a closed campus environment. This paper describes the implementation techniques used in the vehicle modules. Experiments conducted shows that the modules implemented in the vehicle are able to interact within the dynamic environment inside the campus. Even though the vehicle was able to navigate between different destination points within the campus, there are still some unsolved challenges. Damaged/irregular parts from the road are not detectable or classifiable within the current implementation of the perception module. For the drivable area detection module, we can add a prediction module that can track the speed and heading for all the vehicle surrounding objects, using such information we can predict hazardous situations before happening instead of waiting till the hazard is within the path of the vehicle. Successfully tackling such problems will result in a safer ride.

References

1. Buehler, M., et al.: The DARPA Urban Challenge. https://doi.org/10.1007/978-3-642-03991-1, 9783642039904
2. Biber, P., Straßer, W.: The normal distributions transform: a new approach to laser scan matching. In: Proceedings 2003 IEEE/RSJ International Conference on Intelligent Robots and Systems (IROS 2003) (Cat. No. 03CH37453), vol. 3. IEEE (2003)
3. Thrun, S., Leonard, J.J.: Simultaneous localization and mapping. Springer Handbook of Robotics, pp. 871–889 (2008)
4. Redmon, J., Farhadi, A.: Yolov3: An incremental improvement. arXiv preprint arXiv:1804.02767 (2018)
5. Ohta, H., et al.: Pure pursuit revisited: field testing of autonomous vehicles in urban areas. In: 2016 IEEE 4th International Conference on Cyber-Physical Systems, Networks, and Applications (CPSNA). IEEE (2016)
6. Quigley, M., et al.: ROS: an open-source Robot Operating System. In: ICRA Workshop on Open Source Software, vol. 3, no. 3.2 (2009)
7. Zhou, Y., et al.: A fast and accurate segmentation method for ordered LiDAR point cloud of large-scale scenes. IEEE Geosci. Remote Sens. Lett. 11(11), 1981–1985 (2014)

Model-Driven Decision Support System for Broadband Penetration in Nigeria: Smart City Challenge

Cosmas Ifeanyi Nwakanma[1]([⊠]), Achimba Chibueze Ogbonna[2],
Udoka Felista Eze[3], Esther Chiadikaobi Ugwueke[4],
Christiana Chidimma Nwauzor[5], and Joy Okwuchi Chizitere Oguzie[3]

[1] Kumoh National Institute of Technology, Gumi, Korea
cosmas.ifeanyi@kumoh.ac.kr
[2] Babcock University, Ilishan Remo, Ogun, Nigeria
[3] Federal University of Technology, Owerri, Imo, Nigeria
[4] Circuit Pointe, Owerri, Imo, Nigeria
[5] Sabre Travel Network, Port Harcourt, Rivers, Nigeria

Abstract. This paper proposes a novel model used for the development of a decision support system (DSS) to improve broadband penetration which is critical to the global goal of achieving 5G network implementation beyond 2020. Smart cities relied on information and communication technology driven by the rate of broadband provision in each country. Countries with smart cities have high broadband penetration rate as well as improved economic index. The model was developed after an analytical appraisal of factors affecting broadband penetration in a country case study using a 17-year data period (2001–2017). The significant variables were used to develop the DSS. The result of the analysis showed that the model developed achieved 70.2% reliability in terms of its ability to predict broadband penetration.

Keywords: Broadband · Decision support system · Modeling · Regression analysis · Smart city

1 Introduction

Broadband penetration is critical to the full implementation of 5G networks especially in developing countries where the rate remains low. For efficient broadband penetration, three major segments are usually considered such as international network infrastructure, national network infrastructure and last mile connectivity. However, while developed countries like South Korea and the United States of America launched their 5G networks, same cannot be said of several less developed or developing countries due to so many variables affecting broadband penetration in such countries.

A smart city is simply a digital city which can be described as an area or community employing broadband communication infrastructures to connect government and their citizens in a manner that satisfies the aspirations of citizens, businesses, employees and government. More so, a city that can link sensors and software effectively to generate information and manage the information for improve decision

© The Author(s), under exclusive license to Springer Nature Switzerland AG 2021
I. El Dimeery et al. (Eds.): JIC Smart Cities 2019, SUCI, pp. 349–356, 2021.
https://doi.org/10.1007/978-3-030-64217-4_38

making and arriving at smart solutions to every day issues is known as smart city (Hoang et al. 2019). Examples of smart cities are: Songdo, Suwon Gangnam district of Seoul and Seoul (South Korea), Waterloo and Ontario (Canada), Glasgow (Scotland, UK), Calgary (Alberta, Canada), New York City (US) and Singapore to mention but a few. These cities made use of broadband networks and e-services to sustain innovation ecosystem, growth and inclusion (Intelligent Community Forum 2011).

This paper developed a decision support system for broadband penetration using the most populous nation in Africa as a case study - Nigeria. Nigeria is considered here since it is a developing country and it is expected that a model developed can be used to guide broadband penetration targets across other countries sharing similar variables. The country also offers huge potential for investment into next generation network considering its market size and potential for helping to achieve the 5G network implementation. Broadband and its growth is critical to the full realization of all 5 G communication standards and by implication, critical to smart city development in Nigeria. In a former work (Nwakanma et al. 2018), data for the supply side variables were collected and analyzed to ascertain the relationship between broadband penetration and these variables. Some of the variables were not just technical as wrongly believed; rather they were socio-economic factors. This means that modeling broadband penetration, there is a need to recognize these socio-economic dynamics.

According to (Naser and Omar 2017), next generation networks (and smart cities) present a challenge of the need to guarantee high speed that is neither bound by time nor location dependent. Achieving such, demand that all issues of data rate handling in real time, reduction in delay, improved throughput and security are addressed. One of such ways to address these issues is the framework for broadband penetration. This paper proposes the following:

- Development of a model for the prediction of broadband penetration,
- Implement a decision support system which is driven by the model developed,
- Carry out prediction evaluation of the model to ascertain to what level of significance it can predict broadband penetration for a country case study.

The remaining part of the research paper is arranged thus: Sect. 2 focuses on the problem formulation where the need for a model driven DSS was established. In the third section, the overall model and brief description was presented. In Sect. 4, the paper presents the performance evaluation of the proposed model driven DSS. The paper concluded in Sect. 5.

2   Problem Formulation

Broadband penetration is poor in most of the less developed and developing countries. Between 2015, 2016 and 2018, the broadband or telecommunication sector in Nigeria for example contributed about 8.5, 9.8 and 10.43% respectively to the country's gross domestic product (GDP). The outcome of this poor and uncoordinated investment in bradband is evident in Nigeria having only about 30 percent achievement in broadband penetration.

The Federal Ministry of Communication in collaboration with AFRITEXT Initiative had a Nigeria Smart City Summit which gave birth to the Nigeria Smart City Initiative (NSCI) (Kabir 2019). The summit was held on August 8, 2017 in recognition of the need to have smart cities in Nigeria. The outcome of the summit was the fact that smart city in Nigeria will improve her economy and but depends on the deployment of broadband infrastructures since smart city relies on application of information and communication technology (ICT) and smart technologies in the administration, development and management of smart city initiative. Motivated by this, a proposal for a decision support system is presented in this work to guide decision making in broadband deployment to maximize broadband usage considering its attendant cost.

3   Proposed Model

3.1   Model Driven Decision Support System

Model-driven DSS uses statistical, financial, optimization or simulation model as well as parameters provided by the user to guide decision making. A model is simply a mathematical representation that relates variables. This model is used for solving decision problems. Models could be explanatory (such as for forecasting), contemplative (what if analysis) or algebraic (goal seeking or optimization). Model-driven DSS are useful for the following (Galipalli and Madyala 2012):

1) Analytical Capabilities for answering 'what if scenarios',
2) Can be used for deciding which path to take (Goal seek),
3) Can be used to determine what inputs will get you the desired output (solving).

According to (Power and Sharda 2007), a model-driven DSS focuses on access to and manipulation of a statistical, financial, optimization or simulation model. He further argued that model-driven DSS use data and parameters provided by decision makers to aid the analysis of situations even though it is known that Model driven DSS are not data intensive, meaning that very large database is not needed for many Model-driven DSS.

Very recently, (Legato and Mazza 2018) designed a model-driven DSS whose aim was to effectively support the operations manager in a container handling environment. The DSS helps the manager to determine the proper operational policies and equipment management decision to be taken with respect to a proficient interpretation of container handling operations. In (Austero et al. 2018), a model driven DSS for the management of resources during disaster relief was designed. They developed an optimized model which was used in designing the DSS. The drawback however, is that verification and validation of the system was limited to only interview which is not free of bias since interviews are adjudged to be opinion of respondents.

Previously, (You et al. 2017) designed a DSS for the management of post-earthquake emergency, (Kuo et al. 2015) used a mathematical programming approach to develop a support tool to assist in real time response to disaster, while (Guojon and Labreuche 2016) suggested the use of a multicriteria approach to the design of DSS. The major difference in this paper is the use of secondary data for the analysis and development of the model using backward elimination methods of variables based on the p-values of the variables during regression iterations. The advantage is that the model is free of bias and based strictly on a 17-year data trend.

3.2 Mathematical Model of the Proposed System

The model used in this paper is a mathematical model known as multiple regressions model. A simple regression Eq. (1) is stated.

$$Y = b_0 + \sum_{i=1}^{n} (bixi) \tag{1}$$

Where Y is the Investment Decision in broadband proxy by both public and private Investment in Nigeria's Telecommunication Industry. The various independent variables $x_1, x_2 \ldots x_n$ represents the independent variables captured for the model development. At the end of the model analysis and development, the significant independent variables were: tariff, urbanization, corruption index, competition and ease of doing business in the country case study.

3.3 Choice of Programming Language

According to (Power 2014), L.A.M.P are the required tools for building a DSS. LAMP is an acronym for Linux (Operating System), Apache (the web server), MySQL (the database management system) and PHP (the scripting/programming language). These tools make it easy to develop a web-based DSS within a short period. However, this paper adopted the following modification due to access and ease of use:

1) The data was fetched by a web interface created with PHP and JQuery,
2) Database Implementation and Management using MySQL,
3) Microsoft Windows 8,
4) PC Intel Core i 73.5Ghz quad-core processor,
5) 16G RAM,
6) Then a virtual server name "Qlik sense" which is a cloud application running on the web. Qlik Sense is a platform that helps to analyze data and make data discoveries personalized to the user. It aids organization staff to reach decision collaboratively.

3.4 High Level Model of the DSS

The output of the model was used to design the decision support system. Figure 1 is the high-level model (HLM) of the decision support system.

The support system involves a log-in for the user and admin. Three menus provided are dashboard, data warehouse and upload functions. While the upload allows you to add more data, the data warehouse shows the content of the repository. The repository contains the 17-year data of all the dependent and independent variables. Decision makers can view the display of the dashboard and be guided or make informed decisions. Figure 2 and Fig. 3 are examples of screen shot showing ease of doing business Competition and rate of urbanization respectively.

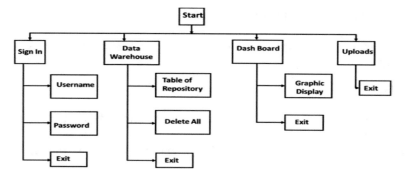

Fig. 1. High level model of the proposed model driven DSS

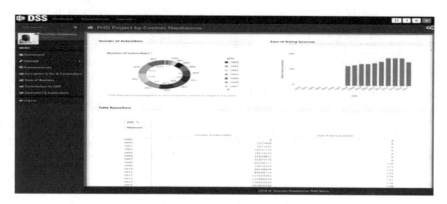

Fig. 2. Sample user interface: ease of doing business

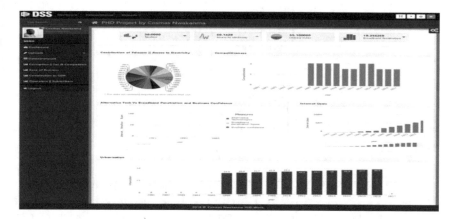

Fig. 3. Sample user interface: competition and urbanization

4 Performance Evaluation

Four conditions are important for the verification and validation of a regression model. They are:

1) Linearity,
2) Independence,
3) 'Homoscedasticity' that is, number of predictors (independent variable) less than number of observations, and
4) that the adjusted R^2 is above 50% within the level of significance less than the confidence interval.

Using the above assumptions or conditions, result summary or analysis is presented in Table 1. It can be argued that the model developed is both verified and valid. Verification entails that the necessary items or components needed to develop the model are present while validation has to do with the significant level of reliability or ability of the model to do what it was designed to do.

Finally, the fourth condition for verification is Homoscedasticity which is automatically met if a regression is linear and well fitted. It is therefore, concluded that the model passed the verification test. Validation of the regression model is based on the R^2 (and/or adjusted R^2) and F-significant change. The R^2 value of 70.2% and adjusted R^2 of 56.7% percent means that the R^2 is above the threshold of 50% and an indication that the regression model explains most of the variability in investment in Nigerian's Telecommunication industry and the fitted model is good.

Table 1. Result of evaluation

Serial no.	Parameters	Computed value
1	Observation	17
2	Significant drivers	5
3	Durbin-Watson	2.843
4	Scatter Plot	Linear
5	Verification	values (1)-(4)
6	R2	0.702
7	Adjusted R^2	0.567
8	F-Sig. Change	0.011
9	Validation	values (6)-(8)

Also, the F-sig change is 0.011, a value less than 0.05 which further validated the significance of the predictors which in this case are broadband penetration drivers. The collinearity statistics shown in Table 2 reveals that the variance inflationary factor (VIF) is less than 5 (corruption (1.069), competition (1.383), Ease of doing business (2.264), Urbanization (3.437) and Tariff (3.600)). This VIF is very good and shows the problem of multicollinearity is eliminated. In all, Backward Elimination method of variable elimination was adopted to reduce the drivers from 16 to 5. This method involves starting with all the variables in the model (in this case, 16 of them) and using the p-values in deciding which variables to remove until few variables are left (Fig. 4).

Table 2. Collinearity statistics

Serial no.	Broadband drivers	VIF	Remarks
1	Corruption	1.069	less than 5.0
2	Competition	1.383	less than 5.0
3	Ease of doing business	2.264	less than 5.0
4	Urbanization	3.437	less than 5.0
5	Tariff	3.60	less than 5.0

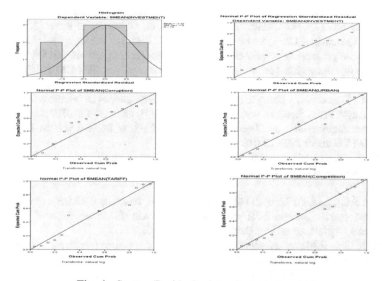

Fig. 4. Scatter (Residual) plots of selected variables

5   Conclusions

The analysis of broadband penetration drivers for investment decision making in Nigeria telecommunication industry was carried out. Five factors were found to be of significant effect. These drivers include: Urbanization, corruption, tariff, competitiveness and ease of doing business. These drivers constituted the component of the regression model that was developed. It was found that the model has an R^2 of 70.2% and reliable for predicting investment decision in Nigeria Telecommunication industry. Several other broadband penetration drivers were found not to be of significant effect such as: GDP, Age, access to electricity, number of subscribers. The model developed was embedded into a decision support system and subjected to verification as well as validation test using regression model analysis (backward elimination of variable approach). The verification and validation test showed a positive level of reliability. Two distinct features of this model driven DSS just like every other DSS are: (1) A model in a model-driven DSS is made accessible to a nontechnical specialist such as a manager through an easy to use interface. (2) A specific DSS in intended for some repeated use in the same or similar decision situation. The situation in the context of this paper is investment decision making in Nigeria telecommunication industry while taking broadband penetration drivers into consideration.

References

Austero, L.D., Brogada, M.A.D., Paje, R.E.J.: Determining resource capacity in disaster relief through a model-driven decision support system. In: Proceedings of IEEE 2018 International Symposium on Computer, Consumer and Control (IS3C), pp. 225–228 (2018)

Galipalli, A.K., Madyala, H.J.: Process to build an efficient decision support system- identifying important aspects of a DSS. MSc. Thesis in Informatics, University of Boras (2012)

Goujon, B., Labreuche, C.: Use of a multi-criteria decision support tool to prioritize reconstruction projects in a post-disaster phase. In: Proceedings of 2015 International Conference of Information Technology for Disaster Management (ICT-DM 2015), pp. 200–2016 (2016)

Hoang, G.T.T., Dupont, L., Camargo, M.: Aplication of decision-making methods in smart city projects: a systematic literature review. Smart Cities 2(3), 433–452 (2019)

Intelligent Community Forum. Intelligent Community of the Year, Internet archive document 22 July 2011–11 October 2018 (2018)

Kabir, M.M.: "Nigeria Smart City Initiatives (NSCI) the Geospatial Perspectives", FIG Working Week 2019. Geospatial Information for Smarter Life and Environmental Resilience, Hanoi, Vietnam (2019)

Kuo, Y.H., Leung, J.M.Y., Meng, H.M., Tsoi, K.K.F.: A Realtime decision support tool for disaster response: a mathematical programming approach. In: Proceedings of IEEE 2015 International Conference on Big Data, pp. 639–642 (2015)

Legato, P., Mazza, R.M.: A decision support system for integrated container handling in a transshipment hub. Decis. Support Syst. 108, 45–56 (2018)

Naser, F., Omar, A.: Technologies for 5G networks: challenges and opportunities. IT Professional 19, 12–20 (2017)

Nwakanma, I.C., Ogbonna, A.C., Asiegbu, B.C., Udunwa, I.A., Nwokonkwo, O.C.: Broadband telecommunication deployment: a supply side analysis of penetration drivers in a developing country case. In: Proceedings of the Tenth International Conference on Construction in the 21st Century (CITC − 10), vol. 10, pp. 342–349 (2018)

Power, D.J.: Using 'big data' for analytics and decision support. J. Decis. Syst. 23(2), 222–228 (2014)

Power, D.J., Sharda, R.: Model-driven DSS: concepts and research directions. Decis. Support Syst. Special Issue Integr. Decis. Support 43(3), 1044–1061 (2007)

You, Y., Li, F., Shao, Z., Zeng, F.: The frame design of prefecture-level post-earthquake emergency decision support system. In: Proceedings of IEEE 2017 International Conference on Cloud Computing, Big Data Analytic ICCCBDA, pp. 538–541 (2017)

Perturbation-Based Analysis of Thin-Walled Steel Tubes Buckling Under Compression: Numerical and Experimental Study

Nicholas Colbert, Mohanad M. Abdulazeez,
and Mohamed A. ElGawady[✉]

Department of Civil, Architectural, and Environmental Engineering, Missouri
University of Science and Technology, Rolla, MO 65401, USA
{nbcwd2,mma548,elgawadym}@mst.edu

Abstract. This paper investigates the development and validation of a numerical model of a thin-walled steel tube under axial compressive loading. The tested specimens consisted of a thin steel tube with a coupler inserted and welded onto each end to reduce buckling due to edge distortion and allow an axial load to be applied when the tube is surrounded by a solid medium. A three-dimensional finite element (FE) model was developed using LS_DYNA. A global imperfection field is used by implementing an imperfection-based perturbation method to accurately trigger the experimentally observed deformation. Validation of the behavior and failure of the model was based upon previous tests performed on four thicknesses of steel with diameter-to-thickness ratios (D/t) ranging approximately from 40 to 120 under monotonic axial compression. The FE results were found accurate for the stress-strain curve up to buckling and in the post-buckling region. It was revealed that the models did precisely match the location and type of local buckling and were able to predict well the buckling (bifurcation) loads.

1 Introduction

Due to the complexity of developing equations that define the buckling and post-buckling behavior of the cylindrical shell it has become common practice to use finite element software to address this case [1–3]. It has been proven that doing so produces accurate results that can replicate the failure mode and stresses occurring in test specimens. Due to the software using geometrically perfect representations of the shells the ideal buckling modes may be created which are represented as sinusoidal waves in the circumferential and axial directions measured in the number of periods that occur [4]. Those fundamental patterns are a part of the classical theory of shells that have been improved upon by adding additional degrees of freedom and changes in geometry [5, 6]. Still, any analytical solution is an approximation which usually overestimates the buckling loads due to localized imperfections which cannot be evaluated by hand calculations [7].

I. El Dimeery et al. (Eds.): JIC Smart Cities 2019, SUCI, pp. 357–366, 2021.
https://doi.org/10.1007/978-3-030-64217-4_39

The finite element program LS_DYNA was used to model the performed tests. The model was then verified against experimental data. The properties used in the model are applicable in the case of compression buckling on the steel. However, due to the sensitivity of its geometry, an imperfection must be introduced to the FE programs using perfect geometries [8]. LS_DYNA has included several methods for applying imperfections perturbations-based method to the simulated models [9, 10]. In this case a linear perturbation using nodal displacement was used. It is also known that refining of the mesh for the shell results in local buckling at progressively lower loads so that a sensitivity analysis must be done to find a sufficiently fine mesh without excessive refining. As the model developed is either unique or previously unpublished it was necessary to research and test all parameters used in its creation with the aid of the published references for LS-DYNA [11–13]. By testing pieces of the model and developing an ever more complex iteration a sufficiently accurate and efficient model was developed.

In the current study, finite element simulations were carried out for thin-walled steel tubes tested previously under axial compression loads. Geometrical imperfections were embedded into a series of explicit FE models to induce buckling modes that obtained during the experimental tests in the steel tubes by updating the wavelength and the magnitude of the imperfection. The aim is to assess the accuracy, robustness, and efficiency of the imperfection-based perturbation method used in this study.

1.1   Experimental Test Program

2   Materials

A total of 4 bare steel tubes with different thicknesses have been investigated under a monotonic quasi-static compressive loading [Table 1]. All specimens had a testing region height of 203 mm (8 inches) and an inner diameter of the steel tube of 108 mm (4.25 inches).

To determine the mechanical properties of the four thicknesses of a1008 cold-formed steel used in the specimens; tension coupons were tested based on ASTM E8/E8M - 16a. the results were illustrated in Fig. 1 and Table 2. It's worth mentioning that the steels for 2.5 mm (0.10 inches) and 0.91 mm (0.036 inches) both had the classic mild steel behavior with the well-defined yield point and considerable ductility [Fig. 1].

Due to the unavailability of the desired steel tubes, the specimens had to be custom fabricated from milled sheets which were shaped with a slip roll and seam welded with a TIG welder. To both allow for the load to be applied and keep the tube's end cross-section circular, a section of steel mandrel two inches long and 4–5/16 inches in outer diameter was made into a coupler and fitted into the tubes. The mandrel had half an inch of the end lathed to fit inside the tube snuggly before the couplers on either end were welded in place [Fig. 2].

Table 1. Geometrical property of the steel tube specimens

Specimen name	Variable	Inner diameter D [mm (inch)]	Steel thickness [mm (inch)]	D/t
S1	Steel tube wall thickness (t_s)	108 (4.25)	2.54 (0.1)	42.5
S2			0.91 (0.036)	118

Table 2. Material properties of the steel tube

ID	Young's modulus [GPa (10^3ksi)]	Yield strength [MPa (ksi)]	Yield strain (%)	Ultimate strength [MPa (ksi)]	Ultimate strain (%)	Rupture strain (%)
S1	218 (31.6)	325 (47.2)	0.34	384 (55.7)	14.5	29.8
S2	217 (31.5)	196 (28.4)	0.29	305 (44.2)	22.1	38.2

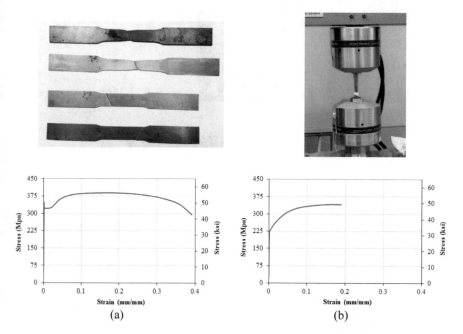

Fig. 1. Axial stress-strain curve for different steel coupon thicknesses in tension test. (a) S1: 2.54 mm (0.1 inches); (b) S2: 0.91 mm (0.036 inches).

3   Test Instrumentation

Displacement was determined by the use of two LVDTs which were set up to record displacement of the top-loading platon. The top platon was meant to be fixed which reduced the difference in readings and allowed accurate displacement data by averaging the LVDTs. Loads were obtained through the testing machine software and its installed load cell. Testing of the specimens was carried out on an MTS 2500 machine under displacement controlled monotonic axial compression loading at a rate of 0.51 mm/min (0.02 in/min) [Fig. 3]. Testing was performed until at least one inch of displacement had been reached. During testing, the top platon was fixed and the bottom was a ball-jointed platon allowing free rotation to represent a pinned connection.

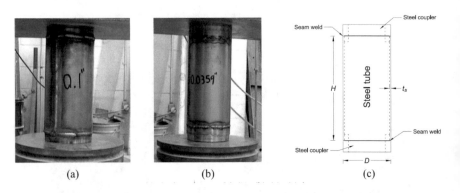

(a)   (b)   (c)

Fig. 2. Specimens (a) S1; (b) S2; (c) layout

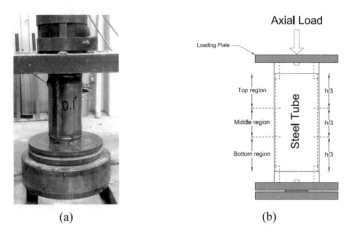

(a)   (b)

Fig. 3. Testing of the bare steel tubes (a) specimen on testing machine; (b) test layout

4   Finite Element Modeling

Geometry

The top and the bottom plate were modeled using solid elements with constant-stress one-point quadrature integration. The average element height, width, and length were 3.2 mm (0.125 inches). The inner steel tubes were simulated using Belytsehko-Tsay four-node shell elements. These shell elements considered bending, membrane, and forces exerted normal to the surface. The element has six degrees of freedom at each node: translations in the nodal X, Y, and Z-directions and rotations about the nodal X, Y, and Z-axes. Stress stiffening and substantial deflection capabilities are included. A sensitivity analysis was conducted to determine the different element sizes. The final model had 2,360 elements and 3,003 nodes. The steel tube elements have an average height and width of 3.18 mm (0.125 inches) and 4.24 mm (0.167 inches), respectively. This size was determined to produce accurate results and proper behavior with a focus on the buckling of the steel tube. The element height in the tube is more significant than the width due to the focus on axial buckling allowing deviation from the 1:1 aspect ratio. The hourglass type and coefficient used in this study were 5 and 0.03, respectively.

Boundary Conditions and Loading

A nodal z-displacement which was applied by defining in a displacement control manner. The bottom node of the base plate was restrained in all directions. While the side nodes of the top plate in the circumferential direction were restrained in the x-y directions only [Fig. 4]. A quarter section of the tested specimens was modeled due to symmetry allowing boundary conditions to replace the continuity in the full model and all the nodes on the surface of the cut were restrained against displacements and rotations. Making these boundaries allows the z-displacement applied to the steel tube to cause the correct buckling behavior at the edges.

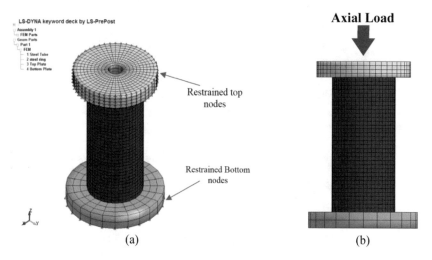

Fig. 4. Simulated model (a) restrained nodes; (b) loads on the column

Material Models

The steel tube was defined using 003-plastic_kinamatic with the option of including rate effects. It is a very cost-effective model and is available for the beam (Hughes-Liu and Truss), shell, and solid elements. The non-tube components were defined using an elastic modulus of 200 GPa (29,000 ksi), the yield stress of 496 MPa (72 ksi), and Poisson's ratio of 0.3. The ultimate stress and ultimate strains were 599 MPa (87 ksi) and 10%. Properties of the steel tube varied in the model due to corresponding variations in the actual material [Table 2] such that two behaviors appeared.

The specimens were tested under quasi-static conditions in axial compression. To apply the load in practice, a coupler was inserted into the steel tube on either end to a depth of 12.7 mm (0.5 inches) and seam welded in place then the weld was ground flush with the surface. This was simplified in the model by incorporating a ring stiffener of 0.5 inches tall and 0.375 inches thick to act as the interior portion of the coupler. The nodes between the ring and tube were merged to simulate the weld. An additional ring of equal thickness to the internal ring and length of 1.5 inches was placed on the internal ring to apply the load. A surface to surface contact was used between both loading rings and the inner rings. This contact type considers sliding and separation between the two surfaces.

Perturbation

Initial perturbations were produced using the *PERTURBATION_NODE keyword command, and the analysis method was single precision explicit using the four-node Belytschko-Tsay shell elements with six degrees of freedom per node. To obtain the failure modes observed in experiments, it was necessary to introduce an imperfection to the steel tube. The real specimens had imperfections due to the combination of roll forming changing internal stresses, forming not producing a perfectly cylindrical shape, residual stresses due to the lengthwise weld, and welding at top and bottom which added residual stresses resulting in a small uneven loading along the edge.

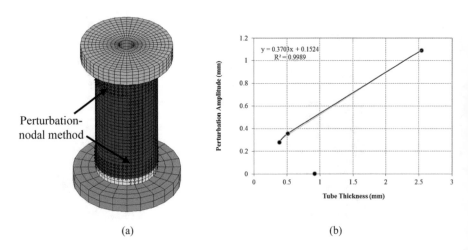

(a)   (b)

Fig. 5. Perturbation (a) perturbation in the model; (b) perturbation amplitude vs tube thickness

In order to incorporate all of those imperfections, a simple nodal perturbation was added which created a radial displacement along the z-axis in the form of a sine wave in the XY-plane [Fig. 5 (a)]. The perturbation was applied to all nodes between the inner rings which matched the experimental data better than having the perturbation only on a small portion of the surface. Shells are extremely sensitive to any imperfections and are even more so at high D/t ratios. In the case of the models described, the perturbation amplitudes respectively for S1, and S2 were 0.043 and 0.003 inches. These results with the exception of S2 are very linear [Fig. 5 (b)]. The exception, S2, failed earlier due to a local buckle at the weld. But its failure mode was very high energy and allowed it to nearly reach its yield stress, hence the lower perturbation.

5 Results and Discussion

FE Validation Results

Figure 6 shows the axial load versus shortening of the experimental and FE analysis conducted on the four bare steel tubes with different thicknesses seen in Table 2. As shown in Fig. 6, the FE results had a good agreement with the experimental results with high accuracy. The FE model predicted the axial load of the steel tube generally by less than a 2% difference and the shortening by 18%. Deformations in the tests and models were in good agreement with two modes seen. Specimen S1 buckled in an axisymmetric elephant's foot buckle and the others failed by non-axisymmetric diamond buckling. Elephant's foot is a single outward buckle and diamond buckling is inward and outward buckles which in an ideal test form a pattern of diamond-shaped buckles around the circumference of the tube (Table 3).

Table 3. Buckling loads and shortening at buckling for finite element and experimental

Specimen name	FE		Exp	
	Buckling load [kN (kip)]	Shortening [mm (inch)]	Buckling load [kN (kip)]	Shortening [mm (inch)]
S1	203.8 (45.8)	2.91 (0.115)	207.6 (46.7)	2.54 (0.100)
S2	58.8 (13.2)	0.76 (0.030)	58.7 (13.2)	0.84 (0.033)

These results are as expected for their D/t ratios since at this length over diameter ratio those modes have been proven to coincide with the tested D/t ratios [14]. The amplitude of the applied perturbation was increased and lead to a lower buckling load making the buckling load very accurate. In contrast, the shortening at the buckling load was not as reliably determined due to their being little control over its location. Due to these factors accuracy will be from highest to least; buckling load, initial stiffness, and shortening at buckling load. As the purpose of the steel tube models is validating the material and geometry for use in a future full model, it makes sense that the buckling load and stiffness be the highest accuracy values as they will be the most important.

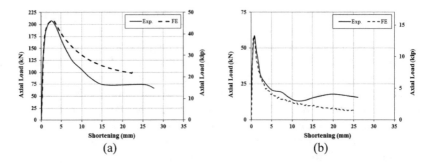

Fig. 6. Experimental versus FE axial load-shortening backbone curve for the tested bare steel tubes (a) S1: 2.54 mm (0.1 inches); (b) S2: 0.91 mm (0.036 inches)

For the bare steel tube with a thickness of 2.54 mm (0.1 inches), the experimental work showed a high-stress concentration at a distance of 32 mm (1.25 inches) from the top [Fig. 7 (a)]. This behavior was similarly displayed in the FE model at a distance of 38 mm (1.5 inches) from the top [Fig. 7 (b)]. Furthermore, the FE model was able to simulate the buckling patterns and mode as obtained in the experiment [Fig. 7 (c and d)]. The steel tube local buckling was uniform around the circumference which means that one wavelength occurred or one period, this represents the first fundamental buckling mode and is commonly referred to as an elephant's foot buckle. The percent differences between the model and experiment were 1.8%, 17.2%, and 2.8% for peak load, shortening at peak load, and initial stiffness respectively.

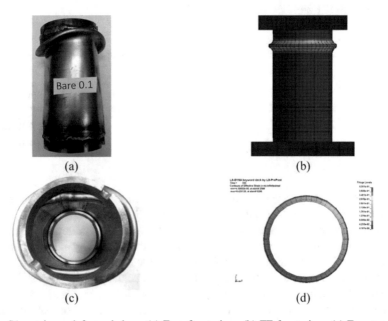

Fig. 7. S1 specimen deformed shape (a) Exp. front view; (b) FE front view; (c) Exp. top view; (d) FE top view

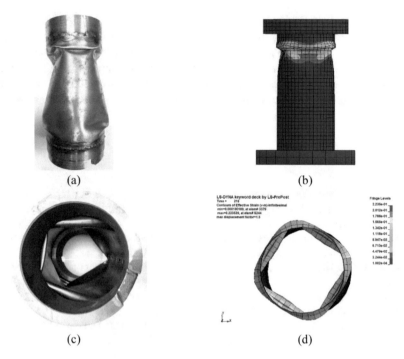

(a)   (b)

(c)   (d)

Fig. 8. Bare tubes deformed shape with 0.91 mm (0.036 inches) thickness (a) Exp. front view; (b) FE front view; (c) Exp. top view; (d) FE top view

For the bare steel tube with a thickness of 0.91 mm (0.036 inches), the experimental behavior was not as predicted due to a local failure at the weld. That caused local buckling at about 19 mm (0.75 inches) from the bottom on the side of the weld [Fig. 8 (a)]. The FE model showed a high-stress concentration at a distance of 38 mm (1.5 inches) from the top [Fig. 8 (b)] very much matching the behavior of the 2.54 mm (0.1 inches) specimen. Due to the local buckling on the side of the weld, an equal but opposite buckle occurred at the top [Fig. 8 (c and d)]. The FE model predicted local buckling uniform around the circumference with a period equal to one which is a circle. The percent differences between the model and experiment were 0.25%, 10.4%, and 16.1% for peak load, shortening at peak load, and initial stiffness respectively. Due to the local failure at the bottom the stiffness was initially lower, a brief stable slope occurred followed by an additional loss of stiffness as the top buckled. That sequence led to a more ductile failure [Fig. 8] and exceeded the expected capacity at higher displacement.

6   Conclusion

This paper investigated the effectiveness of using finite element analysis to model the steel used in the described compression tests. The correct buckling load was achieved with accuracy within a few percent based upon fine-tuning the perturbation card in

LS_DYNA. Axial shortening at the buckling load was within 18.3% difference for all instances or a max shortening difference of 0.37 mm (0.015 inches). Most significant is that the models were able to predict both the location and type of buckling for three of the tests with high accuracy and showed good correlation with part of S2 which had a localized imperfection based failure. Finally, the obtained results indicate that this model is sufficiently accurate to be used as a portion of a more complex composite model.

References

1. Aly, S.S.S.: Buckling assessment of axially loaded cylindrical shells with random imperfections. In: Civil Engineering 1995, Iowa State University, Digital Repository, p. 154 (1995)
2. Haynie, W.T., Hilburger, M.W.: Comparison of methods to predict lower bound buckling loads of cylinders under axial compression. In: Collection of Technical Papers - AIAA/ASME/ASCE/AHS/ASC Structures, Structural Dynamics and Materials Conference (2010)
3. Little, A.P.F., et al.: Inelastic buckling of geometrically imperfect tubes under external hydrostatic pressure. In: Civil-Comp Proceedings, vol. 88 (2008)
4. Hoff, N.J.: The perplexing behavior of thin circular cylindrical shells in axial compression, A.O.O.S. Research, Department of Aeronautics and Astronautics, Stanford University, p. 92 (1966)
5. Flugge, W.: Stresses in shells, p. 1973 (1973)
6. Timoshenko, S., Woinowsky-Krieger, S., Firm, K.: Theory of plates and shells. 2nd edn. McGraw-Hill (1959)
7. Bushnell, D.: Buckling of shells - pitfall for designers, p. 1–56 (1980)
8. Teng, J.G., Hu, Y.M.: Behaviour of FRP-jacketed circular steel tubes and cylindrical shells under axial compression. Const. Build. Mater. **21**(4), 827–838 (2007)
9. He, Y.Z., et al.: Out-of-plane secondary bifurcation buckling behavior of elastic circle pipe arch. In: Key Engineering Materials, Trans Tech Publ. (2011)
10. Rahman, T., Jansen, E.: Finite element based coupled mode initial post-buckling analysis of a composite cylindrical shell. Thin-Wall. Struct. **48**(1), 25–32 (2010)
11. Yan, Q.: Introduction to LS-PrePost (workshops). Livermore Software Technology Corporation (2016)
12. LSTC,: LS-DYNA keyword user's manual, Livermore software technology corporation, p. 2670 (2016)
13. Fanggi, B.A.L., Ozbakkaloglu, T.: Compressive behavior of aramid FRP–HSC–steel double-skin tubular columns. Const. Build. Mater. **48**, 554–565 (2013)
14. Guillow, S.R., Lu, G., Grzebieta, R.H.: Quasi-static axial compression of thin-walled circular aluminium tubes. Int. J. Mech. Sci. **43**(9), 2103–2123 (2001)

Axial Behavior of Concrete Filled Pultrutded FRP Box

Amro Ramadan and Mohamed A. ElGawady[✉]

Missouri University of Science and Technology, Rolla, MO, USA
elgawadym@umsystem.edu

Abstract. Pultruded FRP has been used for decades to save weight, increase corrosion resistance, provide aesthetic beauty and reduce maintenance costs. This paper presents the findings of an experimental study that was conducted to evaluate the axial of six specimens tested under axial loading with different lengths varies from 10 inches to 30 inches using two type of specimens concrete filled box, and reinforced concrete box. The results indicate that the addition on the concrete to the FRP section can improve compressive properties of the section. Also the mode of failure changed from local buckling in case without using concrete to global buckling in case of filling the box with concrete. Design approach were also used to determine the axial capacity of the section filled with reinforced concrete.

Keywords: Pultruded FRP · Axial loading · Experimental results

1 Introduction

Over the past few decades strengthening of reinforced concrete structures using fiber reinforced polymer (FRP) systems has become widely accepted in the construction industry. Lately FRP closed structure shapes also offer potential for use in new construction, especially when used as a confining material for concrete. One such application, which has received much recent attention, involves the use of concrete filled FRP tubes (CFFT) as beams or columns for new construction.

Earlier, the focus was on using FRP sheets or plates for repair of damaged concrete elements by connecting the FRP panels to the concrete elements in order to restore the capacity of the original member and prevent it from further deterioration [1, 2].

Later on, it was more efficient to use the CFFT as a structural element, so researchers started investigating CFFT with different shapes from circular [3–8], rectangular [9, 10], and square [11, 12] to understand the behavior response of the CFFT under axial and bending loading, most the FRP tubes used was based on the directions of the fibers which is either perpendicular to the structural member to provide confinement to the concrete in case of axial loading, or parallel to the member to inhance the bending stiffness of the tested member.

Most of these studies focused on using the filled FRP tubes with bi-directional fibers for compression members [11, 12] and the uni-directional fibers for flexural members[13–17]. However, limited studies have been investigating the use of concrete filled uni-directional FRP as compression element.

© The Author(s), under exclusive license to Springer Nature Switzerland AG 2021
I. El Dimeery et al. (Eds.): JIC Smart Cities 2019, SUCI, pp. 367–373, 2021.
https://doi.org/10.1007/978-3-030-64217-4_40

This study aims to invistage the influence of reinforced concrete infill on the axial behavior of pultruded FRP square columns using uni-directional FRP in the longitudinal direction throught testing six colums divided as three columns with different heights with plain concrete and three with the same height with reinforced concrete infill, and compare the results with the exisiting specifications.

2   Experimental Work

2.1   Test Specimens

A total of six specimens were tested under concentric compression (see Fig. 1). All specimens has 102×102 mm (4×4 in.) square cross section with thickness of 6.35 mm (0.25 in.) with variable height of 51–760 mm (2–30 in.) without concrete, three with height of 10, 20, and 30 in. with plain concrete filled the tube, three with height of 30 in. with reinforced concrete filled the tube (Table 1).

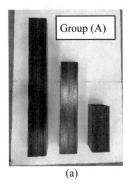

(a)   (b)

Fig. 1. Tested specimens (a) different lengths parameter, (b) reinforced tube group

The specimens in Table 1 were labeled as follows: C indicates columns followed by number of specimen, letter P indicates plain concrete used in specimen while letter E indicates reinforced concrete used.

Table 1. Summary of the tested specimens

	Description	Height (in.)	Concrete Strength N/mm² (ksi)	Reinforcement
Group A	C5-P	254 (10)	35 (5)	–
	C6-P	508 (20)		
	C7-P	760 (30)		
Group B	C8-R	760 (30)	35 (5)	4 #3
	C9-R	760 (30)		
	C10-R	760 (30)		

2.2   Test Setup and Loading Protocol

Figure 2 shows the 2225kN (500 kips) self-sustained MTS testing machine that was used for testing the columns. The load was applied monotonically, at a rate of 0.05in./min, until rupture occurred and the test specimen became unstable.

A test specimen had pin-hinged boundary conditions. The pin was simulated with swivel plate that was connected to the bottom end of the specimen.

Fig. 2. Test setup

2.3   Results and Discussion

Table 2 and Fig. 3 illustrates the axial force versus axial shortening for all the tested columns. Also shown in Fig. 3 are the strengths calculated using the analytical model presented later in this paper. The axial force and shortening were obtained as the force and displacement of the MTS loading machine.

Table 2. Summary of the results for the tested columns

Designation	Axial Capacity	Axial Shortening	Mode of failure
	kN (kips)	mm (in.)	
C5-P	503 (113)	15 (0.58)	RC
C6-P	472 (106)	12.5 (0.49)	RC
C7-P	427 (96)	12.5 (0.49)	RC
C8-R	333 (75)	6.4 (0.25)	RC
C9-R	365 (82)	5.6 (0.22)	RC
C10-R	365 (82)	4.6 (0.18)	RC

[*]RC = FRP rupture at corners

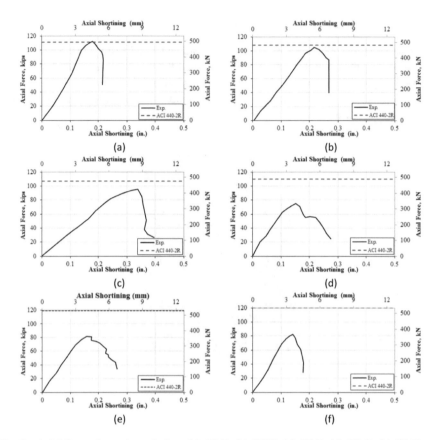

Fig. 3. Axial force shortening relation: (a) C5-P, (b) C6-P, (c) C7-P, (d) C8-R, (e) C9-R, and (f) C10-R

In group (A) for column C5-P, the peak load was 503 kN (113 kips), while for column C6-P, the column failed at peak load of 472 kN (106 kips), for column C7-P, the peak load was 427 kN (96 kips), similarly all the columns failed with the same mode of failure which is rupture of the FRP tube corners.

In group (B) for column C8-R, the peak load was 333 kN (75 kips), while for column C9-R, the column failed at peak load of 365 kN (82 kips), for column C10-R, the peak load was 365 kN (82 kips), similarly all the columns failed with the same mode of failure which is rupture of the FRP tube corners.

3   Analytical Models

3.1   ACI-440

ACI-440-2R (2017), [18] adopted Eq. 1 to calculate the nominal axial strength of nonprestressed FRP members filled with concrete, as a function of the compressive strength of confined concrete, f'_{cc}, gross sectional area of concrete, A_g, specified yield strength of noprestressed steel reinforcement, f_y, and total area of longitudinal reinforcement, A_s.

$$P_n = 0.8 \left[0.85 f'_{cc} (A_g - A_s) + f_y A_s \right] \tag{1}$$

The design approach considers the contribution of fibers on the compressive strength based on the compressive strength of confined concrete; with the contribution of the steel reinforcement.

4 Comparison of Experimental and Analytical Models

Table 3 presents the calculated axial capacities of the tested columns using the analytical model presented in Sect. 3.1. Also presenting the ratio of the peak load – to – the predicted capacity P_n. The analytical model was accurate in predicting the strengths of the CFFT without reinforcement while it was over-predicting the strengths of the reinforced CFFT. The peak strengths measured during the experimental work ranging from (0.9 to 1.02) and (0.68 to 0.69) of those predicted using the ACI-440-2R,[18].

Table 3. Evaluating the axial capacity of the tested columns using analytical model

Designation	$P_{exp.}$ (kips)	ACI 440.2R	
		P_n (kips)	$P_{exp.}/P_n$
C5-P	113	111	1.02
C6-P	106	108	0.98
C7-P	96	107	0.90
C8-R	75	110	0.68
C9-R	82	119	0.69
C10-R	82	120	0.68

5 Conclusions

Six CFFT columns were tested under axial compression. The results of the experimental was compared with the exisiting design specifications of ACI 440.2R (2017), [18]. The research findings led to the following conclusions:

1. Changing the length of CFFT from 254–760 mm (10–30 in.), does not significant effect on the capacity of the columns where it was dropped by 15% without changing the mode of failure.
2. Adding reinforcement to the CFFT weakened the column were the capacity decreased by 17% less than the unreinforced CCFT.
3. Comparing the capacities from the current design specification to the experimental results indicate the ACI 440.2R has well predicted the strength of unreinforced

CFFT with range of predicting from (0.9 to 1.02), while over predicting the strength of steel-rienforced CFFT with range of predicting from (0.68 to 0.69).

Acknowledgements. This research was conducted at the Missouri University of Science and Technology and supported by Mid-America Transportation Center (MATC), The Missouri Department of Transportation (MoDOT), and Missouri Department of Natural Resources (MoDNR). However, any opinions, findings, conclusions, and recommendations presente in this paper are those of the authors and do not necessarily reflect the views of the sponsers.

References

1. Miller, T.C., et al.: Strengthening of a steel bridge girder using CFRP plates. J. Bridge Eng. **6**(6), 514–522 (2001)
2. Honickman, H., Fam, A.: Investigating a structural form system for concrete girders using commercially available GFRP sheet-pile sections. J. Compos. Const. **13**(5), 455–465 (2009)
3. Mirmiran, A., Shahawy, M.: Behavior of concrete columns confined by fiber composites. J. Struct. Eng. **123**(5), 583–590 (1997)
4. Fam, A., Rizkalla, S.H.: Behavior of axially loaded concrete-filled circular FRP tubes. Aci Struct J. **98**(3), 280–289 (2001)
5. Fam, A.Z., Rizkalla, S.H.: Confinement model for axially loaded concrete confined by circular fiber-reinforced polymer tubes. Struct. J. **98**(4), 451–461 (2001)
6. Bakis, C.E., et al.: Fiber-reinforced polymer composites for construction—State-of-the-art review. J. Compos. Const. **6**(2), 73–87 (2002)
7. Youssf, O., ElGawady, M.A., Mills, J.E., Ma, X.: An experimental investigation of crumb rubber concrete confined by fibre reinforced polymer tubes. Constr. Build. Mater. **53**, 522–532 (2014)
8. ElGawady, M.A., Sha'lan, A.: Seismic behavior of self-centering precast segmental bridge bents. Journal of Bridge Engineering **16**(3), 328–339 (2010)
9. Fam, A., Schnerch, D., Rizkalla, S.: Rectangular filament-wound glass fiber reinforced polymer tubes filled with concrete under flexural and axial loading: experimental investigation. J. Compos. Const. **9**(1), 25–33 (2005)
10. Belzer, B., Robinson, M., Fick, D.: Composite action of concrete-filled rectangular GFRP tubes. J. Compos. Const. **17**(5), 722–731 (2013)
11. Ahmed, A., Masmoudi, R.: Axial Response of Concrete-Filled FRP Tube (CFFT) Columns with Internal Bars. J. Compos. Sci. **2**(4), 57 (2018)
12. Youssf, O., ElGawady, M.A., Mills, J.E.: Static cyclic behaviour of FRP-confined crumb rubber concrete columns. Eng. Struct. **113**, 371–387 (2016)
13. Fam, A., Flisak, B., Rizkalla, S.: Experimental and analytical modeling of concrete-filled FRP tubes subjected to combined bending and axial loads. ACI Struct. J. **100**(4), 499–509 (2003)
14. Fam, A., Cole, B., Mandal, S.: Composite tubes as an alternative to steel spirals for concrete members in bending and shear. Constr. Build. Mater. **21**(2), 347–355 (2007)
15. Mohamed, H.M., Masmoudi, R.: Flexural strength and behavior of steel and FRP-reinforced concrete-filled FRP tube beams. Eng. Struct. **32**(11), 3789–3800 (2010)
16. Aslani, F., Gunawardena, Y., Dehghani, A.: Behaviour of concrete filled glass fibre-reinforced polymer tubes under static and flexural fatigue loading. Constr. Build. Mater. **212**, 57–76 (2019)

17. Moustafa, A., ElGawady, M.A.: Strain rate effect on properties of rubberized concrete confined with glass fiber–reinforced polymers. J. Compos. Const. **20**(5), 04016014 (2016)
18. (ACI): A.C.I. Guide for the Design and Construction of Externally Bonded FRP Systems for Strengthening Concrete Structures, in ACI 440-2R-17. American Concrete Institute Farmington Hills, MI, USA (2017)

Compressive Strength of Revibrated Concrete Using Smart Combination of Sawdust Ash from Selected Wood Species to Partially Replace Cement

S. M. Auta[✉], V. N. Agie, and M. Alhassan

Civil Engineering Department, Federal University of Technology,
Minna, Nigeria
samuel.auta@futminna.edu.ng

Abstract. Study into the compressive strength of revibrated concrete using smart combination of sawdust ash from selected wood species to partially replace cement is presented. Apa and mahogany wood species were used to carry out this study. Chemical analysis of the ASDA and MSDA contain the major chemical oxides found in cement which includes SiO_2, Al_2O_3 and Fe2O3. Six (6) concrete cubes of 0% replacement by weight were produced with ordinary Portland cement (OPC), while another Six (6) concrete cubes were cast each using 5, 10, 15, 20, 25 and 30% replacement of OPC with SDA from the specified wood species giving rise to a total of fifty-six (56) cubes produced without revibration. These cubes were cured and subjected to a compressive strength test at 7 and 28 days. The optimum ASDA replacement for cement was attained at 5%, while that of concrete cube containing MSDA was attained at 10%. Another set of concrete cubes numbering eighty-four (84) was produced, fifty-six (56) of which were cast using the optimum percentage from ASDA and MSDA, while the remaining twenty-eight (28) were cast using OPC only. All the eighty-four cubes were revibrated for 1 min 20 s at 10 min intervals to a duration of 1 h after initial vibration. The cubes were cured for 7 days, 28 days and subjected to compressive strength test. The result reveals that the compressive strength of the revibrated concrete cubes increases up to a certain time lag and thereafter decreases to the lap hour of 1 h. The result also reveals that the percentage increase in compressive strength value obtained after revibration for a curing period of 7 and 28 days using 0% (only OPC) increased by 28.01 and 37.31%, that of the 5% ASDA optimum percentage replacement obtained increased by 48.40 and 40.49% and also that of 10% MSDA optimum percentage replacement obtained increased by 33.58 and 40.83% compared to the average value obtained for the corresponding percentage replacement for the non-revibrated concrete cubes. Hence, the study suggests that re-vibration enhances the strength of concrete once done within the plastic stage of the concrete.

Keywords: Compressive strength · Revibration · ASDA · MSDA

© The Author(s), under exclusive license to Springer Nature Switzerland AG 2021
I. El Dimeery et al. (Eds.): JIC Smart Cities 2019, SUCI, pp. 374–384, 2021.
https://doi.org/10.1007/978-3-030-64217-4_41

1 Introduction

Vibration and revibration plays a major role in placing and distribution of high quality concrete. Revibration is the process in which a vibrator is reapplied to fresh concrete at some time after initial vibration (Auta et al. 2015). It is suggested that revibration eliminates defects (honey comb and voids) thereby increasing the compressive strength of the concrete (Krishna et al. 2008). The overall relevance of concrete in virtually all civil engineering practice and building construction works cannot be overemphasized (Adewuyi and Adegoke 2008). Concrete consist of water, cement, fine aggregate, coarse aggregate and admixture if need be, which are mixed in a particular proportion to get a particular strength. However, the construction industry relies heavily on conventional materials such as cement, granite and sand for the production of concrete. The high and increasing cost of these materials has greatly hindered the development of shelter and other infrastructural facilities in developing countries (Olutoge 2010). As the infrastructure of the entire world is kept developing, the construction industry is in need of large amount of raw materials (Olugbenga and Sunmbo 2014). Thus, there is the need to search for local materials as alternatives for construction and building works. Cement as one of the basic ingredient of concrete is one of the main construction materials used widely.

The increasing demand for cement is expected to be met by partial cement replacement. The search of non-conventional local construction materials which may serve as alternative for cement or binder replacement in concrete production led to the discovery and possibility of using agricultural waste and industrial by product as cementitious materials (Obilade 2014). Saw dust also known as wood dust can be define as a loose particles or wood chippings obtained as by-product resulting from the mechanical milling or processing of timber into various shapes and sizes. The resulting ash know as sawdust ash is a form of pozzolan (Obilade 2014). According to American Society for Testing Materials (ASTM, C-618-1978), pozzolana is a siliceous or a siliceous aluminous material which contains little or no cementitious value, but in finely divided form and in the presence of moisture or water, chemically reacts with calcium of moisture at ordinary temperature to form compound possessing cementitious properties. Saw dust is in abundance in Nigeria and other parts of the world. However, its continuous accumulation in recent times can lead to health hazard and environmental problem. It is logical that one way of disposing of it can be its beneficial incorporation into structural concrete system. Cheah and Ramli (2011) investigated the implementation of wood waste ash as a partial cement replacement material in the production of structural grade concrete and mortar,

2 Materials and Equipment

2.1 Materials

Sawdust Ash (SDA): The sawdust from Apa (ASDA) and Mahogany wood species (MSDA) which was used for this research work were sourced from Dei-Dei timber shade in Abuja along Kubwa- Abuja expressway in F.C.T. The saw dusts were brought to Minna and taken to where they were burnt into ash to produce the sawdust ashes (ASDA, MSDA) at 800 °C.

Coarse Aggregate: The gravel (coarse aggregate) which was used for this research work was sourced from Kpankugu in Minna, Niger state. Gravel material with particle size between 10 mm to 14 mm was used as coarse aggregate for this research work. The coarse aggregate was free from impurities such as dust, clay particles.

Fine Aggregate: The sand (fine aggregate) was sourced from Kpankugu in Minna, Niger state, Nigeria. It was free from impurities, organic matter and later sun dried. It conformed to the requirements of BS EN 12620 (2008).

Water: The water which was used for mixing the materials was obtained from borehole which conformed to BS EN 1008 (2002) requirements. Therefore, the water is fit for drinking, free from suspended particles, organic materials and soap which might affect hydration of cement.

Cement: The cement which was used as a binder throughout this project is Dangote cement which is a type 1 ordinary Portland cement (i.e. a general purpose Portland cement suitable for most uses) conforming to BS EN 197-1 (2000) specifications.

2.2 Equipment

The equipment used to carry out this research work includes; electric weighing balance, table vibrator, British standard sieves, stop watch, hand trowel, tamping rod, head pan, Seidner compression machine, bucket and 150 mm × 150 mm × 150 mm metal mould.

3 Experimental Method

3.1 Procedure

The following procedure was adopted in this research work:

Chemical Analysis of ASDA and MSDA: Samples of ASDA and MSDA were taken to a chemical laboratory to determine the oxides composition using the X-Ray fluorescent (XRF) test.

Aggregate Characterization: Physical properties tests of aggregate such as specific gravity, particle size distribution (sieve analysis) and bulk density tests were carried out on the constituent materials used which include fine aggregate, coarse aggregate and ASDA and MSDA.

Production of Concrete Cubes: The concrete mix design was carried out using the absolute volume method for the constituent materials of concrete mix ratio of 1:2:4

with a constant water/cement ratio of 0.5 for the production of concrete cubes. Mixing of the concrete was done manually based on the volume to be use, slump and compacting factor test were carried out on the fresh concrete and the result presented in Table 1. A total of one hundred and sixty (162) concrete cubes specimen of metal mould size 150 mm × 150 mm × 150 mm was produced for this study. The concrete cubes were produced for two procedures, revibrated and non-revibrated. For the non-revibrated, a set of seventy-eight (78) cubes were produced in total, six (6) concrete cubes (control cubes) were produced with pure cement and another six (6) cubes were produced each using 5%, 10%, 15%, 20%, 25% and 30% ASDA and MSDA replacements for cement. However, another set of concrete cubes numbering eighty-four (84) was produce, fifty-six (56) of which were cast using the optimum percentage replacements of ASDA and MSDA for cement. The remaining twenty-eight (28) using OPC were cast and revibrated. The revibration process was carried out for 20 s at 10 min intervals through duration of 1 h after initial vibration. Vibrating table was used to compact the concrete mix in the metal mould. The concrete cubes were de-moulded after 24 h and were cured by method of ponding (immersion of the cubes into water) for a period of 7 and 28 days only. Compressive strength on the concrete cubes was carried out after a 7th and 28th days of curing in accordance with the specification in BS1881: Part 116: (1983) using Seidner compression test machine.

4 Result and Disscussion

The results of the Laboratory tests which include: Chemical composition of ASDA and MSDA, physical properties test of aggregate and test on concrete (fresh and hardened properties) are presented in Tables 1, 2, 3, 4, 5, 6 and 7.

4.1 Chemical Composition of ASDA and MSDA

The chemical composition of the SDA for the specified wood specie used in this study are presented in Table 1. It can be seen that the oxide concentration of the ash using an X-Ray fluorescent (XRF) test showed that its major oxide concentration is CaO, SiO_2, Al_2O_3, Fe_2O_3, MgO and SO_3. Element such as Na_2O, ZnO, Cl, Mn_2O_3, SrO, Cr_2O_3, P_2O_5, TiO_2 and K_2O were detected in trace amount. The analyte concentration of Apa wood specie ash on Table 1 shows that oxides concentration of silicon oxides (SiO_2 = 37.628%), aluminium oxides (Al_2O_3 = 16.529%) and iron oxides (Fe_2O_3 = 7.132%) constitute a total sum of 61.289% of pozzolanic materials which is slightly below 70% but greater than 50%, thus indicating that the Apa wood specie ash falls under class C as specified by ASTM C 618 (1978) for pozzolana based on its lime content while the analyte concentration of Mahogany wood specie ash on Table 4 also shows that oxides concentration of silicon oxides (SiO_2 = 39.873%), aluminium oxides (Al_2O_3 = 18.053%) and iron oxides (Fe_2O_3 = 6.924%) constitute a total sum of 64.814% of pozzolanic materials which is slightly below 70% but greater than 50%, thus indicating that the Mahogany wood specie ash falls under class C as specified by ASTM C 618 (1978) for pozzolana based on its lime content.

Table 1. Chemical composition of SDA from the specified wood species

Element	ASDA	MSDA
	Concentration (Wt %)	Concentration (Wt %)
Na_2O	0.668	0.567
MgO	5.954	5.043
Al_2O_3	16.529	18.053
SiO_2	37.628	39.873
P_2O_5	0.271	0.324
SO_3	1.152	1.160
Cl	0.521	0.530
K_2O	0.412	0.434
CaO	29.167	26.568
TiO_2	0.402	0.449
Cr_2O_3	0.011	0.007
Mn_2O_3	0.042	0.021
Fe_2O_3	7.132	6.924
ZnO	0.109	0.010
SrO	0.046	0.037

4.2 Particle Size Analysis of Fine, Coarse Aggregate and SDAs

The results of particle size analysis of ASDA and MSDA, fine and coarse aggregates used in this study are presented Table 2 and 3. The fineness modulus of MSDA as calculated from Table 2 is 1.51 while that of ASDA as calculated from Table 3 is 1.57, thus implying that the result attained for both wood specie is less than FM of 2.3–3.1 for fine aggregate recommended by American Society for Testing and materials (ASTM) C 33, therefore, ASDA and MSDA are finer than fine aggregate. The uniformity coefficient C_U was calculated as equal to 6.0, while the coefficient of curvature C_C is calculated as 1.63. This signifies that Sand is within specifications of a well-graded sand, which complies with the unified soil classification system (USCS) conforming where $C_U \geq 6$ and $1 \leq C_C \leq 3$ (Arora 2010). The uniformity coefficient C_U was calculated as equal to 1.5, while the coefficient of curvature C_C is calculated as 1.12. Hence the aggregate is well graded gravel because the result complies with the Unified Soil Classification System (USCS) of well graded sand with less than 5% fine has $1 \leq C_C \leq 3$ as stated by Arora (2010).

Table 2. MSDA particle size analysis

Sieve size (mm)	Sample weight retained (g)	Percentage retained (%)	Cumulative percentage retained (%)	Cumulative percentage passing (%)
0.850	0.00	0.00	0.00	0.00
0.600	0.00	0.00	0.00	0.00
0.425	0.00	0.00	0.00	0.00
0.300	22.20	11.10	11.10	88.90
0.150	95.68	47.84	58.94	41.06
0.075	44.39	22.19	81.13	18.87
Pan	36.55	18.27	-	-
Total			151.17	

Table 3. ASDA particle size analysis

Sieve size (mm)	Sample weight retained (g)	Percentage retained (%)	Cumulative percentage retained (%)	Cumulative percentage passing (%)
0.850	0.00	0.00	0.00	0.00
0.600	0.00	0.00	0.00	0.00
0.425	0.00	0.00	0.00	0.00
0.300	24.30	12.15	12.15	87.85
0.150	97.24	48.62	60.77	39.23
0.075	47.58	23.79	84.56	15.44
Pan	28.84	14.42	-	-
Total			157.48	

Bulk Density: The compacted and un-compacted bulk density of the fine aggregate (sand) are 1669 kg/m^3 and 1590 kg/m^3 while that of coarse aggregate (gravel) are 1534 kg/m^3 and 1352 kg/m^3. This correspond to the range of 1200–1800 kg/m^3 specified by BS 812, part 2: 1975 for aggregates. The value for compacted and un-compacted bulk density of SDA from Apa wood specie are 812 kg/m^3 and 650 kg/m^3 while that of mahogany wood specie are 470 kg/m^3 and 449 kg/m^3. The results of bulk density depend on how the samples of materials are closely packed. SDA are loose materials that are light in weight which are not closely packed thereby leading to low coherent bulk density.

Specific Gravity: The specific gravity was found to be 2.62 for fine aggregate and 2.65 for coarse aggregate, these values obtained is within limit of 2.5 to 3.0 for natural aggregate as specified by BS812: part 107, 1995. The specific gravity value of ASDA and MSDA were found to be 2.53 and 2.63, being average of three trials.

Fresh Concrete Properties Test

Slump: The results of slump test for the concrete cubes containing ASDA and MSDA shown in Table 4. It was observed from the slump test result that the slump value reduces as the SDAs content increases upon inclusion in the mix from 0% to 30%, the difference in slump value when cement is partially replaced with SDAs shows the demand of water, hence giving rise to lower workability.

Compacting Factor: The results of compacting factor test from Table 5 shows that the compacting factor values reduce as the SDAs content increases. The compacting factor values reduced from 0.92 to 0.84 as the percentage SDAs replacement increased from 0% to 30% for the concrete cubes containing the ash from the different wood species. Thus, indicating that the concrete becomes less workable as the SDAs percentage increases signifying that more water is required to make the mixes more workable. According to Auta *et al.* (2016), the high demand for water as the SDAs content increases is due to increased amount of silica in the mixture.

Table 4. Slump test result of fresh concrete

Percentage replacement (%)	ASDA slump (mm)	MSDA slump (mm)
0	25	25
5	18	19
10	16	16.5
15	13	14
20	10	11
25	No slump	No slump
30	No slump	No slump

Table 5. Compacting factor test result of fresh concrete

Percentage replacement (%)	Compacting factor value for ASDA	Compacting factor value for MSDA
0	0.92	0.92
5	0.88	0.88
10	0.87	0.87
15	0.87	0.86
20	0.86	0.86
25	0.85	0.85
30	0.84	0.84

Hardened Concrete Properties Test

The compressive strength was test was carried out based on the specification in BS1881: Part 116: 1983 on the hardened concrete.

Compressive Strength: The compressive strength of concrete containing sawdust ash (SDA) from the selected wood species and the effect of re-vibration using the optimum strength from the selected wood species were investigated after a curing period of 7 and 28 days respectively. The summary of the results is tabulated and presented graphically.

Effect of SDA from Selected Wood Specie on the Compressive Strength of Non-revibrated Concrete. The result of the effect of ASDA and MSDA on the compressive strength of non-revibrated concrete is shown in Table 6, it can be observed that the compressive strength value reduces as the percentage SDA from the specified wood specie partially replaced with cement increases 0, 5, 10, 15, 20, 25, and 30%. However, the compressive strength increases as the number of days of curing increases from 7 to 28 days for each percentage SDA from the specified wood specie partially replaced with cement. It was also observed from Table 4. 10 that the concrete cubes containing MSDA tends to show a higher gain in strength than the corresponding concrete mix containing ASDA, thus indicating that SDA from different wood specie tends to affect the compressive strength of concrete. The result also reveals fort concrete cube containing ASDA, the optimum percentage replacement was attained at 5% while that of concrete cube containing MSDA, the optimum percentage replacement was attained at 10% based on the strength obtained.

Table 6. Compressive strength test result obtained for non-revibrated hardened concrete

Percentage replacement (%)	Compressive strength of ASDA (N/mm^2)		Compressive strength of MSDA (N/mm^2)	
	7 days	28 days	7 days	28 days
0	17.74	22.90	17.74	22.90
5	15.06	18.81	16.59	21.24
10	13.77	16.35	15.93	20.03
15	12.05	14.60	13.86	17.68
20	10.16	12.11	10.93	13.86
25	8.88	11.19	9.74	11.85
30	7.46	9.80	8.03	10.01

Effect of Revibration on the Compressive Strength of ASDA and MSDA Concrete. The effectiveness of the research depicted by the effect of revibration on the compressive strength of ASDA and MSDA using the optimum percentage replacements obtained for the non-revibrated concrete cubes. This is clearly shown by the result presented in Table 7.

It is observed from Table 7, that the compressive strength increases at the initial stage of revibration time lag interval from 0th to 40th min giving rise to strength value of 19.07 N/mm^2 to 22.71 N/mm^2 at 7days and 25.69 N/mm^2 to 31.46 N/mm^2 for 28 days of curing but later gradually decreases in strength from 50 min (21.10 N/mm^2,

26.79 N/mm^2 at 7 and 28 days) to 60 min (20.46 N/mm^2, 26.09 N/mm^2 at 7 and 28 days) for the 0% replacement level. It can also be seen from Table 7 where cement was partially replaced with ASDA using the 5% optimum replacement level obtained for the non-revibrated concrete cube that there is also a significant increase in compressive strength from 0th to 40th min giving rise to strength value of 18.75 N/mm^2 to 22.35 N/mm^2 at 7days and 22.35 N/mm^2 to 26.39 N/mm2 for 28 days and later decrease in strength from 50th min (19.46 N/mm^2, 23.88 N/mm^2 at 7 and 28 days) to 60th min (18.25 N/mm^2, 22.24 N/mm^2 at 7 and 28 days). From Table 7, the result concrete cube containing 10% optimum replacement level obtain using MSDA partially replaced for cement shows an increase in compressive strength from revibration time lag interval of 0th to 30th min with strength value of 18.93 N/mm^2 to 21.28 N/mm^2 at 7days and 23.46 N/mm^2 to 28.21 N/mm^2 for 28 days of curing and a gradual drop in strength from 40 min (19.15 N/mm^2, 23.81 N/mm^2 at 7 and 28 days) to 60 min (18.88 N/mm^2, 23.10 N/mm^2 at 7 and 28 days).

Generally, as stated by Auta et al. (2016) the early increase in compressive strength may be attributed to the calcium hydroxide in the OPC and revibration which enhanced densification and volumetric compaction of the concrete, but later re-vibration will debond the chemical compound of C$_3$S which lead to decrease in strength as can be seen from the result obtained. The maximum compressive strength for the 0% was attained at 40 min revibration time lag with strength value of 22.71 N/mm^2 and 31.36 N/mm^2 at 7 and 28 days respectively, that of 5% ASDA was also attained at 40th min revibration time lag with strength value of 22.35 N/mm^2 and 26.39 N/mm^2 at 7 and 28 days while the maximum strength value of 21.28 N/mm^2 and 28.21 N/mm^2 was attained at 30th min revibration time lag for the 10% MSDA. This maximum strength value obtain after revibration seems to have an improved strength compare to the average value obtained for corresponding percentage replacement for the non- revibrated concrete cubes. Furthermore, the result shows that revibration improves the compressive strength of OPC and OPC – SDA concrete cubes which agrees with findings of Krishna et al. (2008), that revibration increases the compressive strength of concrete if carried out within the initial setting time of the concrete.

Table 7. Compressive strength of revibrated hardened concrete

Time interval (Mins)	Compressive strength of Control (0%) replacement level (N/mm^2)		Compressive strength of ASDA 5% optimum replacement level (N/mm^2)		Compressive strength of MSDA 10% optimum replacement level (N/mm^2)	
	7 days	28 days	7 days	28 days	7 days	28 days
0	19.07	25.69	18.75	22.35	18.93	23.46
10	20.27	27.46	19.73	23.24	20.17	25.70
20	21.51	28.51	20.79	24.35	20.66	26.50
30	22.08	30.13	21.64	25.33	21.28	28.21
40	22.71	31.46	22.35	26.39	19.86	25.15
50	21.10	26.79	19.46	23.88	19.15	23.81
60	20.46	26.09	18.95	23.24	18.88	23.10
Mean strength	21.03	28.02	20.24	24.11	19.85	25.13

5   Conclusion

The chemical composition of ASDA and MSDA show that they are of class C Pozzolana, where the sum of $SiO_2+AL_2O_3+Fe_2O_3$ fall under in-between 50%–70% specified by ASTM C 618 – 78 (2005) for pozzolana. The compressive strength generally increases with curing period and decreases with increased amount of ASDA and MSDA as partial replacements for cement.

The optimum compressive strength of concrete cubes were at 10% MSDA and 5% ASDA. It also gave rise to increase of compressive strength of the revibrated concrete cubes up to a certain time lag and thereafter, decreases to the hour lag.

Therefore, maximum strength value obtained after revibration using 0%, 5% ASDA and 10% MSDA seem to have an improved strength compare to the average value obtained for corresponding percentage replacement for the non-revibrated concrete cubes. Revibration has thus been useful to produce smart concrete.

References

Adewuyi, A.P., Adegoke, T.: Exploratory study of periwinkle shells as coarse aggregates in concrete works. J. Appl. Sci. Res. **4**(12), 1678–1681 (2008)

Auta, S.M., Amanda, U., Sadiku, S.: Effect of revibration on the compressive strength of 56 aged RHA-cement concrete. Nigeria J. Eng. Appl. Sci. (NJEAS) **2**(2), 115–121 (2015)

American Standard for Testing Materials ASTM C 618– 78: Specification for fly ash and Raw or Calcined Natural Pozzolana for use as a mineral admixture in Portland Cement Concrete (2005)

Arora, K.R.: Soil Mechanics and Foundation Engineering (Geotechnical Engineering, 7th edn. Standard Publishers Distributors, Delhi (2010)

Auta, S.M., Uthman, A., Sadiku, S., Tsado, T.Y., Shiwua, A.J.: Flexural strength of reinforced revibrated concrete beam with sawdust ash as a partial replacement for cement. Const. Uniq. Build. Struct. **5**(44), 31–45 (2016)

BS EN 12620: Specifications for Aggregates from natural sources for concrete. BSI, London (2008)

BS EN 1008: Methods of test for water for making concrete. BSI, London (2002)

BS EN 197-1: Specification for Portland cement. BSI, London (2000)

British Standards Institution: BS 1881: Part 116 Method for determination of Compressive Strength of Concrete Cubes. British Standards Institution, London (1983)

Cheah, C.B., Ramli, M.: The implementation of wood waste ash as a partial cement replacement material in the production of structural grade concrete mortar. Overview Rev. Art. Resour. Conserv. Recycl. **55**(1), 669–685 (2011)

Elinwa, A.U., Ejeh, S.P., Mamuda, A.M.: Assesing of the fresh concrete properties of self-compacting concrete containing sawdust ash. Constr. Build. Mater. **22**(6), 1178–1182 (2008)

Elinwa, A.U., Mahmmodb, Y.A.: Ash from timber waste as cement replacement material. Cem. Concr. Compos. **24**(2), 219–222 (2002)

Krishna Rao, M.V., Rathish Kumar, P., Bala Bhaskar, N.V.R.C.: Effect of re-vibration on compressive strength of concrete. Asian J. Civ. Eng. (Build. Hous.) **9**(3), 291–301 (2008)

Obilade, I.O.: Use of saw dust ash as partial replacement for cement in concrete. Int. J. Eng. Sci. Invent. **3**(8), 36–40. http://www.ijesi.org/. ISSN (Online): 2319–6734, ISSN (Print): 2319–6726.

Oyedepo, O.J., Akande, S.P.: Investigation of properties of concrete using sawdust as partial replacement for sand. Civ. Environ. Res. **6**(2), 35–42 (2014). ISSN 2225-0514

Olutoge, F.: Investigations on sawdust and palm kernel shells as aggregate replacement. ARPN J. Eng. Appl. Sci. **5**(4), 8–12 (2010)

Costs and Benefits Data Mapping of BIM Laser Scan Integration: A Case Study in Australia

Sherif Mostafa[1($\boxtimes$)], Harold Villamor[1], Rodney A. Stewart[1],
Katrin Sturm[2], Emiliya Suprun[1], and Scott Vohland[2]

[1] School of Engineering and Built Environment, Griffith University,
Brisbane, QLD 4222, Australia
sherif.mostafa@griffith.edu.au
[2] Seqwater, Ipswich, QLD 4305, Australia

Abstract. Building information modelling (BIM) has been increasingly popular in the utility industry due to substantial benefits of cost and time savings, and improved performance and asset management during the operations and maintenance (O&M) phase. With the recent addition of point cloud from laser scanning, the level of details opted BIM 3D models from 'as-designed' to 'as-constructed' which paves the way for more benefits to other stakeholders during the O&M phase. This research identified and developed the cost benefit elements of the laser-scan integrated BIM as part of a case study research project of a water treatment plant (WTP) in Queensland, Australia. The costs elements from stakeholder's perspective are predominantly BIM supporting software, hardware, labour and training. Whereas, the benefits are categorised based on communication, asset, data and risk management elements for the WTP stakeholders. This research evaluated the costs and benefits through 3D modelling, including asset input, of a WTP study section, site surveys and interviews. The research developed an association mapping between all costs and benefits elements for the WTP case selected.

Keywords: BIM · As designed as constructed assets · BIM adoption in the water industry

1 Introduction

Building Information Modelling (BIM) is one of the most promising developments in the architecture, engineering and construction (AEC) industry. It is a digital representation of a physical structure and functions of a project/facility supported by multiple software and technologies; and are managed by AEC professionals and other stakeholders. BIM is currently used by individuals and organisations such as business and government agencies for planning, designing, constructing, operating and maintaining various physical infrastructures from different industries including the water sector [1, 2]. The recent advancement of BIM technologies has opted 3D models from "as designed" to "as constructed". Although construction industries have proven BIM adoption to be beneficial, case studies of the adoption in the water sector have

I. El Dimeery et al. (Eds.): JIC Smart Cities 2019, SUCI, pp. 385–392, 2021.
https://doi.org/10.1007/978-3-030-64217-4_42

demonstrated related benefits such as reduced frequency of site survey, and enhanced client communication and associated costs.

BIM is an object-based organisation of information that represents a building to intelligent objects to carry detailed information which can be extracted, exchanged or networked to support decision-making, in contrast to the traditional method of file-based organisation of information which causes errors and omissions, and leads to financial expenses and delays [2]. However, [1] believes that BIM in a broader spectrum is not just an object but rather a human activity that involves people, system process and software. BIM connects AEC professionals and other stakeholders such as contractors and suppliers contributing and collaborating towards a common data environment (CDE). The CDE consists of project and asset 3D models which involve graphical and non-graphical data, and associated documents [2]. These 3D models are the stereotypical reflection of BIM since it constitutes technological advancement.

BIM 3D models are initially designed to record drawings and documentation during the design and construction phase [1], which is referred as "as-designed" BIM [3, 4]. It is argued that BIM excels mostly during these phases. However, the recent advancement of BIM technologies has opted BIM from "as designed" to "as-built". The latter is referred to as the utilisation of laser scan (LiDAR) technology and associated applications such as point cloud and web-share [5, 6]. For example, FARO Scene software utilises LiDAR equipment, where application includes point cloud generation from images; and extension file conversion to other software [6]. Another application of the software is the ability to share the 3D view (digital twin image) to the selected stakeholders including a webshare cloud option [6]. There are many BIM software products and associate tools either in conjunction or compatible separately on the market, but the majority have the same principle and use. They can be categorised based on the manufacturer, life-cycle stages use, and the main BIM dimensions i.e. 3D-modeling, 4D scheduling, 5D cost, 6D sustainability, and 7D (O&M) [6, 7]. The use of these BIM technologies is also dependent on the activities and specific management system such as the drinking water treatment plant (WTP).

A WTP is a public facility that treats raw water and distributes it as portable water to clients from different sectors such as residential, commercial and industrial. The conventional water treatment processes include coagulation, flocculation and clarification, sedimentation, filtration, and disinfection; and are managed and operated by facility personnel including operators, maintenance planners, process engineers and asset managers. WTP assets include pipes and other equipment within the plant, and piping and instrumentation drawings (P&IDs), structural drawings, asset records/register, warranties and other documentation. In Australia, the water industry continues to recognise the community benefits of water infrastructure and is taking a more businesslike approach, with a number of Water Service Providers being commercialised and some corporatised [8].

The elements of this businesslike approach include a more rigorous infrastructure investment decision-making process with non-asset solutions, full cycle costs, risks and existing alternatives being considered before deciding to construct infrastructure; making the best use of existing infrastructure; ensuring that infrastructure can sustain agreed customer service standards; and a recognition that asset management is a core business function to support service delivery [8]. For this purpose, the Queensland

Government created and recommended the QLD water total management planning (TMPs) [9]. The "TMPs were introduced in 1994 to promote best practice planning and least cost outcomes for water supply and sewerage planning". A TMP quantifies and assesses the condition of assets, prioritises expenditure, and identifies options for cost saving and improvement in ways that are environmentally and financially sustainable" [8, 9]. This research adopts the TMPs' and identified four critical elements for better planning and managing of a water treatment asset. These elements are asset, risk and data management, and communication. This research mapped the BIM integration benefits in regards to these four elements.

BIM adoption has recently been increasing in the water industry due to substantial benefits in cost and time savings; and improved performance and asset management during the construction phase. With the recent addition of point cloud from laser scanning, the level of detail (LOD) and level of information (LOI) of as constructed assets are captured which pave the way for more benefits to other stakeholders during the O&M phase. Although many case studies are elaborating the benefits of using BIM in the construction industry [10], the water industry is also gaining momentum [11]. A case study in Queensland, Australia shows that the Kawana sewage treatment plant upgrade has successfully used BIM for design, construction and operation; through the 3D utility software Autodesk AutoCAD Plant 3D and other BIM software [12]. The case study illustrates the benefits of adopting BIM. Other case studies worldwide have also been reviewed and illustrated similar results [11]. Research reveals some of the workflows of the 3D modelling from laser-scan images to Plant 3D software and other BIM software such as Autodesk Navisworks with the aid from third party software FARO products [13]. Other workflows have also been illustrated in the 3D modelling case studies from other reviews [10]. These workflows will be explored during the 3D modelling process of the WTP case in this research.

2 Research Methodology

The scope of this research includes 3D modelling using point cloud which is extracted from laser-scan images; WTP asset input and validation; and investigation of the WTP stakeholders, for cost and benefit investigation of BIM integration and the creation of a data map for the WTP stakeholders. After a thorough review of relevant literature, the research adopted a mixed-method to allow for a combination of identifying patterns and categorising cost and benefit elements. An in-depth exploration of costs and benefits to relevant stakeholders across the asset life-cycle was conducted. The data collection methods comprised case study reviews and other 3D modelling processes evaluations; utilisation of WTP existing data; structured interview and equipment survey from the WTP site visit and group meetings; and experimentation of 3D modelling processes.

3 Results – 3D Modelling Processes and Case Studies Comparison

The research reveals the costs, benefits and WTP stakeholders of BIM during the data collection mentioned in the previous section. The costs and benefits of the BIM adoption were recorded during the experimentation workflow shown in Fig. 1, reflective of the actual BIM design phase; and literature review through case studies' defined costs and concluding benefits, reflective of the actual O&M and construction phases.

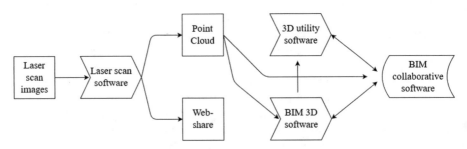

Fig. 1. The workflow of 3D modelling from laser-scan images.

The overall costs include software such as scan processor software (e.g. FARO Scene and Autodesk Recap), BIM software (e.g. Autodesk Revit and Navisworks), and 3D utility software (e.g. Autodesk AutoCAD Plant 3D). Although the research hardware available is web-browsing devices and 8G RAM PC specification, other PC specifications have been recommended for optimal 3D modelling and model use. To quantify the labour and training costs, the intensities have been quantified based on a scale of high, medium and low as illustrated in Table 1.

Table 1. Labour and training scale reflective of the costs identified during the data collection.

Element	Low	Medium	High
Labour	0 to 20 h	20 to 40 h	40 to 80 h
Training	Online tutorials	Same with scale Low plus manual consultations	Same with scale Medium plus expert consultations, e.g. diagnostic blogs, company inquiry request, expert advice

Through experimentation and literature review, the costs of the WTP BIM adoption were observed and recorded during the BIM life-cycle phases i.e. BIM design, O&M, BIM redesign for WTP upgrade and construction for WTP upgrade activities. BIM design and redesign for WTP upgrade activities include laser scanning, laser-scan to point cloud, web-share uploading, and 3D modelling using different BIM software, and utilities, and other tools. The O&M and construction for WTP upgrade activities include data access, input and output, and surveying. Through literature review by comparing benefits from case studies and experimentation by validating benefits from case studies, the benefits include reduced time and costs during all phases, clients' serviceability and improved communication.

4 Discussion – Costs and Benefits Data Mapping

The research reveals that the costs and benefits of BIM adoption can be described based on their respective elements. Costs of point cloud-based BIM or webshare from the laser scan images are predominately software, hardware, labour and training elements as illustrated in Table 2.

Table 2. Cost and source references.

	Software	Hardware	Labour	Training	Source
	S	H	L	T	R
1	Scan processor	PC -8G RAM	High	High	Lit. Review
2	FARO Scene	PC -16G RAM	Medium	Medium	Online Tutorial
3	Recap	PC -16G RAM above	Low	Low	3D Modelling
4	Revit	Web-browsing device	Utility		Site survey
5	Plant 3D		U		Site Interview
6	Navisworks		Internet CXN		Scan Company

These cost elements were identified during the 3D modelling processes experimentation and case study comparisons from all BIM life-cycle phases. Each phase consists of activities and thus, to avoid duplication of costs, the data are presented in matrix form as illustrated in Table 3.

Table 3. Some costs investigated for the different BIM life-cycle phases. (Refer to Table 2)

Activity	Code	Processes/function	Cost element					Source
			Software	Hardware	Labour	Training	Utility	
Experimental BIM design								
Scan	C1	Laser scanning site	AUD $3500 (provided)					R6
P. Cloud	C2	Scanning processes	S2	H1	L2	T1	U6	R1,R2,R3
Webshare	C3	Uploading processes	scan package					R6
O&M and Potential upgrade								
Access	C10	Webshare viewing		H4	L3	T3	U6	R1, R2
Input	C12	Webshare data relay		H4	L2	T2	U6	R1, R2
Output	C15	Webshare print out		H4	L3	T2	U6	R1, R2

Table 3 shows the cost-activity matrix of laser scan processes during the experimental BIM design and fixed prices from the laser scan company. The table also shows the WTP cost-activity matrix during the O&M and potential upgrade found in the literature review. Other WTP costs associated with 3D modelling using utility software and BIM software, are presented in the same manner. It was found that this software has cost impacts when using in conjunction with each other, and could potentially reduce the cost of the processes. Benefits can be organised from the data, asset and risk management, and communication elements based the TMPs' interpretations i.e. Queensland government's recommendation of asset management as a business-like approach as illustrated in Table 4.

Table 4. Some of the LS-I-BIM validated benefits. (Refer to Table 2)

Code	Benefit	Elements	Source
B1	Better design solutions	Data/Asset	R1,R3,R4
B12	Increased client satisfaction	Comm./Risk	R1,R4
B18	Increased profits on projects	Asset/Risk	R1

Table 4 shows some of the potential benefits with respect to the activities illustrated in Table 2. It shows design, client and overall project benefits which were found during literature review and experimentation. Other benefits were also found regarding the O&M and upgrade phases, and the design of BIM models. An accurate design of BIM models is essential since it will become the CDE of all stakeholders. WTP stakeholders were found to have their respective costs and benefits when adopting LS-I-BIM, presented in Table 5.

Table 5. Generalised WTP cost and benefit data map. (Refer to Table 3 and 4)

Stakeholders	Relevance		BIM life-cycle phase				Potential costs	Potential benefits
	Webshare	LS-I-BIM	BIM design	O&M	Upgrade			
					Design	Const.		
3D Team	✓	✓	✓	✓	✓	✓	C1–C14	B1–B5, B8–B9
Facility PER	✓	✓		✓			C10–C18	B2–B9
Managers	✓	✓	✓	✓	✓	✓	C10–C11, C5–C17	B5–B9, B11–B18
Contractors	✓				✓	✓	C10–C11, C15–C18	B2–B9, B11–B18
Clients	✓				✓		C10	B12, B14, B15
Owners	✓				✓	✓	C10–C17	B1–B18
AEC Prof.	✓	✓			✓	✓	C4–C18	B1–B11, B16–B18

Table 5 shows a data map of potential costs and benefits to a generalised WTP stakeholder group. For example, facility personnel include operators, maintenance planners, schedulers and process engineers. These specific WTP stakeholders and respective BIM costs and benefits were identified during the literature review and were validated from the other data collection methods. The data map also shows the relevance of other types of BIM tools such as webshare and the laser scanning BIM functionalities, and during BIM life-cycle phases. It shows that webshare benefits all involved stakeholders. However, the overall relevance of BIM is distributed throughout different phases.

5   Conclusions

This research has evaluated the cost-benefit of laser-scan integrated BIM through experimentation of 3D modelling processes and other data collection and created a data map for WTP stakeholders. Despite an initial increased cost by using multiple software packages, the overall costs are significantly reduced when these are used in conjunction with each other. The benefits were shown to significantly outweigh the costs. The created BIM cost-benefit data map can be beneficial for WTP stakeholders, other water sector facilities, and future researchers interested in laser scanning BIM cost-benefit analysis.

Acknowledgements. The authors would like to thank the Seqwater team for their excellent support and professional insights provided during this research project.

References

1. Eastman, C.M.: ProQuest (Firm). In: BIM Handbook: A Guide to Building Information Modeling for Owners, Managers, Designers, Engineers and Contractors. 2nd edn. Wiley, Hoboken (2011)
2. Garber, R.: ProQuest Ebooks 2014. In: BIM Design: Realising the Creative Potential of Building Information Modelling, Wiley, Chichester (1971)
3. British Standards Institution: (2013) & BIM wiki PAS 1192-2 (2019). https://www.designingbuildings.co.uk/wiki/PAS_1192-2. Accessed 23 Sept 2019
4. Moser, C.: Record models vs. as-built models in a world of complexity (2016). https://www.linkedin.com/pulse/record-models-vs-as-built-world-complexity-cliff-moser. Accessed 28 Sept 2019
5. Cheng, L., Chen, S., Liu, X., Xu, H., Wu, Y., Li, M., Chen, Y.: Registration of laser scanning point clouds: a review. Sens. (Switzerland) 18(5), 1641 (2018)
6. FARO. https://websharecloud.com/index.html#divlogo Accessed 2 Oct 2019
7. Moscardi, L.: BIM software guide (2017). http://www.buildingincloud.net/wp-content/uploads/2017/03/BIM-Software-list.pdf. Accessed 6 Sept 2019
8. Qldwater: Total management planning: asset management-overview (2019a). https://www.qldwater.com.au/TMPs. Accessed 8 Sept 2019
9. Qldwater: Archived: total management plans (2019b). https://www.qldwater.com.au/TMPs Accessed 25 Sept 2019

10. Liu, Z., Zhang, C., Guo, Y., Osmani, M., Demian, P.: A Building Information Modelling (BIM) based Water Efficiency (BWe) framework for sustainable building design and construction management. Electronics **8**(6), 599 (2019)
11. Dodge Data & Analytics: Business value of BIM for water projects (2018). https://www.construction.com/toolkit/reports/business-value-bim-water-projects. Accessed 8 Sept 2019
12. DavidKent, S.S.: Kawana STP upgrade: Using BIM for design, construction and operation (2019). https://www.waternz.org.nz/Attachment?Action=Download&Attachment_id=3383. Accessed 8 Sept 2019
13. Nguyen, H., Radcliffe, I.: Field to finish: point cloud to autocad plant 3D (2015). https://www.autodesk.com/autodesk-university/class/Field-Finish-Point-Cloud-AutoCAD-Plant-3D-2015#handout. Accessed 8 Sept 2019

Author Index

I. El Dimeery et al. (Eds.): JIC Smart Cities 2019, SUCI, pp. 393–394, 2021.
https://doi.org/10.1007/978-3-030-64217-4